Groundwater Monitoring

Water Quality Measurements Series

Series Editor

Philippe Quevauviller
European Commission, Brussels, Belgium

Published Titles in the Water Quality Measurements Series

Hydrological and Limnological Aspects of Lake Monitoring
Edited by Pertti Heinonen, Giuliano Ziglio and Andre Van der Beken

Quality Assurance for Water Analysis
Edited by Philippe Quevauviller

Detection Methods for Algae, Protozoa and Helminths in Fresh and Drinking Water
Edited by Andre Van der Beken, Giuliano Ziglio and Franca Palumbo

Analytical Methods for Drinking Water: Advances in Sampling and Analysis
Edited by Philippe Quevauviller

Biological Monitoring of Rivers: Applications and Perspectives
Edited by Giuliano Ziglio, Maurizio Siligardi and Giovanna Flaim

Wastewater Quality Monitoring and Treatment
Edited by Philippe Quevauviller, Olivier Thomas and Andre Van der Berken

The Water Framework Directive – Ecological and Chemical Status Monitoring
Edited by Philippe Quevauviller, Ulrich Borchers, Clive Thompson and Tristan Simonart

Rapid Chemical and Biological Techniques for Water Monitoring
Edited by Catherine Gonzalez, Richard Greenwood and Philippe Quevauviller

Groundwater Monitoring
Edited by Philippe Quevauviller, A M Fouillac, D J Grath and R Ward

Forthcoming Titles in the Water Quality Measurements Series

Chemical Marine Monitoring: Policy Framework and Analytical Trends
Edited by Philippe Quevauviller, Patrick Roose and Gert Vereet

Groundwater Monitoring

PHILIPPE QUEVAUVILLER
European Commission, Brussels, Belgium

ANNE-MARIE FOUILLAC
BRGM, ORLEANS Cedex 2, France

JOHANNES GRATH
Umweltbundesamt GmbH, Wien, Austria

ROB WARD
Environment Agency – England and Wales, Solihull, UK

**WILEY**

A John Wiley and Sons, Ltd., Publication

Library of Congress Cataloging-in-Publication Data

Groundwater monitoring / Philippe Quevauviller ... [*et al*.].
    p. cm.
  Includes bibliographical references and index.
  ISBN 978-0-470-77809-8
  1.   Groundwater–Pollution–Measurement. 2.   Groundwater–Quality. 3.   Environmental monitoring.
I. Quevauviller, Ph.
  TD426.G715 2009
  363.739′4 – dc22

               2009016232

A catalogue record for this book is available from the British Library.

ISBN   978-0470-77809-8 (H/B)

Typeset in 10/12 Times by Laserwords Private Limited, Chennai, India.
Printed and bound in Great Britain by CPI Antony Rowe, Chippenham, Wiltshire

Contents

Foreword

The assessment and monitoring of groundwater quality has always posed a significant challenge – presenting as it does some special problems. It is by no means a trivial task to know exactly what is going on 'under-our-feet', when it comes to status and trends of the chemical quality of groundwater, bearing in mind that the resource:

- can be distributed over tens to hundreds of metres below ground;
- is characterised by flow regime dynamics with a time-scale ranging from a few years to various millennia;
- will often be threatened by a myriad of potentially polluting activities;
- is normally subject to gradual, often insidious, deterioration under the pressure of contaminant loading from the land surface.

Groundwater quality monitoring has been a neglected aspect of overall environmental surveillance in many countries, both within the European Community (EC) and (even more so) beyond. Despite the major importance of groundwater resources for the economical provision of public water-supply and its key role in sustaining some aquatic ecosystems, many governments have been reluctant to face the significant capital costs and operational logistics associated with dedicated, custom-built, monitoring networks, and have thus placed far too much reliance on the monitoring of drinking water receptors (mainly deep high-yielding water wells). Given the complexity of groundwater flow regimes, such monitoring:

- can be extremely difficult to interpret in terms of identifying and characterising the responsible aquifer pollution processes (and thus specifying remedial and protection measures), because of the major time-lag in the response of deeper groundwater to applied contaminant pressure;
- might be regarded as an 'essentially post-mortem activity' as regards groundwater body protection;
- has often left regulatory agencies uncertain about the seriousness of pollution trends and 'almost blindfold' when it comes to the best approach to protection measures.

The advent of the EC Water Framework Directive (2000) and Groundwater Pollution Protection Directive (2006) is changing all that – since these Directives fully embrace the need for systematic monitoring and periodic assessment of groundwater quality, for the specification of specific management and protection measures and for their effectiveness to be demonstrated through further appropriate monitoring. Thus the appearance of this book could not be more opportune, since it will serve as a detailed guide for water-sector professionals (be they in environment regulatory agencies or in environmental consultancy firms) on the methodology and practice of groundwater quality assessment and monitoring at the level required by these Directives.

The contributors to this book comprise an impressive list of European authors, from the various scientific disciplines and professional functions necessary for the evaluation and management of groundwater quality, who have pooled their experience from different national hydrogeologic and socioeconomic settings. It has been produced largely under the umbrella of the EC-Directorate General for Environment-Groundwater Working Group, and like that group has greatly benefited from the coordinating vision of Dr Philippe Quevauviller together with sound and consistent leadership from Austrian specialists. It is thus an ideal reference work for those undertaking the important fieldwork that needs to be undertaken on this topic.

Parts 2–4 provide in logical sequence:

- an approach to conceptual modelling of the flow regime of groundwater bodies in terms of aquifer typologies and visualisation and an introduction to groundwater pollution processes;

- the characterisation of groundwater pollutant pressures and behaviour of groundwater contaminants;

- groundwater quality standards (in terms of the identification of 'threshold values') and the evaluation of groundwater chemical status and trends;

- complementary methods and tools for groundwater flow and quality evaluation.

Part 5 provides a very useful set of 'case histories' from seven different European countries – whose intention is to illustrate the basic principles and procedures of groundwater quality assessment and monitoring, as required by the EC Directives and described in the preceding chapters. Part 6 then gives an overview of groundwater measurements aspects. The book concludes by providing an insight into stakeholder's involvement in teaching, networking and communication features.

This book is firmly based in sound science, richly illustrated and practically oriented. It will be of considerable interest and direct relevance to all those in the EC and beyond confronted with the challenge of designing and operating programmes of groundwater quality evaluation and pollution protection.

Prof. Dr Stephen Foster[1]
February 2009

---

[1] President of International Association of Hydrogeologists (IAH) 2004-08
Director of World Bank-Groundwater Management Advisory Team (GW-MATE)
Visiting Professor of Groundwater Science, University College, London
British Geological Survey, Honorary Research Fellow

Series Preface

Water is a fundamental constituent of life and is essential to a wide range of economic activities. It is also a limited resource, as we are frequently reminded by the tragic effects of drought in certain parts of the world. Even in areas with high precipitation, and in major river basins, over-use and mismanagement of water have created severe constraints on availability. Such problems are widespread and will be made more acute by the accelerating demand on freshwater arising from trends in economic development.

Despite of the fact that water-resource management is essentially a local, river-basin based activity, there are a number of areas of action that are relevant to all or significant parts of the European Union and for which it is advisable to pool efforts for the purpose of understanding relevant phenomena (e.g. pollutions, geochemical studies), developing technical solutions and/or defining management procedures. One of the keys for successful cooperation aimed at studying hydrology, water monitoring, biological activities, etc. is to achieve and ensure good water quality measurements.

Quality measurements are essential to demonstrate the comparability of data obtained worldwide and they form the basis for correct decisions related to management of water resources, monitoring issues, biological quality, etc. Besides the necessary quality control tools developed for various types of physical, chemical and biological measurements, there is a strong need for education and training related to water quality measurements. This need has been recognised by the European Commission which has funded a series of training courses on this topic, covering aspects such as monitoring and measurements of lake recipients, measurements of heavy metals and organic compounds in drinking and surface water, use of biotic indexes, and methods to analyse algae, protozoa and helminths. In addition, series of research and development projects have been or are being developed.

This book series ensures – and will continue to do so – a wide coverage of issues related to water quality measurements, including the topics of the above mentioned courses and the outcome of recent scientific advances. In addition, other aspects related to quality control tools (e.g. certified reference materials for the quality control of water analysis) and monitoring of various types of waters (river, wastewater, groundwater, seawater) are being considered.

Groundwater Monitoring is the ninth of the series; it has been written by policy-makers and scientific experts in issues related to monitoring groundwater as requested by the

Water Framework Directive and its daughter Groundwater Directive. It offers the reader an overview of technical issues related to groundwater quality assessment and monitoring, as well as case studies illustrating them.

Ph. Quevauviller
Series Editor

Preface

Groundwater is sometimes called 'the hidden asset' – awareness of its existence and its importance is not well known and as a consequence the measures which are required to protect and manage it in an environmental sustainable way are either not taken or are taken too late. Where pollution has occurred and measures are taken too late it may take decades, or longer, until the necessary restoration of quality is achieved. This is due to the slow movement of groundwater (and pollutants) through the ground and the very long residence times. Groundwater is the most abundant source of readily available freshwater in the world making up 97% of all freshwater (excluding glaciers and polar caps). In early times it was thought that the soils and rocks overlying groundwater bodies would provide sufficient protection to groundwater. However, groundwater monitoring, scientific research and investigation have shown that pollutants can penetrate the soil and the unsaturated zone and enter groundwater.

Groundwater protection is covered by several EU Directives covering agricultural and other pressures, which are operated under a common regulatory umbrella, namely the Water Framework Directive (EC 2000/60/EC) and its associated daughter Groundwater Directive (2006/118/EC). It is also considered in the framework of international conventions. In parallel with the establishment of groundwater-related legislation, efforts have been made to better understand groundwater systems, their relationship to other parts of the water environment and the process that control the fate and transport of pollutants. Hydrogeological systems across the world differ greatly due to the complex geological, environmental and climatic variations. They can be extremely complex to understand and hence, the characterisation and assessment of aquifer systems and groundwater bodies can be a very time-consuming process. Improved monitoring is playing a very important role in this by establishing the evidence base to support groundwater protection and management.

The Water Framework Directive imposes EU Member States to undertake wide-scale monitoring programmes for all waters in order to develop river basin management plans and programmes of measures aiming to achieve 'good status' objectives by 2015. With respect to groundwater, these obligations concern chemical and quantitative status objectives. The directive introduces specific requirements in this context, which are often prone to various interpretations. The policy-making and scientific communities, along with industrial stakeholders and NGOs have recognised this and have worked altogether to develop guidance documents reflecting common understanding in relation to the development of the Groundwater Directive, which paves the way for new groundwater quality assessment for the forthcoming decade. This will generate a wide array of collaborations among R&D and policy communities, training activities, educational materials, etc. This book is all about these on-going features. It is very timely in that it is published while the WFD groundwater monitoring programme is fully operational, anticipating a review of monitoring and assessment methods planned in 2012.

The four editors have been striving to collect state-of-the-art information on ground-water quality assessment monitoring from the international groundwater community, providing further stimulation to the work of all parties involved in the huge challenges on the way to a ensure a sound quality assessment of groundwater.

Philippe Quevauviller, Anne-Marie Fouillac, Johannes Grath and Rob Ward

The Series Editor – Philippe Quevauviller

Philippe Quevauviller began his research activities in 1983 at the University of Bordeaux I, France, studying lake geochemistry. Between 1984 and 1987 he was Associate Researcher at the Portuguese Environment State Secretary where he performed a multidisciplinary study (sedimentology, geomorphology and geochemistry) of the coastal environment of the Galé coastline and of the Sado Estuary, which was the topic of his PhD degree in oceanography gained in 1987 (at the University of Bordeaux I). In 1988, he became Associate Researcher in the framework of a contract between the University of Bordeaux I and the Dutch Ministry for Public Works (Rijskwaterstaat), in which he investigated organotin contamination levels of Dutch coastal environments and waterways. From this research work, he gained another PhD in chemistry at the University of Bordeaux I in 1990. From 1989 to 2002, he worked at the European Commission (DG Research) in Brussels where he managed various Research and Technological Development (RTD) projects in the field of quality assurance and analytical method development for environmental analyses in the framework of the Standards, Measurements and Testing Programme. In 1999, he obtained an HDR (Diplôme d'Habilitation à Diriger des Recherches) in chemistry at the University of Pau, France, from a study of the quality assurance of chemical species' determination in the environment.

In 2002, he left the research world to move to the policy sector at the EC Environment Directorate-General where he developed a new EU Directive on groundwater protection against pollution and chaired European science-policy expert groups on groundwater and chemical monitoring in support of the implementation of the EU Water Framework Directive. Since 2008, he has been at the EC DG Research where he is managing research projects on climate change impacts on the aquatic environment, while ensuring strong links with policy networks.

Philippe Quevauviller has published (as author and co-author) more than 220 scientific and policy publications, 80 reports and 6 books for the European Commission and has acted as an editor and co-editor for 22 special issues of scientific journals and 10 books. Finally, he is Associate Professor at the Free University of Brussels and promoter of Master theses in an international Master on water engineering (IUPWARE programme), and he also teaches integrated water management issues and their links to EU water science and policies to Master students at the Universities of Paris 7, Polytech'Lille and Polytech'Nice (France).

List of Contributors

Alice Aureli UNESCO – IHP, 1 rue Miollis, 75015 Paris, France

Bruno Ballesteros Instituto Geológico y Minero de Espana (IGME), Cirilo Amoros, 42, 46004 Valencia, Spain

Eduard Batista Fundación Centro Internacional de Hidrología Subterránea, Provença, 102, 08029 Barcelona, Spain

G. Berthold Hessian Agency for Environment and Geology (HLUG), Rheingaustraße 186, D-65203 Wiesbaden, Germany

Ariane Blum Bureau de Recherches Géologiques et Minières, 3 avenue Claude Guillemin, 45060 Orléans cédex 2, France

Agnès Brenot Bureau de Recherches Géologiques et Minières, 3 avenue Claude Guillemin, 45060 Orléans cédex 2, France

Hans Peter Broers The Netherlands Organisation for Applied Scientific Research (TNO), Built Environment and Geosciences, Princetonlaan 6, P.O. Box 80015, 3508 TA Utrecht, The Netherlands

Enrique Chacon Universidad Politécnica de Madrid, Ríos Rosas, 23, 28003 Madrid, Spain

John Chilton International Association of Hydrogeologists, P. O. Box 4130, Goring on Thames, Reading RG8 6BJ, UK

Patrice Christmann EuroGeoSurveys, 3, rue du Luxembourg, 1000-Brussels, Belgium

Marleen Coetsiers Laboratory for Applied Geology and Hydrogeology, Ghent University, Krijgslaan 281-S8, 9000 Gent, Belgium

Teresa Condesso de Melo Universidade de Aveiro, Departamento de Geociencias, 3810-193 Aveiro, Portugal

Catherine Coxon Geology Department, School of Natural Sciences, Trinity College Dublin, Dublin 2, Ireland

Matt Craig Environmental Protection Agency, Regional Inspectorate Dublin, McCumiskey House, Richview, Clonskeagh Road, Dublin 14, Ireland

Emilio Custodio   Technical University of Catalonia, Dept. of Geotechnics, Gran
                           Capità, s/n Ed. D-2, 08034 Barcelona, Spain

Mette Dahl   Geological Survey of Denmark and Greenland, GEUS, Øster
                           Voldgade 10, 1350 Copenhagen K, Denmark

Donal Daly   Environmental Protection Agency, Richview, Clonskeagh, Dublin
                           14, Ireland

Domenicantonio Di   Tevere River Basin Authority, Via Bachelet, 12, 00185 Roma,
                           Italy

Fiona Dunne   Ecological Consultant, 20 Mount Symon Crescent, Clonsilla,
                           Dublin 15, Ireland

W. Mike Edmunds   Oxford Centre for Water Research, Oxford University Centre for
                           the Environment, South Parks Road, Oxford OX1 3QY, UK

Steve Fletcher   Numphra Consultancy, Higher Numphra, Numphra, St Just, Pen-
                           zance, Cornwall TR19 7RP, UK

Stephen Foster   c/o IAH, P O Box 9, Kenilworth (Warks) CV8-1 JG, UK

Anne-Marie Fouillac   BRGM, Service Métrologie, Monitoring, Analyse, 3 avenue
                           Claude Guillemin – BP 6009 45060 ORLEANS Cedex 2,
                           France

H.-G. Fritsche   Hessian Agency for Environment and Geology (HLUG), Rhein-
                           gaustraße 186, D-65203 Wiesbaden, Germany

Jan Gerritse   Deltares, Subsurface and Groundwater Systems, Geosciences labo-
                           ratories, Princetonlaan 6, P.O. Box 85467, 3508 AL Utrecht, The
                           Netherlands

Johannes Grath   Umweltbundesamt GmbH, Spittelauer Laende 5, 1090 Wien,
                           Austria

Jasper Griffioen   The Netherlands Organisation for Applied Scientific Research
                           (TNO), Built Environment and Geosciences, Princetonlaan 6, P.O.
                           Box 80015, 3508 TA Utrecht, The Netherlands

Juan Grima   Instituto Geológico y Minero de Espana (IGME), Cirilo Amoros,
                           42, 46004 Valencia, Spain

Mark Grout   Environment Agency – England and Wales, Kingfisher House,
                           Peterborough PE2 5ZR, UK

Antoni Gurgui   Departament d'Innovació, Universitats i Empresa, Generali-
                           tat de Catalunya, Passeig de Gràcia, 105, 08008 Barcelona,
                           Spain

Klaus Hinsby   Geological Survey of Denmark and Greenland, GEUS, Øster
                           Voldgade 10, 1350 Copenhagen K, Denmark

Natalya Hunter-Williams   Geological Survey of Ireland, Beggars Bush, Haddington Road,
                           Dublin 4, Ireland.

Paul Johnston   Civil, Structural and Environmental Engineering Department,
                           Trinity College Dublin, Dublin 2, Ireland

Lisbeth Jørgensen	Geological Survey of Denmark and Greenland, GEUS, Øster Voldgade 10, 1350 Copenhagen K, Denmark
Garrett Kilroy	Geology Department, School of Natural Sciences, Trinity College Dublin, Dublin 2, Ireland
Wolfram Kloppmann	Bureau de Recherches Géologiques et Minières, 3 avenue Claude Guillemin, 45060 Orléans cédex 2, France
Neno Kukuric	IGRAC, TNO Princetonlaan 6, PO Box 80015, The Netherlands
Ralf Kunkel	Research Centre Juelich, Institute of Chemistry and Dynamics of the Geosphere (ICG), Institute IV: Agrosphere, Juelich, Germany
Alette Langenhoff	Deltares, Subsurface and Groundwater Systems, Geosciences laboratories, Princetonlaan 6, P.O. Box 85467, 3508 AL Utrecht, The Netherlands
Hélène Legrand	Ministère de l'Ecologie, du Développement et de l'Aménagement durables, Direction de l'Eau – PREA, 20 avenue de Ségur, 75302 PARIS 07 SP, France
Tim Lewis	Entec UK Ltd, Canon Court, Abbey Lawn, Abbey Foregate, Shrewsbury SY2 5DE, UK
Ramon Llamas	Complutense University of Madrid, Dept. of External Geodynamics, 28040 Madrid, Spain
Marisol Manzano	Technical University of Cartagena, Dep. of Mining, Geological and Topographical Eng., P° de Alfonso XIII, 52; E-30203 Cartagena, Spain
Carlos Mediavilla	Geological Institute of Spain, Plaza de España, Torre N, 41013 Sevilla, Spain
Juan Angel Mejia	Instituto de Ecología de Guanajuato, Monte de las Cruces 101 col el monte infonavit 3 Salamanca, Guanajuato, México
Henning Moe	CDM Ireland Ltd, O'Connell Bridge House, 5th Floor, D'Olier Street, Dublin 2, Ireland
Carlos Montes	Autonomous University of Madrid, Dept. of Ecology, 28049 Tres Cantos, Madrid, Spain
Simon Neale	Environment Agency Wales, Head Office, Cambria House, 29 Newport Road, Cardiff, CF24 0TP, Wales
Philippe Negrel	Bureau de Recherches Géologiques et Minières, 3 avenue Claude Guillemin, 45060 Orléans cédex 2, France
Josep Mª Niñerola	Provença, 204–208, 08036 Barcelona, Spain
Áine O'connor	National Parks & Wildlife Service, Department of Environment, Heritage & Local Government, 7 Ely Place, Dublin 2, Ireland
Hélène Pauwels	Bureau de Recherches Géologiques et Minières, 3 avenue Claude Guillemin, 45060 Orléans cédex 2, France
Emanuelle Petelet-giraud	Bureau de Recherches Géologiques et Minières, 3 avenue Claude Guillemin, 45060 Orléans cédex 2, France

Enric Queralt   Comunitat d'Usuaris d'Aigua Subterrània del Baix Llobregat, Avda.
                               Verge de Montserrat, 133, 08820 Prat de Llobregat, Barcelona,
                               Spain

Philippe Quevauviller   Vrije Universiteit Brussel (VUB), Dept. Water Engineering, Bd. du
                               Triomphe, 1060 Brussels, Belgium

Ramiro Rodríguez   Instituto de Geofìsica UNAM, Circuito Institutos Delegación
                               Coyoacan, CU Mexico DF

Stéphane Roy   BRGM, Orleans cedex, France

Manuela Ruisi   Tevere River Basin Authority, Via Bachelet, 12, 00185 Roma, Italy

Jim Ryan   National Parks & Wildlife Service, Department of Environment,
                               Heritage & Local Government, 7 Ely Place, Dublin 2, Ireland

Andres Sahuquillo   Technical University of Valencia, Dept. of Hydraulics, Camino de
                               Vera, s/n., 46071 Valencia, Spain

Andreas Scheidleder   Umweltbundesamt GmbH, Spittelauer Laende 5, 1090 Wien,
                               Austria

Sue Shaw   University of Sheffield, Wetland Research Group, Department of
                               Animal & Plant Sciences, Sheffield S10 2TN, UK

Helen Simcox   Scotland & Northern Ireland Forum for Environmental Research
                               (SNIFFER), Greenside House, 25 Greenside Place, Edinburgh EH1
                               3AA, Scotland

Raya Marina Stephan   UNESCO – IHP, 1 rue Miollis, 75015 Paris, France

M.E. Stuart   British Geological Survey, Maclean Building, Crowmarsh Gifford,
                               Wallingford OX 10 8BB, UK

Kathryn Tanner   Environment Agency – England and Wales, Lutra House, Dodd
                               Way, Walton Summit, Bamber Bridge, Preston PR5 8BX, UK

Cath Tomlin   Environment Agency – England and Wales, Government Buildings,
                               Westbury-on-Trym, Bristol BS20 7FP, UK

Paolo Traversa   Tevere River Basin Authority, Via Bachelet, 12, 00185 Roma, Italy

Bas van der Grift   Deltares, Subsurface and Groundwater Systems, Geosciences labo-
                               ratories, Princetonlaan 6, P.O. Box 85467, 3508 AL Utrecht, The
                               Netherlands

Pieter Jan Van Helvoort   Triqua B.V., Vadaring 7, P.O. Box 132, 6700 AC Wageningen, The
                               Netherlands

Ate Visser   Faculty of Geosciences, Utrecht University, P.O. box 80115, 3508
                               TC Utrecht, The Netherlands

Rob Ward   Environment Agency – England and Wales, Olton Court, Solihull,
                               West Midlands B927HX, UK

Kristine Walraevens   Laboratory for Applied Geology and Hydrogeology, Ghent University, Krijgslaan 281-S8, 9000 Gent, Belgium

Frank Wendland   Research Centre Juelich, Institute of Chemistry and Dynamics of the Geosphere (ICG), Institute IV: Agrosphere, 52425 Jülich, Germany

Bryan Wheeler   University of Sheffield, Wetland Research Group, Department of Animal & Plant Sciences, Sheffield S10 2TN, UK

Mark Whiteman   Environment Agency – England and Wales, Rivers House, 21, Park Square South, Leeds LS1 2QG, UK

Part 1
Groundwater Monitoring in the Regulatory and International Context

1.1

General Introduction: Objectives of Groundwater Assessment and Monitoring

Johannes Grath[1], Rob Ward[2], Andreas Scheidleder[1] and Philippe Quevauviller[3]

[1] *Umweltbundesamt GmbH, Wien, Austria*
[2] *Environment Agency – England and Wales, Olton Court, United Kingdom*
[3] *Vrije Universiteit Brussel (VUB), Department of Water Engineering, Brussels, Belgium*

1.1.1   INTRODUCTION

Groundwater is sometimes called 'the hidden asset'. Awareness of its existence and its importance is not well known and as a consequence the measures which are required to protect and manage it in an environmental sustainable way are either not taken or are taken too late. Where pollution has occurred and measures are taken too late it may take decades, or longer, until the necessary restoration of quality is achieved. This is due to the slow movement of groundwater (and pollutants) through the ground and the very long residence times.

Groundwater Monitoring Edited by Philippe Quevauviller, Anne-Marie Fouillac, Johannes Grath and Rob Ward
© 2009 John Wiley & Sons, Ltd

Groundwater is the most abundant source of readily available freshwater in the world making up 97% of all freshwater (excluding glaciers and polar caps). In early times it was thought that the soils and rocks overlying groundwater bodies would provide sufficient protection to groundwater. However groundwater monitoring, scientific research and investigation have shown that pollutants can penetrate the soil and the unsaturated zone and enter groundwater.

The recognition that groundwater is vulnerable to pollution led to further investigation and research and the subsequent development of groundwater protection policies and strategies. At the European level the Groundwater Directive 80/68/EEC (European Commission, 1980) can be seen as the first formal step towards groundwater protection. This was followed by a Ministerial Declaration, made in The Hague in 1991, that recognised the need for long-term strategic action to protect groundwater quality and quantity.

Groundwater protection is now covered by several EU Directives. These include the Nitrates Directive (European Commission, 1991a), the Landfill Directive (European Commission, 1999), the Plant Protection Products Directive (European Commission, 1991b) and more recently the so-called Water Framework Directive (WFD) (EC 2000/60) (European Commission, 2000) and its associated daughter directive – the new Groundwater Directive (GWD) (2006/118/EC) (European Commission, 2006a).

In parallel with the establishment of groundwater related legislation, efforts have been made to better understand groundwater systems, their relationship to other parts of the water environment and the process that controls the fate and transport of pollutants. Hydrogeological systems across Europe differ greatly due to the complex geological, environmental and climatic variation that exists across Europe. They can be extremely complex to understand and hence, the characterisation and assessment of aquifer systems and groundwater bodies can be a very time-consuming process. Improved monitoring is playing a very important role in this by establishing the evidence base to support groundwater protection and management.

1.1.2   THE ROLE OF GROUNDWATER

Groundwater fulfils many different functions across Europe. The best known and probably the most important in many countries is as a resource for drinking water purposes. It is also widely used for irrigation, food production and industrial purposes. The reliance on groundwater in Europe as a source of drinking water is illustrated in Figure 1.1.1.

As well as use to support human activity, groundwater is also vital for supporting and even enabling ecosystem functions. It is well known that groundwater supports surface water flows and their dependent aquatic and terrestrial ecosystems, but it is less well known exactly how these interactions and processes operate. Further it is even less well known what role groundwater ecosystems play and how significant it is.

The WFD (European Commission, 2000) and GWD (European Commission, 2006a) recognise that groundwater supports ecosystems and contributes in many places to the achievement of surface water ecological objectives. As a result they require Member States to take account of this when assessing groundwater.

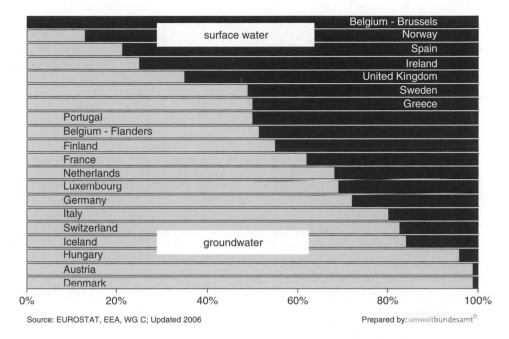

Figure 1.1.1 Share of ground and surface water in the public water supply of Europe.

An assessment performed by all EU Member States in 2004 as part of WFD implementation (Art. 5) provided evidence that approximately 20% of the 7000 groundwater bodies across Europe have associated groundwater dependent terrestrial ecosystems.

1.1.3 GROUNDWATER PROTECTION NEEDS AND OBJECTIVES

The variety of roles that groundwater plays across Europe, from supporting the environment to human activity, can lead to conflicting pressures and priorities and resulting socio-political tensions. In order to address this, the European Commission and Member states of the European Union have recognised that a clear strategy for groundwater protection is vital.

Within the Ministerial Declaration made in the Hague in 1991, the vital importance of groundwater for human health and for safeguarding ecosystems was underlined. Amongst other things, the competent authorities and other groups involved were requested to contribute to the conservation of groundwater as a natural resource in the areas under their control (Müller and Fouillac, 2008).

In order to enact the declaration, it was necessary to understand the protection needs of groundwater *and* the needs of groundwater dependent receptors. The outcome has been a legal framework that is laid down in the WFD (European Commission, 2000) and GWD (European Commission, 2006a) that aims to protect groundwater and the wider water environment. Article 4 of the WFD defines a set of environmental objectives for groundwater (and surface water) which are to be achieved by 2015. The principal goal is to achieve 'good status' for all groundwater (GW) bodies and ensure that there is no future deterioration in status. The definition for good groundwater chemical status is that (WFD Annex V. 2.3.2):

> The chemical composition of the groundwater body is such that the concentration of pollutants do not exhibit the effects of saline or other intrusions (as determined by changes in conductivity) into the groundwater body, do not exceed the quality standards applicable under other relevant Community legislation in accordance with Article 17 of the WFD, and are not such as would result in a failure to achieve the WFD environmental objectives for associated surface waters nor any significant diminution of the ecological or chemical quality of such bodies nor in any significant damage to terrestrial ecosystems which depend directly on the groundwater body.

The other objectives for groundwater set out in the WFD are to: 'prevent or limit' the inputs of pollutants and take measures to reverse significant and sustained upward trends in pollutant concentrations.

Further particular provisions concerning chemical status and the protection of groundwater are set out in the GWD (Art. 4). These include:

- criteria for the assessment of good chemical status and the establishment of environmental quality standards and threshold values for pollutants that are putting groundwater bodies at risk of not meeting their environmental objectives, and

- criteria for the identification of sustained upward trends in pollutant concentration and requirement for trend reversal.

One particular challenge for groundwater protection is the fact that groundwater systems extend across national boundaries. As a result for these transboundary GW-bodies specific regulations are set out in the WFD (Art. 3 and Annex V) and GWD (Art. 3.3 and 3.4). About 110 national and international river basins have been identified within the European Union of which 40 extend across international borders. In each of the 40 cross-border basins bilateral agreements between most countries have established or strengthened cooperation. In six of them – the Danube, Elbe, Meuse, Oder, Rhine and Scheldt – an international cooperation based on multilateral agreements and international commissions is in place. The river basin within Europe that involves the largest number of countries is the Danube River Basin. This extends over 19 countries, both EU Member States and non-Member States, covers an area of about 800,000 km^2 and contains a population of about 81 million people. The International Commission for the Protection of the Danube River (ICPDR) coordinates the activities within this river basin. Groundwater bodies which were identified as being of basin wide importance are dealt with at ICPDR level and the ICPDR supports the harmonisation of WFD and GWD activities including:

- bilateral agreements on approaches and principles (e.g. sampling procedures, network design);

- coordination of conceptual model development;

- revision of risk assessment;

- derivation of groundwater threshold values;

- status and trend compliance assessment;

- monitoring of dependent terrestrial ecosystems and respective assessment criteria;

- exchange of data; and

- QA and QC aspects.

Other international conventions – e.g. the UNECE Convention on the Protection and Use of Transboundary Watercourses and International Lakes (UNECE, 1992) – are addressing harmonisation issues. Harmonisation of the conceptual model of a transboundary GW-body and agreements defining the information exchange are of vital importance. Under the UNECE Convention guidelines have been developed to support the management and monitoring of transboundary GW-bodies.

1.1.4 GROUNDWATER CHEMICAL MONITORING

An essential component in developing and implementing appropriate policy and management strategies for groundwater is having good evidence. This evidence can be obtained from a variety of sources but fundamental to the evidence base is having effective groundwater monitoring.

Groundwater chemical monitoring is needed for various purposes:

- to describe the chemical composition of groundwater;

- to identify the presence and spatial distribution of pollutants;

- to identify trends in pollutant concentrations and natural substances over time; and

- to measure the effectiveness of measures including 'prevent or limit' measures.

Although monitoring can be performed at different scales, all relevant monitoring activities should be considered as complementary and are needed to provide the evidence for characterising all elements of the 'source–pathway–receptor' relationship. At one end monitoring supports the implementation of the 'prevent or limit' obligations by providing the information needed to identify the baseline conditions and then monitor for the impacts of any authorised activities. The original Groundwater Directive 80/68/EEC (European Commission, 1980) addressed the need to 'prevent or limit' input of pollutants and established a role for monitoring in this process (Art. 8, 9 and 10). Although this directive is going to be repealed by 2013, the WFD and the new Groundwater Directive (2006/118/EC) (Art. 6) (European Commission, 2006a) provide comparable provisions. Other European Directives like the Landfill Directive (European Commission, 1999) and the Nitrates Directive (European Commission, 1991a) also require monitoring of a similar nature to support their implementation.

For the WFD the monitoring requirements are focused more on identifying the impacts that potentially affect the objectives at the GW-body scale (Art. 8 and Annex II). The monitoring objective, according to these requirements, is to provide a coherent overview of groundwater chemical status at national and international level. More specifically the monitoring must be able to:

- establish the chemical and quantitative status of groundwater bodies (including an assessment of the available groundwater resource);

- assist in further characterisation of groundwater bodies;

- validate the risk assessments carried out under Article 5;

- estimate the direction and rate of flow in groundwater bodies that cross Member States' boundaries;

- assist in the design of programmes of measures;

- evaluate the effectiveness of programmes of measures;

- demonstrate compliance with drinking water protected areas and other protected area objectives;

- characterise the natural quality of groundwater including natural trends (baseline); and

- identify anthropogenically induced trends in pollutant concentrations and their reversal.

To support the implementation and operation of WFD monitoring programmes, a guidance document has been developed (European Commission, 2006b). This aims to establish consistent and comparable monitoring across the EU.

1.1.5   GROUNDWATER QUALITY ASSESSMENT AND CLASSIFICATION

Of the three principle environmental objectives for groundwater set out in the WFD and GWD (see Section 1.1.3 and Figure 1.1.2) two are new – status and trends. There has previously been no European-wide requirement to classify groundwater bodies or to identify pollutant trends. The third objective – prevent or limit inputs of pollutants – is not new. A similar requirement is already in place through the Groundwater Directive (80/68/EEC).

The objectives for groundwater complement each other and provide the basis for effective groundwater protection and management. For example measures to prevent or limit aim to protect groundwater from current and future pressures and impacts at the local scale, trend reversal addresses existing impacts and status assessment provides the basis for establishing the specific targets required for meeting all the objectives.

Achieving good status means that groundwater quality must meet a number of conditions that are defined in the WFD and GWD. The parameters that must be used in this

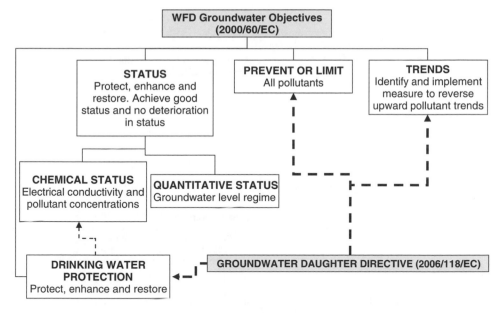

Figure 1.1.2   Groundwater Objectives – WFD and GWD.

assessment are electrical conductivity and concentrations of pollutants. In order to assess status a system needs to be established that can apply to all groundwater bodies. This system needs to be able to allow environmental standards for groundwater to be derived and applied to determine the status of a groundwater body and how much deviation there is from good status if a body is at poor status. An outline of the process involved is shown in Figure 1.1.3.

Because groundwater systems are complex and natural conditions can lead to elevated concentrations of substances that are also pollutants, e.g. metals, and to increase of other parameters such as the conductivity, environmental standards act as triggers for further investigation. This further investigation determines the conditions for good status and whether they are being met in each groundwater body. Because classification of groundwater must take into account the unique features of each groundwater body being assessed environmental standards may differ between bodies for the same pollutants. The standards, how they have been applied and the actions needed to achieve/maintain good status must all be published in River Basin Management Plans – one of the mandatory deliverables for the WFD.

As well as achieving good status, bodies must not be allowed to deteriorate. Preventing or limiting the inputs of pollutants are the mechanisms by which this can be achieved. If properly implemented, measures that prevent or limit inputs of pollutants will ensure that no future deterioration in quality and hence status occurs and, in due course, mean that upward trends are reversed. The prevent or limit objective in the WFD extends the existing Groundwater Directive (European Commission, 1980) regime by extending the scope of the pollutants include to all those that could potentially lead to one or more environmental objectives not being met. It aims to protect at both the groundwater body

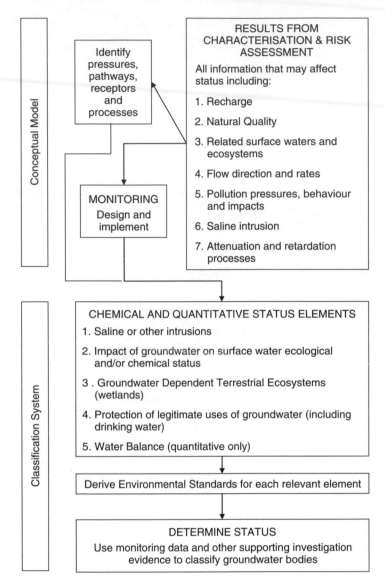

Figure 1.1.3   Groundwater classification process.

scale and local scale to avoid harm to groundwater dependent receptors, e.g. surface waters and wetlands, and to protect groundwater as a water resource.

1.1.6   CONCLUSIONS

The effective protection and management of groundwater require that clear and consistent objectives are established. European (and Member State) legislation now provides these.

The new challenge is how we achieve the objectives and the understanding and methods that need to be developed to support them.

An integrated approach is needed which brings together our understanding of the environment and its needs, the pressures to which it is being exposed and the mechanisms we need to employ to manage the conflicting pressures to achieve a sustainable and protected groundwater resource. This will not be easy but cooperation between and within Member States, and a willingness to deliver environmental improvement, is the key. In time, hopefully, groundwater will no longer be considered as a 'Hidden Asset' but as a 'Treasured Asset'.

REFERENCES

European Commission, 1980. Council Directive 80/68/EEC of 17. December 1979 on the protection of groundwater against pollution, *Official Journal of the European Communities* L 20, 26.1.1980, p. 43.

European Commission, 1991a. Council Directive 91/676/EC of 12 December 1991 concerning the protection of waters against pollution caused by nitrates from agricultural sources, *Official Journal of the European Communities* L 375, 31.12.1991, p. 1.

European Commission, 1991b. Council Directive of 15 July 1991 concerning the placing of plant protection products on the market, *Official Journal of the European Communities* L230, 19.8.1991, p. 1.

European Commission, 1992. Council Resolution of 25 February 1992 on the future Community groundwater policy (92/C 59/02), *Official Journal of the European Communities* No C 59/2.

European Commission, 1999. Council Directive 99/31/EC of 26 April 1999 on the landfill of waste, *Official Journal of the European Communities* L 182, 16.7.1999, p. 1.

European Commission, 2000. Directive 2000/60/EC of the European Parliament and of the Council of 23 October 2000 establishing a framework for Community action in the field of water policy, *Official Journal of the European Communities*, L327, 22.12.2000, p. 1.

European Commission, 2006a. Directive 2006/118/EC of the European Parliament and of the Council of 12 December 2006 on the protection of groundwater against pollution and deterioration, *Official Journal of the European Communities* L372, 27.12.2006, p. 19.

European Commission, 2006b. *Guidance on Groundwater Monitoring. Common Implementation Strategy for the Water Framework Directive.* Guidance Document No. 15. Technical Report – 002 – 2007.

Müller D. and Fouillac A.M., 2008. *Methodology for the Establishment of Groundwater Environmental Quality Standards Groundwater Science and Policy, An International Overview*, RSC Publishing, Cambridge, UK pp. 535–544.

UNECE, 1992. Convention on the Protection and Use of Transboundary Watercourses and International Lakes. www.unece.org/env/water.

1.2

Groundwater Monitoring in International Conventions and Agreements

Raya Marina Stephan

UNESCO – IHP, Paris, France

1.2.1 INTRODUCTION

About two-thirds of the world's population depends for its water supply on groundwater resources (Jousma and Roelofsen, 2004). With a gradually increasing population many of the available groundwater systems in the world are ever more under stress of exploitation and contamination. With these growing problems, the awareness of the need for

---

Groundwater Monitoring Edited by Philippe Quevauviller, Anne-Marie Fouillac, Johannes Grath and Rob Ward
© 2009 John Wiley & Sons, Ltd

sustainable management of the groundwater resources has also increased. Moreover, it has been recognised that successful management needs to be based on sufficient, sound and regular data regarding groundwater resources and their environment, and the stresses upon these systems.

There are still very few conventions and agreements regarding groundwater and transboundary aquifers. The trend seems to be changing, but the move is still very slow. Groundwater monitoring is considered in conventions and agreements of different kinds such as framework conventions and/or agreements on a specific basin and also in very specific legislation.

1.2.2   MONITORING OBLIGATION IN FRAMEWORK INSTRUMENTS

Two main framework instruments include provisions on monitoring. The first one is the UN Economic Commission for Europe Convention on the Protection and Use of Transboundary Watercourses and International Lakes (1992) (UNECE Water Convention).[1] This Convention is a well-established regional instrument, that applies to the 56 Member States of the ECE. These include the countries of Europe, but also countries in North America (Canada and United States), Central Asia (Kazakhstan, Kyrgyzstan, Tajikistan, Turkmenistan and Uzbekistan) and Western Asia (Israel). The second instrument was at the stage of draft articles until recently. At its 60th session (5 May–6 June and 7 July–8 August 2008) the UN International Law Commission (ILC) adopted the draft articles on the law of transboundary aquifers at second reading,[2] meaning that the ILC finished its task and that the draft articles are deferred to the UN General Assembly (GA). The draft articles are now annexed to a UN General Assembly (GA) Resolution A/RES/63/124[3] adopted on 11 December 2008 which recommends to States to make appropriate bilateral or regional arrangements for the proper management of their transboundary aquifers, taking into account the provisions of these draft articles. Both texts include provisions on monitoring. In the draft articles this provision is specific for transboundary aquifers, while in the UNECE Water Convention the provision addresses transboundary waters in general. In the UNECE Water Convention, monitoring is addressed with more details, and working groups were set up on this issue and came out with guidelines.

1.2.2.1   Monitoring in the Draft Articles on the Law of Transboundary Aquifers

As defined in article 1 on Scope, the draft articles apply to:

(a)   utilisation of transboundary aquifers or aquifer systems;

---

[1] Available at http://www.unece.org/env/water/text/text.htm.

[2] http://untreaty.un.org/ilc/reports/2008/2008report.htm.

[3] The Resolution is available on http://www.un.org/ga/63/resolutions.shtml. The Resolution also expresses the decision of the UN GA to include in the provisional agenda of its sixty-sixth session (2011) an item entitled 'The law of transboundary aquifers' with a view to examining, inter alia, the question of the form that might be given to the draft articles (Convention, Protocol, or other).

(b) other activities that have or are likely to have an impact upon such aquifers or aquifer systems; and

(c) measures for the protection, preservation and management of such aquifers or aquifer systems.

Draft article 13 relates to monitoring. In paragraph 1, the article sets an obligation for States to monitor their transboundary aquifers or aquifer systems, and whenever possible this activity should be carried out jointly with the other State concerned, and if required with the assistance of an international organisation. If this is not the case, then article 13 provides that the Aquifer States (or the States sharing a transboundary aquifer) shall exchange the monitored data among themselves. Paragraph 1 sets forth the general obligation to monitor and the sequence of such monitoring activities whether jointly or individually. The draft article adapts to reality and acknowledges that in practice, monitoring is usually initiated individually by a State, and also in many cases by a local government, and develops eventually later into a joint effort with the neighbouring States concerned by the transboundary aquifers. However, the ultimate and ideal monitoring is the joint monitoring based on an agreed conceptual model of the aquifer. Where it is not feasible to monitor jointly, it is important that aquifer States share data on their monitoring activities. In the spirit of this article, States are under an obligation to conduct individual monitoring and share the results with the other aquifer States concerned. This last obligation appears as a minimum requirement under the provision on monitoring.

In its second paragraph, article 13 gives some details about the monitoring activities. It provides that the Aquifer States shall agree or harmonise the standards and methodology they will use for monitoring their transboundary aquifers or aquifer system. It is important that aquifer States agree on the standards and methodology to be used for monitoring or on means to have their different standards or methodology harmonised. Without such agreement or harmonisation, collected data would not be useful for the other State which need to understand the data; in other words a State needs to know the methodology or the metadata in order to be able to understand the data. Article 13§2 provides also that States shall agree on a conceptual model, and on its basis they shall identify key parameters that they will monitor. The paragraph indicates that these parameters should include parameters on the condition of the aquifer or aquifer system such as those listed in draft article 8 paragraph 1 regulating the 'Regular exchange of data and information'. These parameters concern information on the geological, hydrogeological, hydrological, meteorological and ecological nature and related to the hydrochemistry of the aquifers or aquifer systems. Article 13 paragraph 2 adds the requirement of parameters on the utilisation of the aquifers and aquifers system.

As mentioned earlier, draft article 13 adapts to the reality. Joint monitoring is an ideal to be reached. However article 13 acknowledges the difficulties of reaching it. When they are already monitoring the part of the transboundary aquifer on their territory, it can be very difficult for Aquifer States to jointly monitor, or even to harmonize the monitoring, as this sometimes can mean changes in the whole system. A practical solution is the exchange of data, accompanied by the exchange of metadata.

1.2.2.2   Monitoring under the UNECE Water Convention

The Convention addresses transboundary waters in general, and its provisions on monitoring do not include any specificities for groundwater. However, under the Convention, a working group on monitoring and assessment was set up, and a series of guidelines specially dedicated to transboundary groundwaters were published.

Article 4 of the Convention puts in very simple words the general obligation of monitoring, without mentioning if this activity should be joint or not by saying: 'The Parties shall establish programmes for monitoring the conditions of transboundary waters.' Under article 9 on 'Bilateral and Multilateral cooperation', paragraph 2b specifies that one of the tasks of a joint body or mechanism over transboundary waters is 'to elaborate joint monitoring programmes concerning water quality and quantity'. The Convention includes an article dedicated specifically to 'Joint monitoring and assessment' (article 11). This article sets the obligation for the Riparian Parties to establish and implement joint programmes for monitoring the conditions of transboundary waters (paragraph 1). It also requires from the Riparian Parties to 'agree upon pollution parameters and pollutants whose discharges and concentration in transboundary waters shall be regularly monitored' (paragraph 2). And finally it also provides that 'the Riparian Parties shall harmonize rules for the setting up and operation of monitoring programmes, measurement systems, devices, analytical techniques, data processing and evaluation procedures, and methods for the registration of pollutants discharged' (paragraph 4). Article 13 on the Exchange of information between riparian parties establishes for the Riparian Parties the obligation 'to exchange reasonably available data' on monitoring (inter alia).

In its provisions on monitoring, the UNECE Water Convention puts a heavier burden on States than the draft articles as it requires that this activity be undertaken jointly. Under the Water Convention, joint monitoring is an obligation whereas in the draft articles Aquifer States are required to monitor jointly 'wherever possible' (article 13§1). Regarding the parameters to monitor they are broader in the draft articles while in the Water Convention they are limited to the pollution parameters.

Under the Water Convention, a protocol on water and health was adopted in 1999,[4] and entered into force into force on 4 August 2005, after reaching the required number of ratifications. Today 21 countries have ratified this protocol.

According to article 1

> The objective of this protocol is to promote at all appropriate levels, nationally as well as in transboundary and international contexts, the protection of human health and well-being, both individual and collective, within a framework of sustainable development, through improving water management, including the protection of water ecosystems, and through preventing, controlling and reducing water-related disease.

Rather unusually in international law the protocol addresses not only transboundary water resources but also purely domestic ones. The protocol includes some requirements on monitoring such as:

- effective systems for monitoring situations likely to result in outbreaks or incidents of water-related disease (article 4§2e);

---

[4] Available at http://www.unece.org/env/documents/2000/wat/mp.wat.2000.1.e.pdf.

- establishing and maintaining legal and institutional framework for monitoring (article 6§5c);

- achievement of quality assurance (article 14§i).

1.2.3 MONITORING OBLIGATION IN SPECIFIC LEGISLATION

Under this section the obligation of monitoring will be reviewed in two specific texts that are not international agreements or conventions: the European Union Water Framework Directive 'Directive 2000/60/EC of the European Parliament and of the Council establishing a framework for the Community action in the field of water policy' or, in short, the European Union Water Framework Directive (or even shorter the WFD); and the 'United States-Mexico Transboundary Aquifer Assessment Act' adopted by the US Congress on 3 January 2006. These two instruments were included in the present chapter; the first one applies to 27 Member States, and the second one applies to transboundary aquifers on the US Mexico border.

1.2.3.1 The EU Water Framework Directive

This text does not represent an international convention or agreement. It is not part of international law. The EU WFD is part of community law or European law which presents some similarities with international law. European law gives rights and imposes obligations on Member States, but also sometimes on European citizens directly. Member States are primarily responsible for implementing these rules and properly applying them. In this way, European law and hence the rules establishing the European Union have become an integral part of the legal system of the Member States. European law, like international law, is superior to domestic law. And as for international law, the rules of domestic legislation must not contradict its provisions. For these similarities with international law, and because it is a very innovative instrument in water law concerning 27 Member States, the EU WFD is included in this chapter and its important provisions regarding the monitoring of groundwater will be studied. The WFD applies to the protection of all waters including namely inland surface waters, transitional waters, coastal waters and groundwater (article 1), with the objective of reaching 'good status' in 2015 (article 4). In article 8, the EU WFD sets that Member States shall ensure the development of programmes for the monitoring of water status in order to establish a coherent and comprehensive overview of water status within each river basin district (§1). The WFD establishes the river basin district, as the management unit for all waters (article 3). Some EU Member States had already adopted this approach; the new WFD spreads it to all Member States (Stephan, 2008). The WFD defines the river basin district as the area of the whole river catchment with their associated groundwaters and coastal waters (article 2§15). The river basin disctrict can be an international one either among EU Member States, or extending beyond the territory of the Community. In both cases, the WFD requires from States to cooperate; however, when a non-EU Member is concerned, the implementation of the provisions of the Directive is more difficult (Stephan, 2008). These considerations apply for the requirements of monitoring.

For groundwaters, monitoring programmes are defined to cover the chemical and quantitative status (article 8§1), which are the two aspects considered for evaluating the groundwater status (article 2§19 and 20). Monitoring programmes have to be operational in 2006 (article 8§2). According to article 8§3, technical specifications and standardised methods for analysis and monitoring of water status are to be adopted, and if necessary the Commission may issue guidelines (article 20). In view of supporting the implementation of the Directive, the EU Member States, Norway and the European Commission have jointly developed the Common Implementation Strategy (CIS) with the main aim of allowing a coherent and harmonious implementation of the Directive. Under the CIS, a document was produced as a guidance for monitoring under the WFD (CIS, 2003). The report summarises important aspects of groundwater monitoring and includes research and technological developments and examples of practices at the national and regional levels. Annex V of the Directive defines the requirements of monitoring for the quantitative and chemical status. In paragraph 2.2 of Annex V, the WFD sets that the monitoring network has to provide a reliable assessment of the quantitative status of groundwater bodies. The network has to include enough representative monitoring points to allow an estimation of the groundwater level in each groundwater body considering the variations in recharge. A map of the monitoring network has to be included in the river basin management plan, which has to be published in 2009, and then updated every six years (article 13§6 and 7). Member States are requested to provide a map of groundwater quantitative status with the results obtained.

Regarding the monitoring of the chemical status of groundwater, Annex V§2.4 of the WFD requires that the network be designed to provide a coherent and comprehensive overview of groundwater chemical status within each river basin. Monitoring will be of two types: surveillance and operational. The surveillance monitoring programme will be established for each period to which a river basin management plan applies. Its results will be used to establish an operational monitoring programme to be applied for the remaining period of the plan. The WFD indicates a selection of core parameters to be monitored such as oxygen content, pH value, conductivity, nitrate and ammonium. The operational monitoring will be carried out for those groundwater bodies that are identified as being at risk of not meeting the good status objective of the WFD. It will be undertaken at least once a year, for the periods between surveillance monitoring programmes, and at a frequency sufficient to detect the impacts of relevant pressures (Annex V§2.4.3). As for the quantitative status, Member States will prepare a map of groundwater chemical status and include it in the river basin management plan. It is possible for Member States not to separate the maps of both status, quantitative and chemical.

In the EU WFD groundwater monitoring appears as an essential tool in the process of reaching the objective of 'good status'. The Directive installs the obligation for Member States to establish reliable monitoring networks, assessing the essential elements for a continuous evaluation of both aspects defining the status of groundwater, quantitative and chemical status. This obligation is accompanied by the obligation of information, through the publication in the river basin management plan of the map of the monitoring network, and the map of the groundwater status, which is built on the results of monitoring for quantitative and chemical status.

1.2.3.2 The United States-Mexico Transboundary Aquifer Assessment Act

Another specific text is the 'United States-Mexico Transboundary Aquifer Assessment Act'[5] adopted by the US Congress on 3 January 2006. This is also a very special text in the sense that it is a national act with the aim of applying to transboundary aquifers. Its purpose is to direct the Secretary of the Interior to establish a United States–Mexico transboundary aquifer assessment program to systematically assess priority transboundary aquifers. Section 4 defines the objectives of the program as follows:

- to develop and implement an integrated scientific approach to identify and assess priority transboundary aquifers,

- consider the expansion or modification of existing agreements, as appropriate, between the United States Geological Survey, the Participating States, the water resources research institutes, and appropriate authorities in the United States and Mexico,

- and to produce scientific products for each priority transboundary aquifer

The Act already identifies priority aquifers; however, it sets the criteria for the identification of other ones (section 4§c 1 and 2).

Under this Act, monitoring appears as a tool to develop additional data necessary to adequately define aquifer characteristics, and scientifically sound groundwater flow models. The Act mentions the support and expansion of monitoring efforts for achieving these aims. Coordination among and with all entities carrying out monitoring activities with respect to a priority transboundary aquifer is also required (section 5).

1.2.4 MONITORING IN AQUIFER AND RIVER BASIN AGREEMENTS

There are very few agreements on transboundary aquifers; however, groundwater is often included in river basin agreements.

1.2.4.1 Aquifer Agreements

There is only one major and remarkable exception of a treaty dealing exclusively with the management of a transboundary aquifer. It is the Convention on the protection, utilisation, recharge and control of the Franco-Swiss Genevese aquifer, signed between French Communities on one side, and the Canton of Geneva on the other side, and which entered into force on 1 January 2008.[6] This Convention has replaced the Arrangement on the protection, utilisation and recharge of the Franco-Swiss Genevese aquifer (9 June 1977) which

---

[5] Available at http://npl.ly.gov.tw/pdf/5672.pdf.
[6] On file with author.

had been concluded for a period of thirty years. As defined in the preliminary article, the aim of this Convention is to ensure the sustainability of the Franco-Swiss Genevese aquifer, and to guarantee the parties, as much as possible, the capacity of pumping water for drinking water purposes. Like the previous agreement, the new Convention maintains the Commission with the following main responsibilities:

- to define a yearly aquifer utilization programme and to propose all measures for ensuring the protection of the aquifer and to remedy to any pollution;

- to give its technical opinion on all new works or changes in existing ones, and on the withdrawals from the aquifer.

The provisions on monitoring requirements are included in Chapter 4 related to the quantitative and qualitative monitoring of the resource, and in article 10 on the recording and control of withdrawal and levels. According to this article, the control and protection of the resource are organised by each authority on its territory in a concerted manner. The regular readings of the water level and results of the water analysis are transmitted to the Commission, and can be controlled any time at the request of any of the delegations. The readings of the withdrawals in the aquifer will be done by each user and communicated to all the other users and to French and Swiss authorities.

The spirit of dialogue and cooperation has a long history in the management of this transboundary aquifer, and the provisions on monitoring reflect this continuous will. It also reflects the level of trust existing between the authorities of the two countries in the exchange of data, as well as the advance in the process of informing the users.

Another agreement on a transboundary aquifer can be mentioned here, even if it has a more limited scope. It represents nevertheless an example of first cooperation over a transboundary aquifer system (Stephan, 2007). The agreement concerns the Nubian Sandstone Aquifer System (NSAS) shared between Chad, Egypt, Libya and Sudan. In 1992, a Joint Authority was established between Egypt and Libya, and the two other States joined later. The Authority has a wide range of responsibilities such as inter alia collecting and updating data, conducting studies, managing the aquifer on sound scientific bases, rationing the consumption of water if needed and organising scientific workshops in relation to the aquifer (Constitution of the Joint Authority for the Study and Development of the Nubian Sandstone Aquifer Waters, 1992, article 3).[7]

In the frame of the 'Programme for the Development of a Regional Strategy for the Utilization of the Nubian Sandstone Aquifer System' (1998–2001) conducted by CEDARE,[8] the four States two agreements were prepared (5 October 2000) (Burchi and Mechlem, 2005):

- Terms of Reference for the Monitoring and Exchange of Groundwater Information of the Nubian Sandstone Aquifer System;

- Terms of Reference for Monitoring and Data Sharing.

---

[7] On file with author.

[8] Center for Environment and Development in the Arab Region and Europe.

The first agreement relates more to the Regional Information System developed in the frame of the project. In the second agreement, the four riparian countries of the NSAS acknowledge the necessity of maintaining a continuous monitoring of the aquifer in view of its sustainable development and proper management. They also acknowledge the necessity of sharing the monitored parameters of the aquifer among themselves in order to observe its regional behaviour. Therefore the four countries agree to share among themselves the following data:

- Yearly extraction in every extraction site,

- Representative Electrical Conductivity measurements (EC), taken once a year in each extraction site, followed by a complete chemical analysis if drastic changes in salinity is observed.

- Water level measurements taken twice a year.

A map of a proposed regional well monitoring network is annexed to the agreement.

Within the frame of the project, the agreements were signed by the Directors of the water authorities in the four countries, however the agreements did not receive further development nor implementation.

1.2.4.2 River Basin Agreements

While river basin agreements are numerous, not all of them include groundwater in their scope. In this section, some river basin agreements are selected and presented, particularly as they regulate monitoring.

In Europe, river basin agreements were signed referring to the UNECE Water Convention, and more recently to the EU WFD. These two instruments already provide strong provisions for monitoring. The Convention on Cooperation for the Protection and Sustainable Use of the River Danube (Danube Convention) was signed in 1994, and it entered into force on 22 October 1998 (Burchi, 2005), before the adoption of the EU WFD. Therefore it only refers to the UNECE Water Convention. The Convention applies to the catchment area of the Danube river (article 3§1), and it covers groundwater resources as a water resource requiring specific protection measures (article 6). Article 9 relates to monitoring programmes. Under this article the Contracting Parties to the Convention are under the obligation to cooperate in the field of monitoring and assessment. Article 9§1 provides inter alia that the Contracting Parties shall:

- harmonise their monitoring methods;

- develop concerted or joint monitoring systems;

- elaborate and implement a joint program for monitoring the riverine conditions in the Danube.

The exchange of monitoring data is covered under article 12§1 c.

As in the UNECE Water Convention, the Danube Convention stresses the joint character of the monitoring activity.

The Agreement between the Government of the Republic of Hungary and the Government of Romania on cooperation in the field of protection and sustainable use of transboundary waters (2003)[9] refers in its preamble to the three previously mentioned instruments, as the two countries are located in the catchment area of the Danube River. The objective of this agreement is to summarise the rules and obligations, according to which the Contracting Parties proceed in the scope of their cooperation related to transboundary waters, and to define the structural, institutional and economic conditions, upon which they realise their cooperation (article 1). Hungary and Romania are both party to the UNECE Water Convention, to the Danube Convention and are Member States of the EU. The agreement applies to all transboundary waters among the two countries. It establishes the Hungarian-Romanian Joint Commission on Water Issues (article 10). The Commission has many tasks such as:

- regularly evaluating the execution of the provisions of the Agreement;

- establishing environmental aims, ensuring reaching and conserving the good status of surface and groundwater bodies intersected by the boundary;

- coordinating harmonisation of river basin management plans elaborated in accordance with the WFD.

Under this Agreement, Hungary and Romania assure the application of the monitoring provisions prescribed by the WFD for the evaluation of ecological, chemical and quantitative status of surface and groundwaters (article 14§1). As both States are members of the EU, they have to comply with the monitoring obligations of the WFD. These obligations are compatible with the obligations under the UNECE Water Convention which is a framework Convention. It can be expected from this Agreement that both countries cooperate on international river basin districts.

Last, but not least, an example from Europe is the International Agreement on the Schelde River,[10] or Gent Agreement signed in 2002. This Agreement replaces the previous one of 1995. Like the Hungary–Rumania Agreement, the Gent Agreement refers to both the UNECE Water Convention and to the EU WFD. The Agreement covers groundwaters lying in the international Schelde river basin district. The established Commission is responsible for the coordination of monitoring programmes (article 4§3d).

The following examples to be presented come from Africa. The Tripartite Interim Agreement between the Republic of Mozambique, the Republic of South Africa and the Kingdom of Swaziland for Co-operation on the Protection and Sustainable Utilisation of the Water Resources of the Incomati and Maputo Watercourses (Johannesburg, 29 August 2002) (Burchi and Mechlem, 2005) covers groundwaters related to the watercourse. Under this Agreement it is required from the Parties to 'establish comparable monitoring systems, methods and procedures' (article 4§h), 'individually and, where appropriate, jointly'. A list of substances to be monitored has to be established (article 8§1c). In the Convention on the Sustainable Development of Lake Tanganyka (12 June 2003) (Burchi and Mechlem, 2005), groundwater is considered as part of the lake. In article 4 on Cooperation, it is provided that this cooperation includes the exchange of results of the monitoring activities in the lake. This data shall be exchanged through the Secretariat

---

[9] On file with author, draft English translation.

[10] Available at http://www.isc-cie.com/members/docs/documents/4653.pdf.

of the Lake Tanganyka Authority (article 20§1a). The results of monitoring compliance with permits are to be made available to the public (article 19§1b). In the Protocol for Sustainable Development of Lake Victoria Basin (29 November 2003) (Burchi and Mechlem, 2005), groundwaters are included in the Lake Victoria Basin, as in the case of Lake Tanganika. The Partner States are required to monitor within their respective territories, and to adopt standardised equipment and methods of monitoring (articles 16 and 25).

Under these agreements the obligation of monitoring is less demanding that in the previous European agreements. Monitoring is required but can be done individually, and the date exchanged. Only in the Agreement on the Incomati and Maputo watercourses, can the monitoring activity be done jointly 'wherever appropriate'. This last provision (article 4) is similar to the provision of the draft articles on the law of transboundary rivers.

1.2.5 CONCLUSIONS

Monitoring as a scientific requirement for the management of water resources, and especially groundwater resources which are out of sight, has entered the arena of international agreements. In the agreements presented above, it appears that the provisions on monitoring have various level of constraints: monitoring can be undertaken individually with the requirement of exchanging the data among the riparian States, or it is required that monitoring be undertaken jointly, which is more difficult to achieve. In the European continent, legal instruments such as UN ECE Water Convention or the EU WFD are the most developed, and their requirements put a heavy burden on the States. The draft articles on the law of transboundary aquifers represent the last stage in the development of international groundwater law. The article on monitoring presents less constraints. However the draft articles are intended to apply at the global level, and in regions of the world where cooperation over transboundary waters in general and transboundary aquifers in particular is less developed.

REFERENCES

Burchi S. and Mechlem K., 2005. Groundwater in international law, FAO/UNESCO 2005.

European Commission, 2003. Common Implementation Strategy for the Water Framework Directive (2000/60/EC) (2003), Monitoring under the Water Framework Directive, Guidance document n°7.

Jousma G. and Roelofsen F.J., 2004. World-wide inventory on groundwater monitoring, IGRAC, Utrecht 2004, available at http://www.igrac.nl/dynamics/modules/SFIL0100/view.php?fil_Id=56.

Stephan R., 2007. Transboundary Aquifers in International Law: Towards an Evolution. In C. Darnault (ed.), *Overexploitation and Contamination of Shared Groundwater Resources*, Springer, Dordrecht, The Netherlands.

Stephan R., 2008. *The New Legal Framework for Groundwater under the EU Water Framework Directive and Daughter Directive*, forthcoming, ICFAI University Press.

UNECE, 2000. Task Force on Monitoring and Assessment, Guidelines on Monitoring and Assessment of Transboundary Groundwaters.

1.3

Groundwater Monitoring Approaches at International Level

John Chilton

International Association of Hydrogeologists, Reading, United Kingdom

1.3.1 INTRODUCTION

Although in many countries monitoring of groundwater quality in some form or other has been long-established, we have grown accustomed in recent years to the ever-strengthening regulatory framework within which this monitoring is undertaken and the results interpreted and used. This has been particularly so in Europe, starting with the Drinking Water Directive, and then the Nitrates Directive, and coming up to date with the Water Framework Directive and Groundwater Directive. This European regulatory framework has not only provided specific and binding legislation on what chemical substances should be monitored, but increasingly has been supported by guidance on where, when and how the monitoring should be undertaken.

Beyond Europe, the regulatory framework has also developed in recent years. Thus, while the technical objectives of groundwater assessment and monitoring set out in the first part of this chapter are more or less universal – to observe and report on the status and trends of the quality of groundwater – the way in which this is done varies considerably. This section summarises some of the approaches used and highlights important issues arising beyond the European context.

Groundwater Monitoring Edited by Philippe Quevauviller, Anne-Marie Fouillac, Johannes Grath and Rob Ward
© 2009 John Wiley & Sons, Ltd

1.3.2   THE US NATIONAL WATER-QUALITY ASSESSMENT PROGRAMME

The regulatory framework for groundwater quality in the United States has been established over the last 30 years by a succession of federal laws. The Comprehensive Environmental Response, Compensation and Liability Act (CERCLA) which led to the Superfund programme and the Resource Conservation and Recovery Act (RCRA) are targeted at pollution sources, as is the Safe Drinking Water Act. It is the Clean Water Act which provides the federal regulatory basis for the ambient monitoring of groundwater within aquifers (Canter *et al.*, 1987).

In the same way as the Drinking Water Directive did in Europe, these Acts established standards or Maximum Allowable Contaminant Levels (MACs), and it is the responsibility of the US Environmental Protection Agency at federal scale and the state Environment Protection Agencies to review these in the light of increasing scientific knowledge and the development and use of new chemicals. However, the knowledge base from which to judge whether the objectives of this regulatory framework were being met and the investments in protection and remediation of groundwater were sufficient was inadequate. Limitations in consistency and comparability of field sampling procedures and laboratory methods, and in the adequacy of spatial coverage, long-term sampling and the comprehensiveness of parameter selection were all identified (Leahy *et al.*, 1993). Some of the questions being asked by scientists, regulators and the public were: are national water quality goals being met? Is it safe to swim in and drink from our rivers and lakes? Is groundwater polluted? How effective have past actions been? How should limited financial resources be allocated amongst competing water quality problems? For example, what is the balance in extent and importance between point and diffuse sources of pollution? Can and should these limited resources be targeted at specific chemicals and the most sensitive hydrological settings? (Leahy *et al.*, 1993).

The US Geological Survey therefore embarked in 1991 on an ambitious monitoring strategy known as the National Water-Quality Assessment Programme (NAWQA) designed to address these weaknesses. The goals of the programme are to describe the status and trends in quality of the nation's surface waters and groundwater resources and to provide sound scientific understanding of the main natural and human influences and processes affecting their quality. NAWQA is designed to integrate water quality information at different spatial scales and comprises two components; study-unit investigations and national syntheses (Leahy *et al.*, 1993). The study units comprise 60 river basins and aquifers which together account for between 60% and 70% of the nation's water use and population served by public water supply and cover over 50% of the land area (Figure 1.3.1). The similar design of each investigation and the use of standard methods would make comparisons between study units possible, and would also facilitate compilation of regional assessments and national syntheses of priority issues such as acidification, pesticides, nitrate and MTBE.

To ensure that the NAWQA programme was manageable and affordable, intensive assessment activities would be conducted on a cyclical rather than continuous basis, with one third of the study units being intensively studied at any one time (Leahy *et al.*, 1993). Thus, in 1991 NAWQA was established with investigations in 20 study units, with a further 20 being initiated in 1994 and the final 20 in 1997. Each group of study units

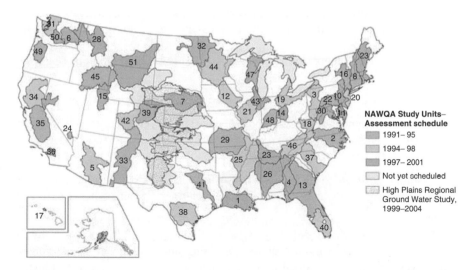

NAWQA assessed 51 hydrologic basins (referred to as "study units") from 1991 through 2001. NAWQA plans to reassess 42 of the 51 study units in the second round of assessments, from 2002 through 2012 (see Gilliom and others, 2001,1st map).

NAWQA Circulars

Available on the World Wide Web at
Water.usgs.gov/nawqa/nawqa_sumr.html

River Basin Assessments

1. Acadian- Pontchartrain Drainages (Circular 1232)
2. Albemarle-Parmlico Drainage Basin (Circular 1157)
3. Allegheny and Monongahela River Basins (Circular 1202)
4. Apalachicola-Chattahoochee-Flint River Basin (Circular 1164)
5. Central Arizona Basins (Circular 1213)
6. Central Columbia Plateau (Circular 1144)
7. Central Nebraska Basins (Circular 1163)
8. Connecticut, Housatonic and Thames River Basins (Circular 1155)
9. Cook Inlet Basin (Circular 1240)
10. Delaware River Basin (Circular 1227)
11. Delmarva Peninsula (Circular 1228)
12. Eastern Iowa Basins (Circular 1210)
13. Georgia-Florida Coastal Plain (Circular 1151)
14. Great and Little Miami River Basins (Circular 1229)
15. Great Salt Lake Basins (Circular 1236)
16. Hudson River Basin (Circular 1165)
17. Island of Oahu (Circular 1239)
18. Kanawha - New River Basins (Circular 1204)
19. Lake Erie - Lake Saint Clair Drainages (Circular 1203)
20. Long Island - New Jersey Coastal Drainages (Circular 1201)
21. Lower Illinois River Basin (Circular 1209)
22. Lower Susquehanna River Basin (Circular 1168)
23. Lower Tennessee River Basin (Circular 1233)
24. Las Vegas Valley Area and the Carson and Truckee River Basins (Circular 1170)
25. Mississippi Embayment (Circular 1208)
26. Mobile River Basin (Circular 1231)
27. New England Coastal Basins (Circular 1226)
28. Northern Rockies Intermontane Basins (Circular 1235)
29. Ozark Plateaus (Circular 1158)
30. Potomac River Basin (Circular 1166)
31. Puget Sound Basin (Circular 1216)
32. Red River of the North Basin (Circular 1169)
33. Rio Grande Valley (Circular 1162)
34. Sacramento River Basin (Circular 1215)
35. San Joaquin-Tulare Basins (Circular 1159)
36. Santa Ana Basin (Circular 1238)
37. Santee River Basin and Coastal Drainages (Circular 1206)
38. South-Central Texas (Circular 1212)
39. South Platte River Basin (Circular 1167)
40. Southern Florida (Circular 1207)
41. Trinity River Basin (Circular 1171)
42. Upper Colorado River Basin (Circular 1214)
43. Upper Illinois River Basin (Circular 1230)
44. Upper Mississippi River Basin (Circular 1211)
45. Upper Snake River Basin (Circular 1160)
46. Upper Tennessee River Basin (Circular 1205)
47. Western Lake Michigan Drainages (Circular 1205)
48. White River Basin (Circular 1150)
49. Willamette Basin (Circular 1161)
50. Yakima River Basin (Circular 1237)
51. Yellowstone River Basin (Circular 1234)

Figure 1.3.1   NAWQA study units, 1991–2001 (after Hamilton *et al*., 2004).

was to be intensively studied for three years followed by six years of lower intensity activity designed to maintain the information necessary for identifying and assessing trends.

The groundwater components of the work in these study units would comprise:

- study-unit surveys which would provide a broad overview of groundwater quality;

- land-use studies designed to assess the quality of recently recharged, shallow groundwater associated with the most important combinations of land use and hydrogeological conditions;

- flowpath studies designed to characterise the spatial and temporal distribution of water quality in shallow groundwater systems and the evolution of groundwater quality with time.

Such a strategic, long-term commitment to monitoring and assessment of the aquatic environment was unusual and implied an equally long-term commitment of funding to make it work. In practice, it has been difficult to maintain funding against competing demands through successive national political administrations, but the NAWQA programme has indeed prospered. During its first decade from 1991 to 2001, assessments were completed in 51 study units (Figure 1.3.1), and the results drawn out into regional

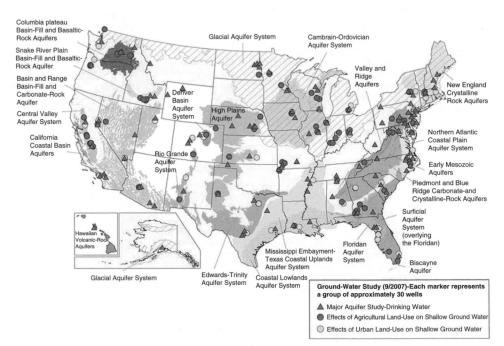

Figure 1.3.2 Location of regional assessments in principal aquifers and groundwater land-use studies (after Lapham *et al.*, 2005).

and national syntheses as planned. The findings showed that in streams and groundwater there was widespread pollution in agricultural and urban areas by nutrients, trace elements, pesticides, VOCs and their degradation products (Hamilton *et al*., 2004).

The programme is characterised by a very high level of open reporting aimed at both technical and non-technical readers. The results are made widely available through traditional USGS technical report series, circulars and fact sheets. The latter are readily obtained from the website http://water.usgs.gov/nawqa from which much of the monitoring data itself can also be accessed.

During the second decade (Cycle II) from 2002 to 2012, the NAWQA programme has been reassessing 42 of these study units (Hamilton *et al*., 2004), returning to 14 units in 2002, 14 in 2004 and 14 in 2007. These return assessments will fill some critical gaps in the characterisation of water quality, and build on the earlier assessments that link water quality to natural and human factors. The baseline data obtained in the first surveys, the lower level of monitoring activity in the intervening period and the new assessments will be used to provide information on trends in water quality, which will be a key focus of the work in Cycle II. During the second decade there will be a major focus for groundwater on regional assessments of quality in 19 major aquifers (Lapham *et al*., 2005), which together account for about 75% of the estimated withdrawal of groundwater for drinking water supply. This will be combined with land-use studies to improve the understanding of the processes by which agriculture and urbanisation impact on groundwater quality (Figure 1.3.2).

1.3.3   OTHER NATIONAL PROGRAMMES

Approaches to groundwater quality monitoring vary greatly, reflecting the broad range of national regulatory frameworks, institutional settings and economic circumstances. Most other countries with national groundwater quality monitoring programmes use existing abstraction boreholes as sampling points, largely for reasons of cost and practicality. This is often not ideal from the point of view of regulatory agencies and others who require information about the quality status and trends of groundwater within the natural environment as part of the overall hydrological cycle. The monitoring is often undertaken by water utilities or other operators primarily to assess groundwater quality in relation to drinking water standards, industrial process requirements, irrigation demands or other uses. In these circumstances, the selection of monitoring sites, frequency of sampling and choice of parameters is made primarily to meet the operator's objectives and the design and implementation of the monitoring is in his hands.

If the dominant groundwater use and management concern is not drinking water but agriculture, then the picture can be very different. In the Lower Indus basin in the Sind and Punjab Provinces of Pakistan, for example, groundwater quality monitoring has been established for many years to provide information for managing groundwater and soil salinity in what is one of the largest irrigation command areas of the world. A restricted range of parameters – electrical conductivity, sodium and chloride – related to salinity are measured by the Central Monitoring Organisation in repeated regular survey rounds of many hundreds of operating irrigation boreholes to assess the status and trends of groundwater salinity and the effectiveness of control measures in Salinity Control and Reclamation Projects (SCARPS). In contrast, routine monitoring of drinking water supplies as yet lags far behind.

In the absence of anything else, the monitoring data are often used to provide information about the groundwater environment, but the continuity of sustained monitoring cannot always be assured, and the sharing of data between operator and regulator is not always guaranteed. In other countries, monitoring is undertaken by regulatory agencies, whose personnel regularly (and by agreement with the operator) visit the pumping boreholes to take water samples. This provides the regulator with much greater control over frequency of sampling and choice of parameters, and often also greater confidence in the analytical results obtained.

As an example, the Swiss Federal Office of the Environment has a national groundwater quality monitoring network NAQUASPEZ of just over 500 sites. These meet selection criteria which provide adequate coverage of the range of geological and hydrogeological conditions, altitudes and land uses within the country. Of the 508 sites, 260 are springs, 243 are abstraction wells or boreholes and only 5 are piezometers or observation wells. Almost all (483) of the sampling sites are used for potable water supply and operated by utilities or communes, but sampled by the federal authority.

One consequence of dependence on existing groundwater abstraction sites is that these are likely to have been located to meet water the demand from nearby communities. Their distribution is skewed towards centres of population, intensive agriculture and industry, and the resulting information will be representative of these potential impacts. Moreover, within this existing distribution of sites and land uses, the limited resources of staff and budget available for monitoring are likely to be focused by water managers and supply operators on the largest and most important supplies and those which are most vulnerable to pollution, or already show signs that pollution is occurring. An assessment of groundwater quality is likely to be much more indicative of human impacts and less able to provide a balanced picture quality of the overall groundwater resources of the country. This situation is common and, for example in New Zealand (anon, 2007) most of the 1100 groundwater monitoring sites are shallow abstraction boreholes located in agricultural, urban and suburban areas. There are in turn few in the country's large unpopulated regions, and the resulting national picture of, for example, nitrate in groundwater reflects this distribution (anon, 2007).

1.3.4   DRINKING WATER GUIDELINES AND STANDARDS

Outside of the European Union and the US, groundwater quality monitoring results are compared either with national standards or the Drinking Water Guidelines of the World Health Organisation (WHO). The former are (as within the EU) fixed and usually legally binding obligations for compliance. They often adopt or closely follow EU or US EPA values, but sometimes take into account local social, economic, political and environmental considerations in setting more relaxed values or targets. In contrast, the WHO guideline values are advisory and risk-based, taking account of the potential human health impacts, and are intended as a basis for national governments and regulatory agencies

Table 1.3.1   Drinking water guidelines and standards for selected inorganic parameters of direct health significance.

Parameters	WHO Guidelines 3rd edition,	EU Directive 98/83/EC	US EPA Office of Water, 2003	Notes
arsenic (μg/l)	10	10	10	Reduced from 50 μg/l
boron (mg/l)	0.3	1.0	–	
cadmium (μg/l)	3.0	5.0	5.0	
chromium (μg/l)	50	50	100	
copper (mg/l)	2.0	2.0	1.3	
cyanide (μg/l)	70	50	200	
fluoride (mg/l)	1.5	1.5	4.0	Varies with climate
lead (μg/l)	10	10	15	
nitrate (mg/l as NO_3)	50	50	45	
nitrite (mg/l as NO_2)	3.0	0.5	3.3	

to develop standards and regulations that are appropriate to their own socio-economic conditions and water quality situations.

An example of the consequence of the varying approaches is that for pesticides the EU Drinking Water Directive adopted a standard of 0.1 μg/l for all compounds, whereas the US EPA and WHO have taken a toxicity-based approach. The values for the most toxic organo-chlorine pesticide compounds are less than the 0.1 μg/l of the EU, and for less toxic compounds range from 2–10 μg/l for the majority to 20–40 μg/l for less toxic compounds (MacDonald *et al.*, 2005). The most important inorganic parameters with health significance are shown in Table 1.3.1, noting that natural constituents of health concern such as fluoride and arsenic are widespread in groundwater. These need to be properly incorporated into groundwater quality monitoring programmes where they are known to occur or where the geological setting is likely to give rise to their occurrence.

Clearly from the point of view of the WHO, the assessment of risks to health is the greatest priority because of the potentially devastating consequences of waterborne infectious diseases (MacDonald *et al.*, 2005) and the protection of microbiological quality remains the highest priority. In many developing countries, however, even where groundwater supplies a significant proportion of both rural and urban populations, the individual abstraction wells or boreholes are numerous and scattered. Treatment of the abstracted water is rarely possible and the emphasis is instead on adequate wellhead protection measures to ensure that the quality of water remains good. Routine and regular chemical and microbiological monitoring of perhaps thousands of small supplies is almost invariably beyond the capabilities of the responsible local institutions. In these circumstances, water quality assessment and management has to be based on a combination of regular sanitary inspections of the type proposed by Lloyd and Helmer (1991) and microbiological sampling, backed up by the capacity to improve the protection of the supplies which are shown to be most at risk (Howard *et al.*, 2003; MacDonald *et al.*, 2005).

1.3.5   GROUNDWATER QUALITY ASSESSMENTS WITHIN THE UN SYSTEM

While the overall objectives remain broadly the evaluation of status and trends in quality, the need to assess at the global, continental or regional scale and the desire to make state of the environment reports for a broad range of stakeholders impose new difficulties. These include the conceptually simple but difficult practical issue of the comparability of data from a wide range of institutions and countries, and the more fundamental issue of data aggregation over ever larger areas.

The Global Environment Monitoring System (GEMS) Water Programme was established by WHO, UNESCO and UNEP in 1978 to act as a focal point for the collection of information on the quality of the world's freshwater systems. Hosted by the National Water Research Institute in Canada and now managed within UNEP, GEMS Water has been steadily building up its water quality database by receiving information for selected monitoring sites from national institutions, and providing summary reports and maps and contributing to global environmental assessments. While it has a component of quality control and improvement of laboratory capacity activities and training and guidance in monitoring, its success is dependent on the submission of data from within national programmes. In that sense it is similar to other endeavours such as those of the European Environment Agency and the International Groundwater Assessment Centre (IGRAC).

Efforts to develop global assessments of this type, however, need to take account of the simple hydrological facts of life. It is reasonable to take the water quality at downstream monitoring sites in river basins to be broadly representative of the aggregate quality of surface water in the catchment draining to this point. Groundwater occurrence is complex and has a depth dimension which is often problematic to evaluate, and the timescales of groundwater movement and residence times are much longer than surface waters. It is clear from examples later in the book that aggregation of data for groundwater and aquifers is not so simple, and it is not reasonable to use data submitted from a single or even a few monitoring points to be representative of groundwater quality. Groundwater quality changes more slowly than surface water with time, but varies much more spatially and vertically over short distances. This presents an almost insurmountable barrier to valid global assessments for groundwater based on assembling a database of this type. Contributions to global assessments for groundwater instead have to rely on existing published case studies selected to be as representative as is feasible, or solicit new case studies on each occasion. As a consequence, groundwater quality remains poorly described at levels above the national scale. While this is of concern to some stakeholders, it is not actually too serious, as actions to protect and manage groundwater quality are generally required at national scale and below.

REFERENCES

Canter L.W., Knox R.C. and Fairchild D.M., 1987. *Groundwater Quality Protection*. Lewis Publishers, Chelsea, Michigan.

Environment New Zealand, 2007, Chapter 10 Freshwater. Ministry of the Environment, ME 848. Wellington, New Zealand; also available at www.mfe.govt.nz.

Hamilton P.A., Miller T.L. and Myers D.N., 2004. Water quality of the nation's streams and aquifers – overview of selected findings, 1991-2001. *USGS Circular* 1265.

Howard G., Pedley S., Barrett M., Naulbenga M. and Johal K., 2003. Risk factors contributing to microbiological contamination of shallow groundwater in Kampala, Uganda. *Water Research*, **37**, 3421–9.

Lapham W.W., Hamilton P.A. and Myers D.N., 2005. National Water Quality Assessment Programme – Cycle II: Regional Assessment of Aquifers. USGS Fact Sheet 2005–3013.

Leahy P.P., Ryan B.J. and Johnson A.I., 1993. An introduction to the US Geological Survey's National Water Quality Assessment Program. *Water Resources Bulletin*, **29** (4), 529–32.

Lloyd B. and Helmer R., 1991. *Surveillance of Drinking Water Quality in Rural Areas*. Longman, Harlow, UK.

MacDonald A.M., Davies J., Calow R. and Chilton P.J., 2005. *Developing Groundwater: A Guide for Rural Water Supply*. ITDG Publishing, Bourton-on-Dunsmore, UK.

Part 2
Conceptual Modelling and Network Design

2.1

Conceptual Modelling and Identification of Receptors as a Basis for Groundwater Quality Assessment

Cath Tomlin[1] and Rob Ward[2]

[1] *Environment Agency – England and Wales, Government Buildings, Westbury-on-Trym, Bristol, United Kingdom*
[2] *Environment Agency – England and Wales, Olton Court, West Midlands, United Kingdom*

2.1.1   LEGISLATIVE REQUIREMENTS FOR PROTECTING AND ASSESSING GROUNDWATER QUALITY

The legislative requirements for the protection of groundwater in Europe are now established in several European Directives (European Commission 1980, 2000 and

Groundwater Monitoring Edited by Philippe Quevauviller, Anne-Marie Fouillac, Johannes Grath and Rob Ward
© 2009 John Wiley & Sons, Ltd

2006) – the original 1980 groundwater directive (80/68/EEC), and more recently the Water Framework Directive 2000/60/EC (WFD) and its daughter directive on groundwater, 2006/118/EC (GWD).

The purpose of the 1980 directive is to prevent the pollution of groundwater by dangerous substances – so called 'listed substances'. It does this by requiring Member States to take the necessary steps to:

1. *prevent* the introduction into groundwater of certain substances – List I; and

2. *limit* the introduction into groundwater of other substances – list II – so as to avoid pollution of this water by these substances.

The WFD updates these requirements by establishing a broader framework for the protection of the aquatic environment. This includes inland surface waters, transitional waters, coastal waters and groundwater and requires Member States to manage the water environment in an integrated way. The WFD, and its daughter directive on groundwater, sets out several environmental objectives for groundwater quality that Member States must aim to meet. Some of these are new and others improve upon the regime set out in the original groundwater directive (80/68/EEC). The environmental objectives for groundwater quality in the WFD are to:

- prevent or limit the input of [all] pollutants into groundwater;

- protect, enhance and restore all bodies of groundwater[1] with the aim of achieving good chemical status;

- prevent the deterioration in status of all bodies of groundwater;

- implement measures to reverse any upward trends in pollutant concentrations.

There are also other directives that indirectly protect groundwater by requiring controls on activities that could present a risk to groundwater quality if left uncontrolled. For example, the aim of the landfill directive (99/31/EC) is to prevent or reduce as far as possible negative effects on the environment from the landfilling of waste, during the whole lifecycle of the landfill. This includes pollution of groundwater. The IPPC directive (96/61/EC) created an integrated approach to establishing pollution prevention from stationary 'installations', and its main objective is to achieve a high level of protection of the environment through measures to prevent or, where that is not practicable, to reduce emissions to air, water and land from certain activities.

2.1.2   THE IMPORTANCE OF CONCEPTUAL MODELS/UNDERSTANDING

Aquifers and groundwater bodies are complex three-dimensional systems. In order to be able to properly assess and manage the risks to groundwater quality from activities on the ground surface, it is important to understand the groundwater system – how it behaves,

---

[1] A groundwater body is defined as a distinct volume of groundwater within an aquifer or aquifers (European Commission (2003), WFD CIS Guidance Document No. 2 Identification of Water Bodies). Note that not all groundwater is necessarily within an aquifer.

and how it interacts with the aquatic environment at the surface. Predicting the behaviour of contaminants as they pass from sources through to receptors is made easier by having an understanding of the physical, chemical, and biological processes that affect them.

This is known as a Conceptual Understanding or Conceptual Model, and has been defined as

> the set of assumptions that represents our simplified perception of the real system which is to be mathematically modelled (Bear and Verruijt, 1987).

It is this understanding that is the key to managing groundwater systems at all scales of assessment, be that assessing good chemical status across a large groundwater body, or assessing the impact of a pollution incident on a single industrial site.

It is important to note that a conceptual model does not have to be a mathematical or analytical computer simulation, although in complex situations this degree of assessment may be needed. In most cases, a relatively simple descriptive and/or visual representation of the three-dimensional groundwater systems and its relationship to sources, pathways, and receptors is all that is required. The term Conceptual Understanding is therefore preferred. Examples of how groundwater systems can be presented are provided in Chapter 2.3.

A conceptual understanding is therefore a simplified representation of the groundwater system it can include information about geology, groundwater flow paths, background groundwater quality, pollution sources (volumes and concentrations in any discharge), discharge mechanisms, receptors, plus any likely fate and transport processes that may act on the pollutants. The information brought together to form this 'picture' can be tested to see if the assumptions made about how the real groundwater system works are correct. The conceptual understanding can then be refined and improved.

Producing a conceptual understanding is therefore an iterative process of testing and development as more is learned about the site (see Figure 2.1.1). The degree of detail

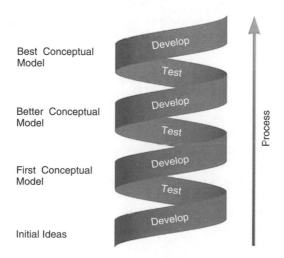

Figure 2.1.1   The conceptual model spiral (from Environment Agency 2008a).

needed for any given assessment will differ, with a groundwater flow numerical model being the most detailed level of conceptual understanding that might be needed for some complex groundwater assessments.

The importance of conceptual models or a conceptual understanding of the groundwater system in groundwater quality assessment is outlined in WFD Common Implementation Strategy Guidance Document No. 17 (European Commission, 2007b). This document identifies the importance of the relationships between sources, pathways and receptors in deciding whether pollution has occurred or will occur from any given activity.

It states that the key considerations to include in a conceptual model are:

1. the physical and chemical nature of the discharge or source of contamination (installation or contaminated part of the subsurface);

2. the physical and chemical characteristics of the aquifer;

3. the subsurface processes, e.g. dilution and degradation, that act on the pollutant as it moves down towards the water table or moves within the groundwater flow;

4. the location of all the receptors and their relationships to groundwater flow; and

5. the environmental standards (for water quality) that apply to the receptors and by which harm can be measured.

Only once a conceptual understanding of the situation has been produced is it then possible to undertake any groundwater quality assessment.

2.1.3   GROUNDWATER QUALITY ASSESSMENT AND RECEPTOR PROTECTION

The term groundwater quality assessment in relation to EU groundwater protection and management, applies to several different activities:

- assessing risk to groundwater from activities as part of the regulation of that activity (for the purpose of protecting groundwater and complying with the prevent or limit objective of the WFD/GWD);

- assessing the chemical status of a groundwater body for the purposes of the WFD;

- assessing whether a groundwater body is subject to a rising trend in pollutant concentrations for the purposes of the WFD.

Each of these requirements relies on predicting the impacts on different types of receptors. The term 'receptor' includes all legitimate uses of groundwater, both active and passive. Passive receptors include rivers receiving base flow from groundwater; active receptors include pumped boreholes.

In certain circumstances, groundwater itself can also be a receptor, rather than a pathway to other receptors (see following section on prevent or limit).

2.1.3.1 Assessment for Preventing or Limiting Inputs of Pollutants into Groundwater

The 'prevent or limit' objective in the WFD/GWD aims to protect all groundwater (not just groundwater bodies) from unacceptable inputs of pollutants. It is the most important of the objectives in the WFD/GWD as it is preventative in nature – complying with this obligation should ensure that pollution of groundwater does not occur in the first place. The full scope of the objective is described in Article 6 of the GWD.

Preventing Inputs of Hazardous Substances

Member States must take all measures necessary to prevent inputs into groundwater of any hazardous substances. Hazardous substances are those that are toxic, persistent and liable to bio-accumulate, and other substances which give rise to an equivalent level of concern.

Understanding the preventive part of the objective is relatively straightforward. Hazardous substances must not enter groundwater, and so in this case groundwater itself is the receptor. Therefore any assessment of the potential effects of such a discharge has to be measured at the water table.

In practice, however, it can be very difficult to prove that there is no (zero) discharge and whether the measures to prevent inputs have been successful or not. This is for two main reasons:

- most laboratory analytical techniques have minimum levels of detection and so are unable to determine that a substance is not present (concentration is zero); and

- the groundwater monitoring point often cannot be sited directly adjacent to the discharge point and therefore measure the maximum impact on groundwater quality.

A more pragmatic approach has to be taken to decide what constitutes 'prevent'. European Commission guidance on Inputs recognises this by stating that it is sometimes not technically feasible to stop all inputs of hazardous substances (European Commission, 2007b). Technical regulatory guidance for England and Wales also interprets this requirement in a pragmatic way by identifying discharges as unacceptable if there is 'discernible entry' into the saturated zone (Environment Agency, 2008b). Discernible entry is determined where detection is at or above the lowest practical levels of laboratory analytical detection.

Despite these difficulties the protection of groundwater quality is a fundamental requirement and so the principle of 'best endeavours' should be used. In other words, all measures that are practicable and cost-effective should be undertaken to prevent an input of hazardous substances entering groundwater. These measures could be a combination of source control (e.g. treatment of effluent), removing the pathway to groundwater (e.g. lining a landfill), or maximising the pathway and hence exploiting the full attenuation capacity within the unsaturated zone (e.g. making the discharge point as shallow as possible).

The GWD itself recognises that some discharges are impossible to control. For this reason, certain activities and circumstances are exempted from this requirement in Article 6.

Limiting Inputs of Non-hazardous Substances

The other part of the 'prevent or limit' objective applies to non-hazardous pollutants. For pollutants which are not considered hazardous but are considered to present an existing or potential risk of pollution, Member States shall take all measures necessary to limit inputs into groundwater so as to ensure that such inputs do not cause deterioration [in status] or significant and sustained upward trends in the concentrations of pollutants in groundwater.

The original requirement in 80/68/EEC was to ensure that such discharges did not cause pollution. Thus, in order to meet the requirement that the regime established by the GWD should be no less protective than the regime established back in 1980, this new requirement has to be interpreted as a requirement to prevent pollution.

In order to comply with this, Competent Authorities of Member States need to have a clear understanding of the basis for judging 'pollution'.

Pollution is defined in the WFD as:

> the direct or indirect introduction, as a result of human activity, of substances or heat into the air, water or land which may be harmful to human health or the quality of aquatic ecosystems or terrestrial ecosystems directly depending on aquatic ecosystems, which result in damage to material property, or which impair or interfere with amenities and other legitimate uses of the environment.

For pollution to occur, there needs to be some actual or likely harmful effect resulting from human activity on the named receptors. Receptors included in the definition of pollution in the WFD are:

- human health, the quality of aquatic ecosystems or terrestrial ecosystems directly depending on aquatic ecosystems;

- material property;

- amenities and other legitimate uses of the environment.

This list covers all passive and active uses of groundwater, but does not specifically include groundwater itself. The assessment of whether an activity will pollute groundwater therefore has to judge whether the activity will cause harm to any of these receptors.

2.1.3.2   Assessment of Groundwater Chemical Status

The GWD (Articles 3 & 4) and the WFD (Article 4 & Annex 5) require Member States to assess the chemical status of each groundwater body, and classify them as either good or poor. This assessment involves comparing groundwater quality monitoring data to threshold values and groundwater quality standards, and where they are exceeded, undertaking further investigation to see whether the conditions for good status are being met.

The definition of, and therefore the conditions for, good chemical status specified in the WFD/GWD are limited to only a few receptors and specific circumstances. The receptors included within the definition of good chemical status are:

- drinking water abstractions (subject to the Drinking Water Directive, 98/83/EC);

- groundwater dependent terrestrial ecosystems (wetlands that are fed by the groundwater body);

- surface water bodies.

This list of receptors does not include all types of abstractions, nor groundwater itself. However, in achieving good status, Member States also have to ensure that groundwater abstraction is not leading to sustained alteration of flow direction that is resulting in saline or other intrusions, nor that the general quality of the body as a whole has not been significantly impaired. This last criterion is a large-scale assessment of the quality of the water contained within the groundwater body.

The assessment of status is carried out for the whole of a groundwater body. In most cases this will be a large area. However there will be a range in the size of groundwater bodies across Europe. For example in England and Wales, the size of groundwater bodies varies from just a few square kilometres to several thousand square kilometres. In other parts of Europe, groundwater bodies may be even larger.

Many inputs of pollutants into groundwater only have a very localised effect as the pollutant(s) may be subject to dilution and attenuation along the flow path to a receptor. These inputs may still result in pollution but it is restricted to only a small area. Because the scale of the impact is small relative to the size of the groundwater body this pollution may not lead to a status failure. However, the more widespread the pollution becomes, the more likely it is that the groundwater body will be at poor status. Therefore under the WFD/GWD it is quite possible to have small-scale impacts within a groundwater body and for it still to be at good chemical status. These events should be investigated (and remedied if necessary) via prevent or limit measures.

2.1.3.3 Assessment of Upward Pollutant Trends

The WFD/GWD contains a requirement for Member States to identify significant and sustained upward trends in pollutant concentrations caused by human activity. The assessment only has to be carried out in groundwater bodies found to be at risk. Member States also have to implement measures to reverse such trends in order to progressively reduce pollution of groundwater.

In order to understand the receptors involved in the assessment, it is important to understand what is meant by 'significant'. The meaning of 'significant' is described in European Commission guidance on trend assessment (European Commission, 2009). This document states that for a trend to be significant it must be both environmentally *and* statistically significant. The definition of an environmentally significant trend is one that would:

> ' ... lead to the failure of one or more of the WFD's environmental objectives if not reversed'.

So a statistically significant trend with a very shallow slope that will not cause pollution (according to the definition of the Directive) or lead to a failure of status is not environmentally significant in terms of the requirement in the directive. This means that when undertaking the trend assessment, the receptors that need to be included are all those covered by both the prevent or limit objective and the status objective. These are all legitimate uses of groundwater.

2.1.4   COMPLIANCE POINTS

Whether undertaking an assessment of chemical status, or regulating to ensure that pollution of groundwater is prevented, it is essential to have an understanding of the groundwater system in order to properly identify all receptors. This should include an understanding of the hydrogeological flowpaths and interactions with surface water. In addition, it is necessary to understand the hydrochemistry in order to recognise the potential scale of the impact on the identified receptors.

Whatever the nature of the assessment of groundwater quality, once all receptors have been identified it is then possible to set appropriate compliance points at which to measure or monitor the situation. Setting compliance points, and the relevant compliance criteria, requires the development of a conceptual understanding of the relationships between sources, pathways and receptors.

According to the European Commission guidance on Inputs (European Commission, 2007b), there are two types of compliance point:

1.   a theoretical point within a model for calculating an acceptable discharge concentration or restoration target, e.g. the required level of clean-up at a contaminated site;

2.   a physical monitoring point (e.g. an observation borehole) for the purpose of measuring compliance with a permit or a clean-up regime.

The guidance document only looks at groundwater quality assessments related to preventing or limiting inputs to groundwater. It is worth noting therefore that there is a third type of compliance point – those physical monitoring points used for the purpose of assessing compliance with chemical status or trend objectives.

For status, and for most trend assessments, these compliance points will be monitoring points that are part of Member State operational monitoring networks put in place for the WFD (European Commission, 2007a). Monitoring points for assessing compliance with permits on the other hand are likely to be close to the discharge. They are specifically put in place to measure the impact of the activity, and assess compliance with environmental permit conditions. This type of monitoring is known as 'prevent and limit monitoring'

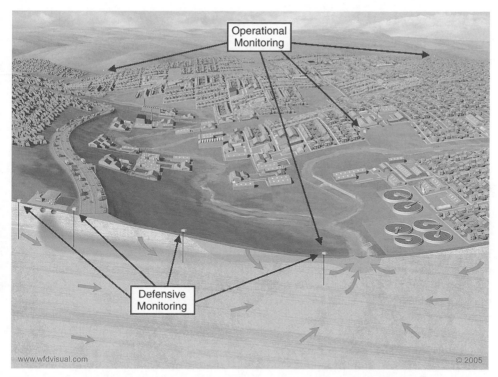

Figure 2.1.2 Operational (status) and defensive monitoring (from UKTAG 2008). Base image copyright WFDVisual.com 2005 (representing SNIFFER, the Environment Agency and the Geological Survey of Ireland).

(European Commission, 2007a) or 'defensive monitoring' (UKTAG, 2008). Where large inputs may impact on status, the defensive monitoring points may also be part of the operational monitoring network. This is illustrated in Figure 2.1.2.

Compliance values are set at compliance points and can vary according to the assessment being undertaken. Compliance values must be set in order to determine:

- for prevent or limit – an acceptable concentration in groundwater which will ensure that an environmental standard is not exceeded at a receptor (and ensure the receptor is adequately protected);

- for status assessments – a threshold value or groundwater quality standard (exceedance of which may indicate a failure of one or more of the criteria for good status, subject to further investigation);

- for trend assessments – a relevant starting point for trend reversal (derived from a threshold value or groundwater quality standard)

Further description of Compliance Values can be found in European Commission guidance document on Inputs (European Commission 2007b) and Chapter 4 of the present book.

2.1.5   THE ROLE OF GROUNDWATER MONITORING

The role and importance of groundwater monitoring are intrinsically linked to all parts of the groundwater assessment process. Conceptual understanding must include an assessment of perceived pressures on groundwater quality. This can only be corroborated by evidence of actual impacts, and groundwater monitoring can supply this evidence. Monitoring therefore fulfils one of the basic data needs for the conceptual understanding of the groundwater system. As more monitoring is undertaken, this further supports the testing and refinement of the conceptual understanding.

Groundwater monitoring also plays a fundamental part in the assessment of groundwater quality, as without such monitoring, the impacts on groundwater quality would not be able to be determined. It also provides a benchmark for assessing future change.

The requirement to monitor groundwater chemistry/quality is set out in Article 8 of the WFD. This is in the context of providing the necessary information to assess whether WFD environmental objectives are being met. In other words, the monitoring must provide sufficient information to undertake an adequate assessment of characteristics and risks to groundwater, assess chemical status of groundwater bodies and establish the presence of significant and sustained upward trends in pollutant concentrations. Therefore these operational monitoring networks have to be representative of the quality of the whole groundwater body.

As discussed, there may also be the need for additional specific monitoring points or programmes aimed at assessing the impacts of point source discharges (defensive monitoring). These monitoring points have to be representative of the local groundwater conditions.

In order to achieve the correct design for monitoring networks, Member States must have a clear understanding of the environmental conditions and human activities likely to impact on the ability to achieve these objectives. Monitoring programmes can only therefore be effectively designed using information contained within the conceptual understanding of the groundwater system and with clear understanding of the information required from the programmes (European Commission, 2007a).

2.1.6   CONCLUSIONS

The effective implementation of EU-wide requirements for groundwater protection and management requires clearly defined objectives and strategies for achieving them. An adequate conceptual understanding of the groundwater system and its relationship to other parts of the environment and human activities that may affect it is essential.

In developing and refining the conceptual understanding (or model) of the system, information and data are required. Much of this can only be obtained through monitoring. Monitoring helps to characterise the system and allows measurement of change over time as well as compliance with environmental objectives.

It is therefore essential that the relationship between each component part of the overall process for groundwater quality assessment, management and protection is understood and continued efforts made to maintain and refine this understanding.

REFERENCES

Bear Jacob and Verruijt A. 1987. *Modelling Groundwater Flow and Pollution*, Kluwer Academic Publishers Group, UK.

Environment Agency, 2008a. Groundwater Protection: Policy and Practice. Part 2 – Technical Framework.

Environment Agency, 2008b. Operational Instruction 429_05, Framework for Assessing the Risk of Pollution to Groundwater from Permitted Discharges.

European Commission, 1980. Directive on the Protection of Groundwater against Pollution Caused by Certain Dangerous Substances, *Official Journal of the European Communities*.

European Commission, 2000. Directive 2000/60/EC of the European Parliament and of the Council of 23 October 2000 Establishing a Framework for Community Action in the Field of Water Policy, *Official Journal of the European Communities*.

European Commission, 2003. WFD CIS Guidance Document No. 2 Identification of Water Bodies.

European Commission, 2006. Directive 2006/118/EC of the European Parliament and of the Council of 12 December 2006 on the Protection of Groundwater against Pollution and Deterioration, *Official Journal of the European Communities*.

European Commission, 2007a. WFD CIS Guidance Document No. 15 Monitoring Guidance for Groundwater.

European Commission, 2007b. WFD CIS Guidance Document No. 17 Guidance on Preventing or Limiting Direct and Indirect Inputs in the Context of the Groundwater Directive 2006/118/EC.

European Commission, 2009. WFD CIS Guidance Document on Trend Assessment.

Scottish and Northern Ireland Forum for Environmental Research 2005. Groundwater Concepts Visualisation Tool. http://www.wfdvisual.com

UKTAG, 2008. UK Technical Advisory Group on the Water Framework Directive, Application of Groundwater Standards to Regulation.

2.2

Aquifer Typology, (Bio)geochemical Processes and Pollutants Behaviour

Hélène Pauwels[1], Wolfram Kloppmann[2], Kristine Walraevens[3] and Frank Wendland[4]

[1,2] *Bureau de Recherches Géologiques et Minières, Orléans cédex, France*
[3] *Laboratory for Applied Geology and Hydrogeology, Ghent University, Gent, Belgium*
[4] *Research Centre Juelich, Institute of Chemistry and Dynamics of the Geosphere (ICG), Institute IV: Agrosphere, Juelich, Germany*

Groundwater Monitoring Edited by Philippe Quevauviller, Anne-Marie Fouillac, Johannes Grath and Rob Ward
© 2009 John Wiley & Sons, Ltd

2.2.1   INTRODUCTION

Knowledge, both on the intrinsic properties of potential pollutants and on the mineralogical and biogeochemical characteristics of the aquifer through which they move, is essential regarding their fate within groundwater. Such information is consequently an indispensable prerequisite for any strategy of groundwater protection against pollution as required by the Groundwater Directive. Determination of threshold values, for example, to be established for pollutants or groups of pollutants, should take into account, among others, the behaviour of each substance including its origin, natural or man-made, its dispersion tendency and persistence potential, all three characteristics depending on the aquifer properties.

This chapter focuses on a scheme to establish an aquifer typology developed in the frame of the European BRIDGE project. Such a typology may serve as a basis for referencing on the one hand the chemical composition of groundwater, in particular the so-called natural background level, and, on the other hand, the transfer and fate of pollutants. It can be used not only to assess the 'good groundwater chemical status', to derive threshold values for contaminants or to monitor the evolution of groundwater quality but also to guide the investigations needed for environmental risk assessment in cases where the contaminant concentrations exceed relevant quality standards or threshold values.

Particularly, it relates the expected variability of water quality to the fate of pollutants in groundwater bodies, as many parameters in the concerned aquifer types show characteristic natural variations. In many cases, establishing the relative contribution of a substance from natural and anthropogenic sources constitutes a real challenge. The proposed scheme for an aquifer typology provides overall geochemical baselines and thus gives, for several classes of pollutants, first hints to potential ambiguity with respect to their natural or anthropogenic origin.

2.2.2   PURPOSE AND DESCRIPTION OF PROPOSED TYPOLOGY

2.2.2.1   Aquifer Typologies for Hydrogeological Mapping

Hydrogeological maps are based on aquifer typologies and UNESCO elaborated an international legend for hydrogeological maps as soon as 1984. The wide use of Geographic Information Systems (GIS) has favoured the fast development, during the last decades, of thematic mapping including hydrogeological maps. However, aquifer typologies applied for establishment of these maps are primarily based on the physical or hydrodynamic properties of aquifers. Generally, hydrogeological maps provide information on groundwater resources (porosity, permeability, productiveness), groundwater flow, recharge and discharge zones, extension of transboundary aquifers, occurrence of low permeability covers, location of productive wells or springs, relationship between ground and surface waters, areas of multi-layered aquifers, or even on groundwater vulnerability. Resulting maps can be drawn for small-scale up to large-scale areas, from countries up to world-wide scale (Gilbrich, 2000; Struckmeier *et al*., 2006; IGRAC, 2005; Schürch *et al*., 2007). They provide an overview of the groundwater resource and are valuable tools for water managers. However, an aquifer is not only a water container, but can actually be considered as a (bio)geochemical reactor, where the chemical composition of water and the concentration of any pollutant evolve over time as the result of water-rock-biomass interaction processes. Natural chemical composition of groundwater depends on rain water chemistry, evapotranspiration before recharge and various reactions of groundwater with minerals, original fluids and gases, with or without bacteria involvement. A crucial role is played by the aquifer matrix with its specific chemical and mineralogical composition but also by the duration of water–rock interaction given the slow kinetics of many geochemical reactions. This natural chemical baseline may be perturbed by anthropogenic activities, mainly through the input of pollutants, which in turn may interact with the aquifer matrix, both within the vadose or the saturated zone, leading to extremely wide variations of the chemical composition of groundwater. If a large range of thematic hydrogeological mapping already exists (Novoselova, 2004) improvements are nevertheless required, as far as vulnerability to pollution is concerned. Existing typologies focus on the capacity of vertical and lateral transport of pollutants exclusively based on physical properties of the aquifer whereas it is also essential to take account of the hydrogeochemical condition of groundwater.

2.2.2.2 Aquifer as a (Bio)geochemical Reactor for Pollutants

Groundwater chemistry depends on water–rock interaction processes but may be impacted by human activities, through inputs both into the vadose zone and into the saturated parts of the aquifer. Before setting up a typology adapted to the fate of pollutants, it is worth reminding ourselves of the potential physico-biochemical processes occurring during their transport, which may induce a significant modification of pollutant concentrations in groundwater. Substantial attenuation of contaminants can occur through filtration, settling of suspended particulate matter and diffusion, which will depend on physical (matrix properties) and hydrodynamic characteristics of the aquifer. As for volatilization, it depends mainly on pollutants properties and not on those of the aquifer or groundwater except if it occurs after a transformation of a nonvolatile species into a volatile species – as for example transformation of nitrate (NO_3) into nitrogen gas (N_2).

(Bio)-geochemical processes governing the fate and behaviour of pollutants during their transfer into groundwater include dissolution/precipitation, sorption, complexation, ion exchange, abiotic or biotic degradation (Table 2.2.1). They depend on the nature of the aquifer (mineralogical, chemical composition, organic matter content, occurrence of bacteria) and on the nature of groundwater which in turn depends on composition of the aquifer. These processes can either induce the immobilization of an anthropogenic pollutant or, on the contrary, increase its concentration through leaching of the aquifer matrix.

Table 2.2.1 Dominant processes affecting the fate and transfer of pollutants and potential impact on groundwater composition.

Process	Description and dependence	Potential impact on pollutant levels
Precipitation	Depends on GW composition and occurs when the solution attains over-saturation with respect to a pollutant bearing solid phase	Can cause the immobilisation of pollutants, in particular through co-precipitation, e.g. with hydroxides or carbonates
Dissolution	Human activity can modify GW chemical composition which in turn induces under-saturation with respect to a pollutant-bearing solid phase thus promoting its dissolution	Can increase the concentration of matrix-specific pollutants
Complex formation	Depends on the GW chemical composition, can be favoured by different dissolved compounds such as Cl, SO_4, HCO_3, organic matter,...	Can increase the solubility of pollutants
Sorption	May occur on organic matter, mineral oxides, clay minerals	Delays the transfer of the pollutant
Cation exchange	– Pollutants are exchanged with cations initially sorbed on aquifer mineral surfaces, particularly on clays – Inversely, a harmless cation can be exchanged with a naturally occurring hazardous/undesirable compound initially present on clay minerals	– Can cause immobilisation of the pollutant or a delay if sorption and pollutant input are reversible – May increase the concentration of naturally occurring undesirable compounds in GW.
Abiotic degradation	– Chemical transformation, such as hydrolysis or redox processes – Can occur on man-induced pollutants – Can involve naturally occurring species in GW or in the aquifer	– Can decrease concentration of the pollutant – Can transform a pollutant into hazardous metabolites – Can contribute to leaching of naturally occurring hazardous species previously immobilized within the aquifer matrix (e.g. through pH increase or sulphide oxidation accompanying heterotrophic denitrification)

Table 2.2.1 (*continued*).

Process	Description and dependence	Potential impact on pollutant levels
Biodegradation	– Microbially mediated redox reactions that degrade/transform the pollutants or naturally occurring species – Generally higher reaction rates compared to abiotic degradation	– Can decrease concentration of the pollutant – Can transform a pollutant into hazardous metabolites – Can contribute to leaching of naturally occurring hazardous species previously immobilized within aquifer matrix

2.2.2.3 Typology Adapted to the Investigation of the Fate of Pollutants in Groundwater

To take account of the different possible chemical and biochemical reactions involving pollutants, we propose a scheme for a typology composed of 2 levels of parameters (Table 2.2.2):

- primary parameters which relate to the origin of compounds and include, for example, lithology and salinity – they lead to the definition of 10 basic units;

- secondary parameters which relate to processes and include, for example, hydrodynamics, redox conditions, geological age and particularities of the aquifer material such as occurrence of organic matter, oxides and sulphide minerals.

Table 2.2.2 Parameters of an aquifer typology adapted to the groundwater chemical composition and fate and behaviour of pollutants in groundwater.

Ten basic units	Secondary parameters	Merged groups
Limestone	Redox conditions	Carbonate group
Chalk		
Sands and gravels	Particular occurrences:	Unconsolidated group
Marls and clays	Organic matter, clays, oxides,	
Sandstones	sulphide minerals	Sandstone group
Crystalline basement		Hard rock group
Schist	Geological age	
Volcanic rocks		
Evaporites	Hydrodynamics (recharge,	Evaporites
Saline influence	residence time, leakage) and physical properties	Saline influence

As a basis for the chemical composition of groundwater, this typology with its ten basic units is listed in Table 2.2.2. It does not take into account the productiveness of the aquifer types and may include aquifers of very low productivity.

These criteria for a typology have proven to be comprehensive with respect to the chemical composition of most current groundwaters (Wendland *et al.*, 2008). Statistical examination of the chemical composition of groundwater from 16 European aquifers (up to 3300 water samples per aquifer), belonging to three different aquifer types, highlighted the comparable composition (pH, major elements such as HCO_3, Na, Mg, Ca...), represented by the 50th percentile, for aquifers of the same type whereas significant differences have been observed from one type to another. Such a consistency between aquifers of the same type and variability between aquifers of different types is illustrated for pH and calcium (Figure 2.2.1).

However, the variation of groundwater chemistry within each aquifer makes it necessary to consider additional secondary parameters that are more or less independent of the primary parameters.

The ten basic units of the typology provide actually a basis for the determination of natural levels of contaminants in groundwater, the so-called Natural Background Level, of either major or minor/trace elements (see companion chapter by Blum *et al.*). When appropriate, the 10 basic units may be merged into six groups as shown in the right-hand column of Table 2.2.2. Cumulative frequency plots for several major and trace elements in groundwater indicate clear differences between the four main groups which are carbonates, unconsolidated, sandstone and hard rock groups (Griffioen *et al.*, 2006).

Besides lithology, saline influence is another parameter which drives the overall chemical composition of groundwater. Most of aquifer-forming sediments were deposited in a marine environment. The presence of connate seawater constitutes an inherent part of the aquifer characteristics. Although such 'fossil' seawater will be gradually washed out by groundwater circulation, mainly triggered by rainwater recharge, its influence may persist over long geological periods, and determine the general composition of groundwater in the aquifer. Therefore, saline influence is considered as a primary parameter in the developed aquifer typology: as long as the saline influence persists, it is overruling the lithological determinants. Under saline influence, the general groundwater composition, e.g. in a limestone, will be comparable to the one in a sand.

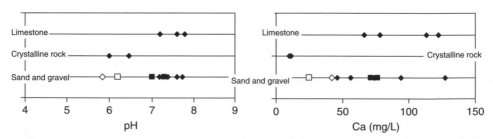

Figure 2.2.1   50 percentile of pH and Ca values of groundwater of European limestone, crystalline and sand and gravel aquifers. Within the sand and gravel aquifers: Squares refer to glacial deposits whereas diamonds to fluviatile deposits, filled symbols refer to carbonate bearing aquifers whereas open symbols to carbonate-free rocks.

2.2.2.4   Secondary Parameters as an Aid to Assess Groundwater Composition Variability

When necessary, each unit of the left-hand column may be further differentiated by applying the secondary parameters of the typology. For example, the ten basic units of the proposed typology are not discriminatory with respect to the residence time of groundwater, whereas residence time and then interaction time of groundwater in the aquifer is a key factor of water–rock interaction processes. Almost all basic units, have a common range of residence times from decades to millennia. In some cases, this range is exceeded, either towards very short or very long residence times, up to 100,0000 years or more. Residence time can be assessed through the knowledge of hydrodynamic conditions or the use of environmental tracers (CFC, tritium, ^{14}C, . . .). For example, a detailed typology has been implemented for drawing the map of hydro-geochemical units of outcropping aquifers in several European states (Wendland *et al*., 2008). This typology is consistent with the variability of, among others, pH and Ca in groundwater from sand and gravel aquifers (Figure 2.2.1). A contrast is indeed highlighted between glacial sand and gravel deposit (squares) and fluviatile deposits (diamonds) related to longer and shorter residence time of groundwater respectively (Figure 2.2.1). A similar difference is observed between carbonate-bearing (solid symbol) and carbonate-poor (empty symbols) aquifers, which could represent another breakdown criterion.

As illustrated for pH and Ca in groundwater from sand and gravel aquifers, the impact of interaction time or any other secondary parameter on the chemical composition of groundwater can be apprehended through a statistical investigation applied to the overall aquifer. The secondary parameters proposed for a typology are not only useful to evidence variability among aquifers but to apprehend the evolution of chemical composition along the flow path within a given aquifer. For instance, recharge or interaction time increasing along the flow path, can lead to a characteristic chemical evolution of the groundwater. Examples are the freshening of saline aquifers or the accumulation of specific elements in groundwater as illustrated in the following.

Recharge of a saline aquifer by rainwater leads to freshening driving the overall chemical composition of groundwater and inducing an evolution of groundwater chemistry along the flow path. Actually, freshening processes of a saline aquifer result in gradual dilution of the interstitial water, followed by cation exchange, in which calcium ions from the recharge water are exchanged with the cations of marine origin adsorbed to the clay minerals. The resulting release of the dominant marine cation sodium into the solution leads to a sodium-bicarbonate watertype, typical for freshening conditions. This reaction may cause high sodium concentrations of around 500 mg/L in fresh water (and much higher in brackish water), which exceed the EC acceptable drinking water level of 200 mg/L. In a subsequent stage of cation exchange, marine potassium and magnesium are released to groundwater from clay surfaces, where they are again replaced by calcium. Also ammonium is desorbed from the clay minerals, rising in the groundwater. Yet, ammonium is not present in appreciable concentrations in seawater, but is formed in the interstitial solution at the sea bottom, from the decomposition of organic matter. As a result of these exchange reactions, potassium and ammonium may rise well above the recommended drinking water levels of 12 mg/L and 0.5 mg/L respectively (Walraevens *et al*., 2007).

Fluoride is a typical example of an element whose accumulation in groundwater is favoured by interaction time. If fluoride is essential to enamel formation and dental health, when consumed in excess it provokes dental fluorosis, and even skeletal fluorosis not to say crippling fluorosis depending on the excess level. Consequently, the tolerance limit in drinking water has been fixed at 1.5 mg/L by the WHO. Fluoride is commonly a minor component in groundwater and is mainly derived from the weathering of rocks, either from dissolution of F- minerals such as fluorite (CaF_2) or fluoroapatite ($Ca_5(PO_4)_3F$) or leaching of fluorine occurring in substitution for hydroxyl positions in silicate minerals, as for example in biotite or amphibole. Actually, fluorite mineral (CaF_2) presents a low solubility and, depending on the concentration of both calcium and fluoride in ground-water, it may precipitate limiting the fluoride concentration in groundwater. Calcium is therefore a key component regulating fluoride concentration, and the lower the calcium, the higher the fluoride can rise, if present in the aquifer matrix. Any processes involving a decrease in calcium concentration favour the occurrence of high fluoride concentration. Such a decrease can occur through ion-exchange (substitution of Na by Ca on the mineral surface) during the circulation of groundwater within the aquifer. Such a situation is illustrated along the groundwater flow path in several European aquifers, independently of their lithological characteristic. For example, F concentration in groundwater of the Jurassic Limestone aquifer of Southwest France exceeds 2 mg/L and the increase of fluoride concentration is accompanied by an increase of Na/Ca (Figure 2.2.2). Another example is provided by the Palaeozoic Basement Aquifer in the western part of Flanders (Belgium).

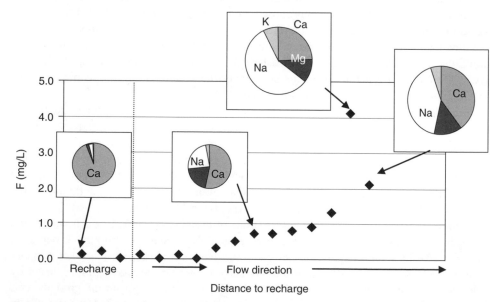

Figure 2.2.2 Evolution of fluoride concentrations in groundwater from a Jurassic limestone aquifer (southwestern France) according to the distance from the recharge zones. Pie charts illustrate the increase of sodium (Na) and decrease of calcium (Ca) concentrations with fluoride increase. Modified from Geosciences, la revue du BRGM pour une Terre Durable, issue no.5, March 2007, pp 68–73, www.brgm.fr.

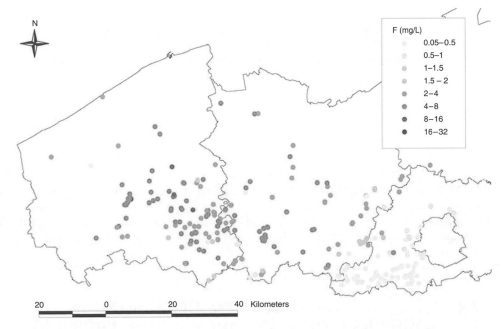

Figure 2.2.3 Distribution of fluoride concentration in the Palaeozoic Basement Aquifer in Flanders.

Freshening conditions in this marine aquifer have caused the soft sodium-bicarbonate type to be the dominant water type in some parts of the aquifer. The observed fluoride concentrations in this aquifer range from 0.05 to 30.9 mg/L (Figure 2.2.3). Low concentrations are found in the south-eastern part of the area (mostly in the province of Flemish-Brabant). Here, the drinking water level is not exceeded. Those low concentrations are limited to the $CaHCO_3$-watertype. The elevated calcium concentrations limit the rise of F through precipitation of the mineral fluorite. The high fluoride concentrations are found in very soft waters, of the $NaHCO_3$- and the NaCl-watertype. The calcium concentrations in these areas are below 10 mg/L, allowing for the fluoride concentrations to rise. Finally, in both cases, the older the water, the higher the fluoride concentration highlighting the slow kinetics of fluoride release.

2.2.3 TYPOLOGY TO ASSESS THE BEHAVIOUR AND FATE OF POLLUTANTS AS A FUNCTION OF GROUNDWATER CHEMICAL COMPOSITION

A first insight into the chemical composition of groundwater is provided by the knowledge of the aquifer lithology or any saline influence, namely through the primary criteria of the proposed typology. Chemical composition of groundwater is a key factor of (bio)geochemical processes impacting the fate and transport of many types of pollutants as it drives processes such as precipitation or sorption. All parameters are important and we can expect that the behaviour of many pollutants presents contrasts between aquifer types just because of differences in pH. This is particularly true for metals and metalloids, since some of them behave as anions (e.g. Se, As, Mo, . . .), whereas others (e.g. Cu, Pb, Zn, Cd) behave as cations, which signifies that adsorption rate decreases or increases with increasing pH. As early as 1995, Allard highlighted the fact that the range pH 5 to pH 7, most current in groundwaters, is particularly critical with respect to the sorption processes of Cu, Pb, Zn and Cd.

In Europe, the main threat for groundwater quality is diffuse nitrate and pesticide pollution from agriculture. The knowledge of the chemical composition of groundwater, and in particular pH, is also important in this domain. In groundwater, pesticides are subject to degradation as well as to adsorption onto clays and organic matter. But a significant and increasing share of the active substances used in Europe are ionizable pesticides and the formation of acidic metabolites is common during degradation. Due to their partial ionization within the range of pH in soils and groundwater, the adsorption of ionisable compounds has been recognized to be strongly influenced by the pH (Kah and Brown, (2006)).

2.2.4   EXAMPLES OF REACTIONS OCCURRING IN AQUIFERS UNDER ANTHROPOGENIC PRESSURE

One of the main difficulties in environmental studies is to establish the relative contribution of substances from geogenic and anthropogenic sources. The proposed parameters for an aquifer typology constitute a precious aid for determining the natural background level (see companion chapter, Blum *et al.*). However, even if the origin of a pollutant is natural (geogenic), its concentration may be strongly impacted by anthropogenic activities. Several situations have been reported throughout Europe where groundwater–aquifer interactions are modified following either human induced pollution or some change of hydrodynamics (quantitative status of groundwater) leading to a release of contaminants. Such situations are illustrated below. These examples illustrate the importance of the secondary parameters within our aquifer typology. They furthermore show how human activities modify one or several of the bio-geochemical processes described in Table 2.2.1 and how this impacts fate and transfer of chemical compounds in groundwater.

2.2.4.1   Denitrification and Leaching of Sulphate and Undesirable Elements in Hard-rock Aquifers of French Brittany

Nitrates from pollution of agricultural origin are considered as one of the main threats for groundwater quality in Europe. However, denitrification, namely nitrate reduction to gaseous species, enables removing nitrate from polluted groundwater. This process, generally microbially mediated, is mainly dependent on the presence of oxidizable compounds in the matrix, constituting the total reduction capacity (TRC; Pedersen *et al.*, 1991) of the rocks. TRC is mainly related to the content of organic matter, sulphides or ferrous iron. The presence and reactivity of these compounds are independent of lithology and denitrification can occur in limestones, chalk, sand and gravels, as well as in crystalline or volcanic aquifers. However, TRC is linked to the aquifer's depositional conditions and geological age and hence depends on the secondary parameters listed in Table 2.2.1. For example, in Flanders, it was shown that in Quaternary and young Tertiary, mainly terrestrial deposits, nitrate reduction is accompanied by the oxidation of organic matter, while in older, mainly marine sediments, the persistent organic matter apparently is less reactive, and nitrate reduction occurs together with pyrite oxidation (Eppinger, 2008).

Brittany in western France is characterized by intensive agriculture, which has developed over several decades and lead to high nitrate concentrations in surface- and shallow groundwater. Such a situation induces environmental concerns and may conduct local authorities to close down individual wells or to prohibit further use of the concerned groundwater bodies for drinking water purposes. The entire area is classified as vulnerable zone regarding the European Nitrate Directive. Its aquifers fall in the merged Hard rock group (Table 2.2.1) combining crystalline rocks and schists. Hundreds of points with indications of denitrification in groundwater have been revealed though, with sulphide minerals (pyrite) as main factor of the TRC. Actually, the observed denitrification processes do not attain completely NO_3^- free groundwater. The persistence of nitrates despite denitrification is mainly due to two factors:

- depletion of TRC, namely depletion of available pyrite, at least in some parts of the aquifer and mainly in shallow parts;

- variability of the denitrification rate. Actually, the half-life time (time after which the concentration of NO_3 is divided by 2) of denitrification by pyrite oxidation can vary from a few days to several years (Pauwels *et al.*, 1998).

Besides, secondary pollution may result from denitrification, depending on the electron donor: oxidation of organic matter may lead to acidification and increased mineral dissolution (increasing water hardness). Oxidation of sulphides releases sulphates and possibly arsenic and heavy metals. In French Brittany, increase of sulphate concentration linked to denitrification by oxidation of sulphide minerals has been evidenced, although in some cases these sulphates are removed from water through precipitation of new sulphate phases (Pauwels *et al.*, 2000). Anyway according to the stoichiometry of the reaction, the elimination of 50 mg/L of NO_3, the thresholds value of the nitrate directive, causes an increase in SO_4 of only 52 mg/L, which is relatively low compared with current quality standards for sulphates. However, in specific cases, it has been observed that denitrification could indirectly lead to much higher levels of SO_4. This is particularly illustrated at the Ploemeur site, where groundwater is exploited for drinking water supply of about 20,000 inhabitants. In recent years, SO_4 concentrations have risen sharply up to 100 and even 500 mg/L in observation piezometers located around the pumping well. For several years, denitrification caused the precipitation of iron hydroxide. But, the high and recent increase of sulphate concentration is partly interpreted as the re-mobilization of accumulated iron (III) to oxidize the sulphides, similarly to processes occurring under mining environment (Tarits *et al.*, 2006).

The sulphide minerals contain trace elements that may be released into groundwater during denitrification. The relationship between occurrence of denitrification and presence of metals and metalloids has received little attention compared to the large number

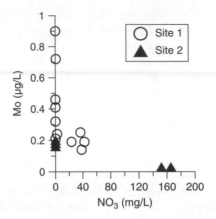

Figure 2.2.4   Mo concentration vs NO_3 at 2 sites where denitrification prevails.

of investigations on denitrification. However, as illustrated for Molybdenum at the Naizin and Lopérec sites (Figure 2.2.4) where denitrification prevails, abatement of nitrate concentration is inversely correlated to the trace element concentration. According to the chemical composition of iron sulphides and redox conditions prevailing during denitrification or subsequently to denitrification (when TRC is exhausted), mobilization of metals or metalloids with higher toxicity than Mo, such as Co, Ni, Zn is also noticed.

From the standpoint of sustainability of water resources and protection of groundwater, denitrification is a process which cannot be ignored. The two main reasons are the potential secondary pollution by sulphate and trace metals, even if their concentrations do not exceed quality standards, and the non-sustainability of the process. Denitrification lifespan must be tackled and compared to the residence time of groundwater as continuous nitrate reduction exhausts TRC, and the aquifer's potential to eliminate nitrate. In the worst case, nitrates increase could be observed even if agricultural practices were unchanged or even ameliorated to lower nitrogen loads.

2.2.4.2   Boron Mobilisation through Groundwater Exploitation of the Coastal Alluvial Cornia Aquifer (Italy)

The coastal alluvial Cornia Aquifer in Southern Tuscany combines several mechanisms of pollutant release listed in Table 2.2.1: man-induced salinization due to overpumping and seawater intrusion, salinity-triggered cation exchange processes as well as desorption of undesired trace elements from the clayey fraction of the aquifer material where they were fixed by previous geothermal activity (Bianchini *et al*., 2005, Pennisi *et al*., 2006, 2009). According to the aquifer typology, the Cornia Aquifer falls in the 'unconsolidated group' combining 'sand and gravels' and 'marls and clays' with 'saline influence' and, as a secondary parameter, the particular occurrence of geothermally impacted clays.

This aquifer is a demonstration of how the geochemical background may become a major regional water management problem when it is influenced by quantitative human impact on the groundwater body (pumping). The target compound of all geochemical studies conducted on this groundwater body is boron. Boron concentrations are already high in the upstream part of the aquifer and exceed, in the coastal band, where saline intrusion occurs, 8 mg/L (Figure 2.2.5). This concentration is around two times higher than seawater concentrations (4.5 mg/L), ruling out seawater boron as unique contaminant source, and exceeds the EU limit for drinking water by a factor of 8. Several geochemical and isotopic studies of both the aquifer sediments and the groundwater in and around the Cornia aquifer allow, at present, to set up the following scenario of processes leading to the extremely high boron concentrations in the lower part of the aquifer:

- **Geothermal activity:** In the upper part of the catchment lies one of the most active geothermal fields of Europe, the Larderello field. Natural outflow of geothermal steams and fluids occurred up to the nineteenth century, when they where captured to produce boric acid and, later, electric energy. These outflows contributed to the surface runoff and, probably, to groundwater flow over thousands of years at least. The alluvial sediments of the Cornia river basin contain high concentrations of boron (up to 144 mg/kg) but also remarkable anomalies of arsenic and antimony (Pennisi *et al*. 2009). Those trace elements are characteristic for geothermal fluids and are clearly related to the clayey fraction of the Cornia aquifer material (except for As). They may have been synsedimentarily fixed or adsorbed after sedimentation.

- **Desorption:** In contact with the present-day low-boron runoff, the fixed boron is mobilized which leads to the high concentrations (>1 mg/L) in the upper basin where groundwaters are mainly of Ca-HCO$_3$ type.

- **Pumping along the coastline** induced seawater intrusion and consequently high chloride concentrations in a strip of around 4 km along the shoreline with maximum values of 7 g/L in some pumping wells. Seawater mixes with local groundwater giving rise to Na-Cl dominated waters.

- **Ion exchange:** With respect to conservative mixing of seawater and groundwater, the coastal groundwaters show a clear deficit in monovalent cations (Na, K) but an equivalent excess of bivalent cations (Ca, Mg) so that, in some spots, the predominant ions are Ca-Cl. This cation exchange is triggered by the pumping-induced seawater intrusion. Simultaneously to cation exchange, boron is released to a much higher extent than in the upstream basin so that the final concentrations largely exceed seawater values. Predominant boron species are the negative ion borate and uncharged boric acid so that the exchange mechanism will not be the same as for cations but the influence of chloride rich groundwater on boron desorption is clearly evidenced.

The impact of this complex combination of geochemical mechanisms, partly natural, partly triggered by human activities in the river basin, on regional water management is evident. As the contaminant boron stems from the aquifer material itself, source control is not an option: No direct human input of boron exists that could be stopped. Even if the saline intrusion could be mastered through control of the groundwater abstraction near the shoreline, boron concentrations would still exceed the drinking water limit. In fact, the fresh groundwater inflow from the upstream part of the basin is already highly contaminated through simple desorption of boron from the aquifer material.

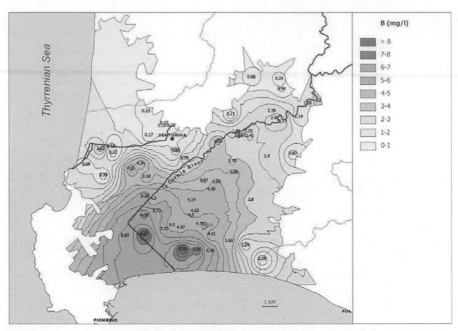

Figure 2.2.5   Geochemical map showing the boron concentrations in groundwaters of the Cornia plain alluvial aquifer (modified from Bianchini *et al*., 2005) (See Plate 1 for a colour representation).

Due to overpumping, the Cornia Aquifer is, at its present state, clearly of 'bad status' according to the WFD terminology. Though, given the high natural boron background values, it will never reach a status better than 'poor quality-good status' (see companion chapter by Dietmar Müller *et al*.): the overall geochemical background is too low to match drinking quality standards. This situation is further strained by the fact that almost known standard water treatment techniques, except high pH reverse osmosis with boron specific membranes, are inefficient with respect to boron (Kloppmann *et al*., 2008).

2.2.4.3   Increasing Arsenic Content in the Neogene Aquifer (Flanders) due to Groundwater Exploitation

The Neogene Aquifer occurs in the northeast of Flanders where it forms an important groundwater body with large amounts of potable water. The deposits consist mainly of sands, deposited in shallow marine to continental conditions. The sediments contain locally glauconite, lignite, vivianite and shells. A large part of the aquifer in the east and in deeper parts is decalcified. In the west and towards the north, layers contain calcite. The Boom Clay forms the base of the aquifer. In the south the aquifer is unconfined, towards the north the presence of a Pleistocene clay layer makes the aquifer semi-confined. A large amount of groundwater is extracted from the Neogene Aquifer for drinking water and industrial, agricultural and household purposes. An abstraction of more than 300 million m^3 per year from the aquifer is permitted. A regional groundwater flow model has been composed by Coetsiers *et al*. (2005). In this model only large groundwater extraction wells of more than 30,000 m³ per year were taken into account. The calculated piezometric head shows clearly influences from the extensive pumping activities. Depression cones are present where large groundwater abstraction wells are operating.

Lowering of the water table by pumping activities can influence the redox conditions prevailing in the surroundings of the production well. The occurrence of arsenic in groundwater is strongly influenced by the redox state of the aquifer. Arsenic may be released to groundwater by the oxidation of pyrite in the oxic zone. In this oxidizing environment arsenic can be sorbed onto the precipitated Fe(III)-hydroxides. In more reducing conditions arsenic is released while iron hydroxides are reduced. In the sulphate reducing zone arsenic can be removed from the groundwater due to incorporation into the formed iron sulphides. The shifting of redox zones by long-term pumping can thus influence the arsenic levels found in groundwater.

At several pumping stations, arsenic is seen to increase. In Figure 2.2.6 the evolution of arsenic and iron is given for a pumping station at Ravels for the period between 1990

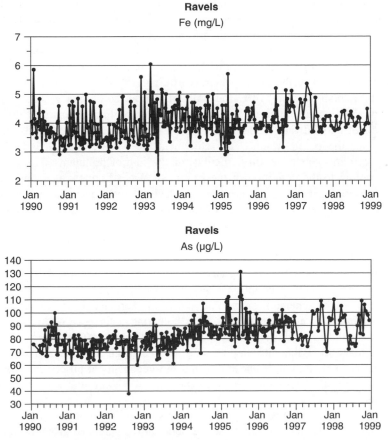

Figure 2.2.6 Arsenic and iron evolution in a production well at Ravels in the Neogene Aquifer in Flanders (Belgium) (Van Camp and Walraevens, 2008).

and 1999 (Van Camp and Walraevens, 2008). The pumping well extracts groundwater from 100 to 210 m deep. Although the arsenic-values show short time fluctuations, a linear fit of the series reveals an increase of around 20% over 10 years. The increase in arsenic is coupled to an increase in iron. Sulphate concentrations on the other hand are below detection limit in this production well, indicating sulphate reducing conditions. At other pumping stations, the increases in iron and arsenic are accompanied by increasing sulphate. The amount of arsenic dissolved in groundwater can reach more than 100 µg/L. Although the sources of arsenic and iron are natural, their increasing concentrations over time indicate that the pumping activities cause a shift in redox conditions, influencing the distribution of these species in groundwater.

2.2.5   CONCLUSIONS

The proposed scheme of a typology provides a general framework for understanding the natural chemical composition of groundwater as well as the fate and transport of pollutants, which represent a key challenge in the implementation of criteria for the assessment of good groundwater chemical status. It gives support to the related tasks such as threshold value derivation or monitoring issues.

Compared to previous hydrogeological typologies, the criteria proposed here consider the aquifer as a biogeochemical reactor where a great number of reactions drive the natural chemical composition of the groundwater. Those natural processes may furthermore be perturbed by human activity. All European aquifers can be classified according to their lithology or saline status. Such a classification can be the first indicator of the chemical composition of groundwater and of the potential behaviour of pollutants. Secondary parameters for the typology allow its refinement for a better knowledge of both the natural background levels and, in particular, of the fate of pollutants. The groundwater bodies according to the definition of the European Water Framework Directive, can be part of a non-homogeneous aquifer system and even comprise several aquifer levels. In both cases, this implies heterogeneities of the groundwater chemical composition and of the conditions governing the fate and transfer of pollutants. The two levels of criteria for the typology adapted to biogeochemical processes allow taking into account such heterogeneities.

Non-point source pollution of agricultural origin is a major threat for groundwater resources as it implies direct pollution of groundwater through transfer of nitrates or pesticides. The proposed scheme for a typology allows to apprehend their fate and transport. Secondary parameters also give hints to the potential occurrence of indirect pollution which may be linked to the reactivity of species like nitrate (e.g. trace metal release from denitrification). Pollutants may be of geogenic origin but their concentrations in groundwater are potentially enhanced through bio-geochemical processes. The resulting concentrations must be differentiated from the natural background level of elements in groundwater.

Overexploitation of resources may also induce secondary pollution, which implies that overexploitation is not only a threat for the quantitative status of groundwater, but can contribute to the deterioration of its chemical status. As illustrated above through the examples of arsenic or boron, such a pollution depends on the secondary parameters of the proposed typology. Impact of overexploitation on both pollution and deterioration can be expected to worsen in the coming decades due to increasing water demand or recharge deficiency in response to climate change and consequently cannot be neglected.

Acknowledgments

The content of this chapter has been initiated in the framework of the EU-Specific Targeted REsearch Project 'BRIDGE' (Background cRiteria for the Identification of Groundwater thrEsholds). The proposed scheme for a typology has been firstly discussed during a meeting attended by 19 European Institutes. The study on boron in the Cornia basin was conducted within the framework of the 5th FP European research project BOREMED (contract EVK1-CT-2000-00046) co-financed by the European Union. The views expressed in the chapter are purely those of the writers and may not in any circumstances be regarded as stating an official position of the European Commission.

REFERENCES

Allard B., 1995. Groundwater. In B. Salbu and E. Steinnes (eds), *Trace Elements in Natural Waters*, CRC Press, Boca Raton, US, 51–76.

Bianchini G., Pennisi M., Cioni R., Muti A., Cerbai N., Kloppmann W., 2005. Hydrochemistry of the Cornia alluvial aquifer (Tuscany, Italy). *Geothermics*, 34: 297–319.

Coetsiers M., Van Camp M., Walraevens K., 2005. Influence of the former marine conditions on groundwater quality in the Neogene phreatic aquifer, Flanders. In: Araguas L, Custodio E, Manzano M (eds), *Groundwater and Saline Intrusion, Selected papers from the 18th Salt Water Intrusion Meeting*, *Cartagena 2004*: 499–509.

Eppinger R., 2008. Mobility and degradation of nitrate in cenozoic aquifers in Flanders and their application to determine vulnerability of shallow aquifer systems (in Dutch). Ghent University – Laboratory for Applied Geology and Hydrogeology. PhD Dissertation.

Gilbrich W.H., 2000. International hydrogeological map of Europe. Feature article, *Waterway* 19, Paris, p. 11.

Griffioen J. *et al.*, 2006. State of the art knowledge on behaviour and effects of natural and anthropogenic groundwater pollutants relevant for the determination of groundwater threshold values. Final reference report- Deliverable 7- BRIDGE project.

IGRAC, 2005. International Groundwater Resources Assessment Centre – www.igrac.nl

Kah M., Brown C.D., 2006. Adsorption of ionisable pesticides in soils. *Rev. Environ. Contam. Toxicol.* 188: 149–218.

Kloppmann W., Vengosh A., Guerrot C., Millot R., Pankratov I., 2008. Isotope and ion selectivity in reverse osmosis desalination: Geochemical tracers for man-made freshwater. *Environ. Sci. Technol.*, 42: 4723–4731.

Novoselova L.P., 2004. Analysis of methods of hydrogeological mapping. *Water Resources* 31(6): 610–16.

Pauwels H., Kloppmann W., Foucher J.C., Martelat A., Fritsche V., 1998. Field tracer test for denitrification in a pyrite-bearing schist aquifer. *Applied Geochemistry* 13(6): 767–78.

Pauwels H., Foucher J.-C., Kloppmann W., 2000. Denitrification and mixing in a schist aquifer: influence on water chemistry and isotopes. *Chemical Geology* 168: 307–24.

Pedersen J.K., Bjerg P.L., Christensen T.H., 1991. Correlation of nitrate profiles with groundwater and sediment characteristics in a shallow sandy aquifer. *J. Hydrol.* 124: 263–77.

Pennisi M., Bianchini G., Kloppmann W., Muti A., 2009. Chemical and isotopic (B, Sr) composition of alluvial sediments as archive of a past hydrothermal outflow. *Chem. Geol.* in press.

Pennisi M., Bianchini G., Muti A., Kloppmann W., Gonfiantini R., 2006. Behaviour of boron and strontium isotopes in groundwater-aquifer interaction in the Cornia Plain (Tuscany-Italy). *Applied Geochem.* 21: 1169–83.

Schürch M., Kozel R. and Jemelin L., 2007. Hydrogeological mapping in Switzerland *Hydrogeology Journal* 15: 799–808.

Struckmeier W.F., Gilbrich W.H., Gun Jvd, *et al*., 2006. WHYMAP and the Groundwater Resources Map of the World at the Scale of 1:50 000 000. Special edition for the 4th World Water Forum, Mexico City.

Tarits C., Aquilina L., Ayraud V., Pauwels H., Davy P.H., Touchard F., Bour O., 2006. Oxido-reduction sequence related to flux variations of groundwater from a fractured basement aquifer (Ploemeur area, France). *Applied Geochemistry*, 21(1): 29.

Van Camp M., Walraevens K., 2008. Identifying and interpreting baseline trends. In: W.M. Edmunds & P. Shand (eds.), *Natural Groundwater Quality*. Blackwell Publishing, Oxford, pp. 131–54.

Walraevens K., Cardenal-Escarcena J., Van Camp M., 2007. Reaction transport modelling of a freshening aquifer (Tertiary Ledo-Paniselian Aquifer, Flanders-Belgium). *Applied Geochemistry* 22: 289–305.

Wendland F., Blum A., Coetsiers M., *et al*., 2008. European aquifer typology: a practical framework for an overview of major groundwater composition at European scale. *Environ. Geol*. 55: 77–85. DOI 10.1007/s00254-007-0966-5.

2.3

Visualising Groundwater – Aiding Understanding Using 3-D Images

Donal Daly[1], Steve Fletcher[2], Natalaya Hunter-Williams[3], Simon Neale[4] and Helen Simcox[5]

[1] *Environmental Protection Agency, Dublin, Ireland*
[2] *Numphra Consultancy, Cornwall, United Kingdom*
[3] *Geological Survey of Ireland, Dublin, Ireland*
[4] *Environment Agency Wales, Cardiff, Wales*
[5] *Scotland & Northern Ireland Forum for Environmental Research (SNIFFER), Edinburgh, Scotland*

2.3.1   BACKGROUND

Humankind has much to lose if the relationship with the geological environment is not managed effectively. It is essential that decision-makers use geoscientific information and expertise in land use and environmental planning as a means of achieving the maximum long-term benefit for all. The required information and expertise are now generally available, but are frequently not utilised adequately.

Groundwater Monitoring Edited by Philippe Quevauviller, Anne-Marie Fouillac, Johannes Grath and Rob Ward
© 2009 John Wiley & Sons, Ltd

The perception by many hydrogeologists of the groundwater environment has long been that, for the wider community, it is 'out of sight and out of mind'; this is often used as both a reason and excuse for the failure to get proper consideration of the 'underground environment'.

Hydrogeologists automatically see 'the world', whether a local area or the planet, in 3-D. This is not necessarily automatic for other stakeholders, whose 'mind picture' of the world is usually in 2-D, often perhaps with a vague 'underground' dimension that varies from 'hell's fire' to a general knowledge that rocks, minerals and groundwater are present. This 2-D vision is inadequate as a basis for environmental decision-making.

To a large degree, the knowledge and judgement of environmental issues among non-hydrogeological specialists and, in particular, among the general public, are based on what they perceive as 'common-sense knowledge' – knowledge deriving from their own personal experiences. In dealing with a situation that cannot usually be 'seen' or experienced, developing this 'common-sense knowledge' so that it includes an understanding of groundwater is a major challenge. However, it is a challenge that must be confronted and overcome. In addition, the threat to groundwater may be exacerbated as people in rural areas often have a mystical view of groundwater as springing or issuing from the ground at a point, without connecting it to what is happening in the surrounding area.

Essentially, there is only one solution to these situations – effective, convincing communication of relevant groundwater concepts and issues to diverse technical and non-technical audiences. As groundwater is hidden from view and as the concepts are often complex, success in achieving this solution is a fundamental requirement for ensuring that the results of groundwater assessments and monitoring are used properly in environmental planning and protection, and in implementing the Water Framework Directive.

Linguistic commentary, whether written or verbal, is the most commonly used means of communication of scientific issues. Conceptual models, which describe how real groundwater and surface water systems behave, are increasingly being used not only as the basis for numerical models, but also as a way of representing relevant groundwater aspects, such as flow, velocity, recharge, pollutant attenuation, the interaction between surface water and groundwater, etc. However, for most people, visual images are more effective than 'words' in developing 'common-sense knowledge', 'mind pictures' and 3-D conceptualisation. Therefore, there is a need for hydrogeologists to use good quality graphical communication material, with emphasis on 3-D images rather than cross sections, to aid the process of successfully communicating groundwater concepts.

2.3.2   PROVIDING A SOLUTION

The Water Framework Directive (WFD) considers the water environment as a continuum, and aims to manage our waters in a holistic and integrated way. Successful implementation of the Directive requires specialists from a wide range of scientific areas – hydrogeology, ecology, hydrology, hydrochemistry, engineering, environmental science, etc. – to work together to integrate both their specialist areas and the different components of the hydrological cycle. Public information and consultation is also a critical part of the implementation process. In considering these issues, the WFD UK-Ireland Water Framework Directive Groundwater Task Team concluded that, as good, relevant images are not available, graphical communication material should be produced to explain key concepts and hydrogeological settings, for use in presentations, publications and on websites. As a response, the Scotland & Northern Ireland Forum for Environmental Research (SNIFFER) commissioned and part-funded development of suitable graphics packages, which would be used by funding partners, such as the Environment Agency (England and Wales), the Geological Survey of Ireland, Environmental Protection Agency (Republic of Ireland) and SNIFFER member organisations. The resulting graphics material is now available to be downloaded, free-of-charge, from a specifically commissioned website: www.wfdvisual.com.

2.3.3   OBJECTIVES OF THE GROUNDWATER CONCEPTS COMMUNICATION PACKAGE

The main aims of the groundwater concepts communication material were (Daly *et al.*, 2006):

- to produce an initial 'library' of images, representing three or four common hydro-geological situations with a variety of pressures operating on them;

- to develop a dedicated website, that will enable easy download of images.

The images available for downloading were intended to:

- assist in enabling an improved understanding by non-hydrogeologists of groundwater, and its relevance to water and river basin management;

- provide a visual representation of the main topographical and hydrogeological set-tings, and how groundwater bodies function, in Britain and Ireland;

- illustrate the main pressures impacting on groundwater, surface water and ecosystems;

- emphasise the role of groundwater as a potential pathway for contaminants to reach surface water;

- allow a presenter to generate sufficient appropriate images to tell a hydrogeological 'story'.

The intended audience for the graphics consists of the following groups:

- 'non-specialists' such as planners, water managers and the general public;

- relevant environmental scientists, e.g. agronomists, ecologists, hydrologists,

- technical hydrogeologists.

2.3.4   CREATION OF THE IMAGE LIBRARY

The high-quality graphics are designed to illustrate conceptual models of hydrogeological systems that are important in the British and Irish contexts by using 3-D 'cutaways', with various land and surface water features shown on the ground surface, and a vertical cross-section through the subsoils (drift) and aquifers beneath the ground. However, while aimed mainly at users in Britain and Ireland, it is considered that the images may also be appropriate for hydrogeological settings in northern Europe and other areas with similar climate and geology. A river is shown on every image, and emphasis is placed on the interconnection and interdependencies between groundwater, surface waters and ecosystems.

Currently, the image library consists of:

1. a 3-D catchment block model of an overall upland-to-lowland catchment (i.e. part of a River Basin District), showing the basic components of the hydrological cycle, and other aspects such as groundwater body delineation (Figure 2.3.1);

2. more detailed topographic/land use settings located within the catchment, with three scenarios currently represented:

    o   upland rural (Figure 2.3.2);

    o   lowland rural (Figure 2.3.3);

    o   urban (Figure 2.3.4).

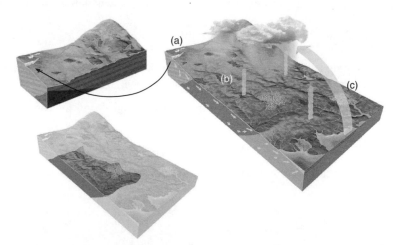

Figure 2.3.1   Block diagram showing a typical upland to lowland River Basin District (RBD) catchment, and the basic components of the hydrological cycle. Notice that groundwater flows through several aquifer types. It discharges to rivers, lakes, wetlands and the sea. The river basin comprises (a) upland areas, see Figure 2.3.2; (b) rural lowland areas, see Figure 2.3.3; and (c) urban lowland areas, see Figure 2.3.4. The insets show (top) the upland rural slice and (bottom) a groundwater body delineated using low flow aquifer boundaries and a topographic and groundwater flow divide that coincides with a topographic boundary within the basin. www.wfdvisual.com copyright 2005, SNIFFER

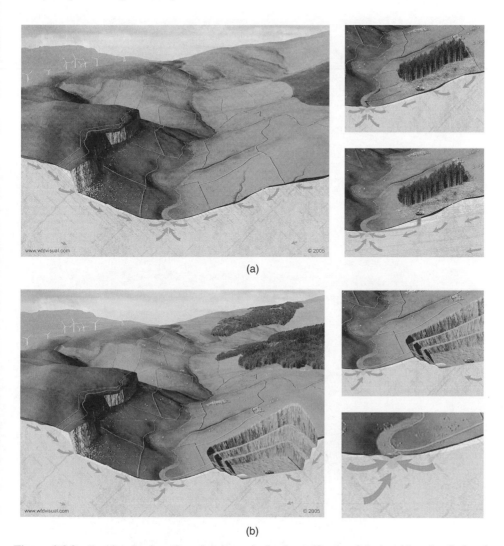

(a)

(b)

Figure 2.3.2 Rural upland setting showing a bedrock aquifer overlain by thin subsoil. Landscape elements include wind turbines, upland bog, grazing sheep, drystone walls, upland forestry. (a) Groundwater flow in a low transmissivity fissured aquifer. The insets show forestry development on a low transmissivity aquifer (top) and a productive sandstone aquifer (bottom). Note the different groundwater-surface water interactions, and overland flow of sediment-laden water to the river via drainage ditches (top) and flow of fuel to the river via groundwater (bottom). (b) Quarrying in a hard rock aquifer showing ponding in the base of the quarry as a result of the high water table. The insets show the discharge to the river of sump water contaminated by a fuel/oil leak. There is no pollution plume in the groundwater, as groundwater in the fissured bedrock aquifer is flowing upwards into the quarry. (c) (next page) Sheep dip contamination of a small stream via overground and subsurface pathways. Note the influence of the water table depth on the fate of the pollution. www.wfdvisual.com copyright 2005, SNIFFER

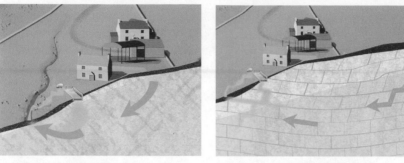

Fissured, low transmissivity aquifer   High transmissivity fractured limestone aquifer

(c)

Figure 2.3.2   (*continued*).

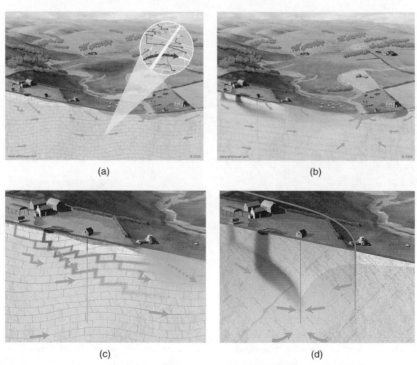

(a)   (b)

(c)   (d)

Figure 2.3.3   Rural lowland setting showing a bedrock aquifer overlain by subsoil (drift). The subsoil thickness can be varied to represent different degrees of aquifer protection. Both point and diffuse pressures may be represented. Receptors are ecosystems and drinking water supplies. (a) Highly vulnerable Chalk aquifer supporting bog, wetland and river ecosystems. Point and diffuse pressures are shown. Blowout shows representation of fracture/pore scale flow in a dual porosity system. (b) Highly vulnerable Sandstone aquifer supporting river flow. The pollution arising from the pressures shown in (a) impacts upon ecosystems and drinking water sources. (c) Point (fuel tank, slurry pit) and diffuse (fertiliser) pollution of a highly vulnerable Chalk aquifer shown in close up. The pollution plumes are migrating towards the river and wetlands under natural groundwater gradients. (d) Point (fuel tank, slurry pit) pollution of a highly vulnerable Fractured aquifer shown in close up. The pollution is drawn towards the pumping borehole and 'captured' within the zone of contribution (shown by red line). www.wfdvisual.com copyright 2005, SNIFFER.

(e)   (f)

Figure 2.3.3 (*continued*). (e) Fractured rock aquifer that is protected from slurry pollution by the thick, low permeability subsoil (drift) overlying it. The contaminant pathway is overground. (f) Migration of septic tank effluent towards a private borehole and the river in an extremely vulnerable limestone aquifer. www.wfdvisual.com copyright 2005, SNIFFER

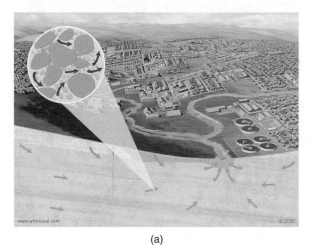

(a)

Figure 2.3.4 Urban lowland setting showing a bedrock aquifer overlain by subsoil (drift). The subsoil thickness can be varied to represent different degrees of aquifer protection. Typical urban point pressures are represented. Receptors are ecosystems and drinking water supplies. (a) Extremely vulnerable Sandstone aquifer underlying an urbanised area. Groundwater discharges to the river flowing through the town. The blowout shows a representation of groundwater flow at the pore scale.

(b)

(c)

Figure 2.3.4   (*continued*). (b) Migration of leachate from an unlined landfill to a borehole abstracting groundwater from a highly vulnerable Limestone aquifer. Note that flow into the aquifer from the river is being induced by pumping. The insets show (top) the migration of leachate to the river under the natural groundwater gradient in a highly vulnerable fractured bedrock aquifer, and (bottom) the natural containment of landfill leachate by thick, low permeability subsoils (drift) underlying the landfill. (c) Migration in a highly vulnerable Sandstone aquifer of hydrocarbons leaking from petrol station fuel tanks and from beneath a gasworks. Contamination of groundwater by light and dense non-aqueous phase liquids (LNAPLs and DNAPLs) are represented beneath the gasworks. Note that the DNAPL plume follows the stratification in the sandstone. Concentrations of petrol and DNAPL derivatives are high in the zone above the water table due to volatilisation of the hydrocarbons. www.wfdvisual.com copyright 2005, SNIFFER.

Table 2.3.1 Aquifer types currently represented in groundwater visualisation images.

Aquifer type and transmissivity (T)	RBD setting	Type of flow
1a Fissured rock. Low T & low productivity. 1b 1a plus weathered zone at top of rock.	Upland & Lowland	Fissure flow, concentrated at top of rock, along permeable faults; topography controlled, short flow paths.
2 Pure Limestone. High T and high productivity.	Upland & Lowland	Fissure flow; flat water tables, long flow paths.
3 Porous sandstone. High T & high productivity.	Upland & Lowland	Primarily intergranular with some fissure flow. Unconfined and confined, low hydraulic gradients, long flow paths.
4 Chalk. High T & high productivity.	Lowland	Mainly fissure flow, with diffusive flow through pores; flat water tables, long flow paths.

The topographic/land use settings summarised in Table 2.3.1 represent many of the key scenarios found in different regions of Ireland and Britain. Representation of typical pressures and impacts associated with the different land uses (Table 2.3.2) in each of the three settings is created in Photoshop®, using a series of interchangeable 'wallpapers' and layers. By adding or removing layers, a conceptual model can be built up to tell a particular story. A narrative can be created by using a series of increasingly-detailed images, with features emphasised by highlights and zooms.

The conceptual model can be considered within the source-pathway-receptor schema. For example, consider the contamination of an aquifer (pathway and receptor) due to a farm fuel tank and slurry pit leak (source) (Figure 2.3.3 (c), (d) and (e)). The contaminant pathways are mainly subsurface but, as shown, can also be overland if contaminants are prevented from infiltrating to the subsurface in low vulnerability settings (or zones of upward groundwater flow). The receptors are mainly ecosystems (rivers, groundwater-dependent ecosystems such as fens, lakes, etc.), but also include drinking water boreholes.

The graphics are generated using Adobe PhotoShop®, and saved as jpeg image files so that they can be imported and viewed by a wide range of computer applications. All the images are designed according to a standard style, and are named using the same convention. The names include an indication of the topographic/land use setting, aquifer type, aquifer vulnerability, pressures, and contamination types.

2.3.5 IMAGE DELIVERY

The images are available free of charge on the WFDVisual website (http://www. wfdvisual.com) for anyone who wishes to use them for educational or explanatory purposes (detailed conditions of use are available on the website). There are currently

Table 2.3.2 Currently-represented topographic/land use settings within a catchment, hydrogeology, and pressures and impacts associated with the different land uses.

Setting	Landscape elements[1]	Aquifer types[2] and vulnerability	Activity/Pressure[3]	Hazards	Pathway(s)[4]	Receptors
Rural upland	Windfarm, forest, bog, sheep grazing, drystone walls, scree, small streams, upland path, soil layer	Aquifer types: 1a & b, 2, 3 Vulnerability: Extreme	Forestry[P,D] Quarrying[P] Farming[P] Peat cutting	Fuel spills, sediment Fuel spills, sediment Sheep dip leaks/spills -	Over- and underground Over- and underground Over- and underground -	Streams and groundwater -
Rural lowland	Trees, farm buildings, grazing cattle & sheep, hedgerows, wetlands – fen, bog	Aquifer types: 1b, 2, 3, 4 Vulnerability: Extreme, High, Low	Rural housing[P] Farm buildings[P] Livestock grazing[D] Crop growing[D] Land-spreading[D]	Septic tank effluent Fuel tank leaks/spills, slurry pit leaks/overflow Nutrient loading Fertiliser application, agri-chemicals Nutrient loading	Underground Over- and underground Underground Underground Underground	River, wetlands, groundwater, water supply (borehole)

Urban lowland	Urban area blocks, radio masts, river culvert, sewage treatment works, wetland	Petrol station	Fuel tank leaks/spills	Underground	River, groundwater, water supply (borehole)
	Aquifer types: 1b, 2, 3, 4	Gasometer	Tank leakage/spills	Underground	
	Vulnerability: Extreme, High, Low	Landfill	Leaching from unlined/leaking	Underground	
		Quarry			
		Motorway	Road runoff	Underground	
		Colliery (mine)	Acid Mine Drainage, water table drawdown	Underground	

Notes: (1) Layers can be turned on or off in images. (2) See Table 2.3.1 for more details. (3) P = point pressure, D = diffuse pressure. (4) Pathway depends on aquifer vulnerability and water table elevation.

over 1000 images available, covering many permutations of the currently available topographic and land use setting, pathway characteristics (e.g. aquifer type, vulnerability), pressures relevant to the setting, and receptor type (e.g. river, wetland, borehole).

2.3.6 FUTURE PLANS

The initial drive in producing these images was to explain the more common hydrogeological concepts and situations. However, it has become apparent that the WFD has generated the need to explain increasingly more complex concepts as part of its implementation. The initial drawings are to be supplemented by new ones currently being produced. These will comprise the following four new settings:

- riparian river valley;
- high relief river valley area;
- estuary flood plain;
- lowland/upland karst;
- various ad hoc drawings showing detailed monitoring boreholes and borehole construction.

They will be used to illustrate the following concepts:

- wetland functioning;
- land drainage;
- estuary flood plain mechanisms – winter flooding, summer irrigation, water meadows;
- groundwater pumping near streams – contours, flowlines, zones of contribution to boreholes;
- effect of low, high and excessive groundwater abstraction;
- 3-D leachate plumes using complex cutaways;
- coastal saline intrusion under pumping and non pumping conditions;
- rivers connected to and separated from underlying aquifers;
- winter and summer water levels in chalk and poorly permeable aquifers, perennial stream heads;
- regional monitoring networks and monitoring groundwater using streams as a surrogate.

As previously, all the diagrams will be available with four different geology 'wallpapers' and with different groundwater monitoring arrangements.

2.3.7　CONCLUSIONS

1.　At the heart of effective river basin management lies the concept of an *integrated* approach to water bodies – groundwater, surface water, transitional and coastal waters – at the river basin scale; this is an important driver for the WFD.

2.　Successful integration needs to be based on a good conceptual understanding of the water flow regime, which includes 3-D visualisations of the conceptual models.

3.　Hydrogeolgists have a responsibility to communicate effectively and give the 3-D 'eye sight' (and insight) to surface water-oriented specialists, planners and the general public. This requires the following:

- giving priority to effective communication;

- giving priority to producing conceptual models (simplified representations of water systems, catchments, groundwater bodies);

- using 3-D diagrams to get the message across; preferably 'realistic' rather than schematic.

4.　This 'one picture is worth ten thousand words' philosophy provides the basis for the WFDVisual website http://www.wfdvisual.com

Acknowledgements

The initial development of the graphics was undertaken by Entec UK Limited; the website was produced and developed by Freeman Christie and Urban Element. John Dooley of the Cartography Section, Geological Survey of Ireland, produced most of the graphics currently in the image library.

REFERENCE

Daly D., Fletcher S., Hunter-Williams T., Martin G. and Neale S., 2006. Visualising groundwater – a tool to aid implementation of the WFD. *Proceedings of European Groundwater Conference 22–23 June 2006 in Vienna*; organised by Umweltbundesamt, Austria.

Part 3
Groundwater Pollutants and Other Pressures

3.1

Occurrence and Behaviour of Main Inorganic Pollutants in European Groundwater

Pieter-Jan van Helvoort[1], Jasper Griffioen[2] and W. Mike Edmunds[3]

[1] *Triqua B.V., Vadaring 7 Wageningen, The Netherlands*
[2] *The Netherlands Organisation for Applied Scientific Research (TNO), Built Environment and Geosciences, Utrecht, The Netherlands*
[3] *Oxford Centre for Water Research, Oxford University Centre for the Environment, Oxford, United Kingdom*

Groundwater Monitoring Edited by Philippe Quevauviller, Anne-Marie Fouillac, Johannes Grath and Rob Ward
© 2009 John Wiley & Sons, Ltd

3.1.1   INTRODUCTION

This chapter presents a state-of-the-art survey of the hydrogeochemistry of inorganic groundwater solutes that are considered as common groundwater pollutants according to the European Groundwater Directive and the Priority Substances List. The review focuses on groundwater conditions mainly within Europe but other examples from outside Europe are also considered. Each substance is discussed here in a standard format as follows: 1. mobility in groundwater environments, 2. natural occurrence addressing both general background concentrations as well as environments with elevated concentrations, and 3. anthropogenic contaminant sources. For every compound, a summary table presents the geochemical controls and the associated environment in terms of geology, climate and anthropogenic activities. Pollution cannot be defined without knowledge of the natural baseline concentrations and in a related publication the natural background concentrations for many inorganic constituents have been defined for a series of European reference aquifers (Edmunds and Shand, 2008a).

3.1.2   ALUMINIUM

3.1.2.1   The Mobility of Aluminium in Natural Waters

In the pH range of most natural waters, aluminium is not mobile. However, in acidic waters (pH < 5) aluminium can be mobilised by dissolution of primary alumino-silicate minerals and the accelerated weathering of clay minerals (e.g. kaolinite). The main control on aluminium solubility is usually gibbsite. Aluminium concentrations in acidic waters containing considerable amounts of sulphate could also be limited by the formation of Al-SO_4 minerals such as alunogen or alunite and hydroxysulphates such as basaluminite and jarosite (Nordstrom, 1982). In some cases, Al solubility could be limited by variscite precipitation ($AlPO_4.2H_2O$).

3.1.2.2   Natural Occurrence of Aluminium in Groundwater

Compared to its high abundance in the crust, Al concentrations are very low in natural waters due to the extreme low solubility of Al-bearing minerals at circumneutral pH. Concentrations in groundwater are strongly pH dependent (Nordstrom, 1982; Fest *et al*., 2007) and it is most unlikely that high aluminium will be found in well-buffered carbonate-containing lithologies. The median value in groundwaters from 25 European reference aquifers was 4 µg/l (Shand and Edmunds 2008). Hitchon *et al*. (1999) report a background concentration in the ppb range. Publications from the UK (Edmunds *et al*., 2003), Norway (Frengstad *et al*., 2000) and Germany (Kunkel *et al*., 2004) report median values in the range of 5–43 µg/l for pristine aquifers with quite diverse lithology. Concentrations up to the mg/l level occur in acidic, unconfined aquifers, especially consisting of unconsolidated materials and at shallow depths (e.g. Kjöller *et al*., 2004). Fast dissolution of alumino-silicates, clay minerals, and/or gibbsite in unbuffered soils or aquifers at low pH is the responsible process. In case of Al mobilisation, Al^{3+} may also take part in cation exchange reactions. Groundwater acidification leading to enhanced Al mobility could have several causes, including the release of acidity coinciding with the oxidation of sulphides or ammonia in the subsurface, or infiltration of acid rain.

3.1.2.3 Anthropogenic Sources of Aluminium in Groundwater

High groundwater aluminium concentrations are almost without exception linked to acidic conditions which may have either natural or anthropogenic causes. Industrial emissions, especially long-range pollution from fossil fuel combustion has caused acidification of shallow groundwaters (Edmunds and Kinniburgh, 1986) with concomitant Al mobilisation, although with emission reductions in Europe and elsewhere, rainfall acidity is now becoming less of a problem. Apart from atmospheric deposition, groundwaters may be affected by acid mine drainage or leachates from colliery wastes. Aluminium concentrations in mine drainage can reach 100s of mg/l in combination with other metals and extreme low pH's (<2). Fest *et al*. (2007) found that DOC-complexation enhanced the solubility of Al in groundwater below agricultural areas that contained several tens of mg DOC/l (Table 3.1.1).

Table 3.1.1 Sources and sinks processes of aluminium in groundwater.

	Natural environment	Anthropogenic/polluted environment
Source process		
Weathering of Al-silicates	No specific environment	
Dissolution of Al-oxides, kaolinite, clay minerals	Possibly acidic thermal waters (pH < 4)	Acidified aquifers, especially unconfined, non-calcareous aquifers only to shallow depth
		Aquifers affected by acid mine drainage or leachates from colliery wastes
Sink process		
Precipitation of secondary Al-silicates, Al-oxides, or Al-sulphates	Virtually all groundwaters with pH > 4	Virtually all groundwaters with pH > 4
Adsorption to DOC	DOC rich waters	DOC rich waters (agricultural leachate)

3.1.3   ARSENIC

3.1.3.1   The Mobility of Arsenic in Groundwater

The fate of arsenic in groundwater is made complex as a result of the large variety of arsenic species found under different redox and pH conditions (Smedley and Kinniburgh, 2002). Whereas most toxic elements occur as cations, arsenic occurs mainly as oxyanions, which tend to desorb as the pH increases from acid to near-neutral. At lower pH, As(V), and to a lesser extent As(III) can adsorb to a variety of aquifer minerals, including Al-(hydr)oxides, Fe/Mn-hydroxides and clay minerals. The most important sorbents are iron-(hydr)oxides (Fe-oxyhydroxide and goethite), which have strong sorption sites and are abundant as grain coatings in many aquifers. The adsorption of As is enhanced in the presence of freshly precipitated metal hydroxides, and decreases with ageing of mineral surfaces. Weaker adsorption of arsenic is expected when competing anions such as phosphate, bicarbonate and silicate, are present in the groundwater.

The mobility of arsenic is also controlled by precipitation/dissolution reactions. Arsenic occurs as impurities in several minerals, but iron (hydr)oxides and metal sulphides (pyrite) are the most important. Arsenic forms arsenopyrite (FeAsS) under strong reducing conditions (sulphate reduction) either in aquifers associated with large amounts of fresh organic matter, or ore mineralisation in hydrothermal systems (where the temperature is typically $>100\,^\circ C$). Under oxidising conditions and low pH, arsenate is likely to be incorporated in iron hydroxides. Organic forms of arsenic may be present in groundwater, but are generally negligible (Chen *et al.*, 1995).

Harvey *et al.* (2002) state that the input of fresh organic carbon or its degradation products may quickly mobilise arsenic due to carbon-driven reduction of Fe-hydroxides and/or desorption by carbonate ions. Desorption from Fe-hydroxides by competitive dissolved organic matter has been demonstrated by Bauer and Blodau (2006).

3.1.3.2   Natural Occurrence of Arsenic in Groundwater

Background levels of As in groundwater are mostly less than $10\,\mu g/l$, and sometimes substantially lower. High-As groundwater areas have been found in many regions of the world (e.g. Argentina, Chile, Mexico, China, West Bengal, Vietnam) as well as in Europe (Hungary, Spain, The Netherlands). In 25 European groundwaters it was found that the median concentration of arsenic was only $0.5\,\mu g/l$ with an overall range of $<0.05 - 79\,\mu g/l$ As (Shand and Edmunds, 2008). Naturally high arsenic is mainly related to oxidising or reducing environments where As has been adsorbed onto secondary Fe minerals and is notably high in regions with young, poorly mineralised sediments. In Europe high As concentrations have been described for example from the Neogene aquifer in Belgium (Coetsiers and Walraevens, 2008) where a median value of $6\,\mu g/l$ As and a maximum of $52\,\mu g/l$ As are recorded.

The concentrations of As may vary within a groundwater body as a result of time-dependent hydrochemical processes, as observed in the UK East Midlands aquifer (Smedley and Edmunds, 2001). In this aquifer, anthropogenic impacts are limited to the near outcrop and for As are negligible. The residence time of flow is around 30 kyr. Arsenic concentrations build up with time within the oxic section of the aquifer and then are less mobile in the anoxic groundwaters due to low Eh and pH-controlled surface reactions with iron oxides.

High natural arsenic concentration in groundwater are also found in closed basins in arid or semi-arid areas, or in strongly reducing aquifers of alluvial and deltaic sediments with rapid burial of large amounts of sediment together with high fresh organic matter contents (Smedley and Kinniburgh, 2002). Both environments contain geologically young sediments and to be in flat, low-lying areas where groundwater flow is sluggish. Historically, these are poorly flushed aquifers and any As released from the sediments following burial has been able to accumulate in the groundwater, even when the As content of the aquifer materials is not exceptionally high (in the range of 1–20 mg/kg).

There appear to be two distinct triggers that can lead to the release of As on a large scale (Smedley and Kinniburgh, 2002). The first one is the development of high pH (>8.5) and oxidising conditions in semi-arid or arid environments as a result of the combined effects of mineral weathering and high evaporation (salinity). In such environments, As(V) predominates and arsenic concentrations are positively correlated with those of other anion-forming species such as HCO_3^-, F^-, H_3BO_3, and $H_2VO_4^-$. The high pH leads either to desorption of adsorbed As from metal oxides, or it prevents them from being adsorbed.

Under strongly reducing conditions at near neutral pH typically found in young (Quaternary) alluvial aquifers, arsenic is mobilised by reductive dissolution of Fe and Mn oxides. Iron (II) and As(III) are relatively abundant in these groundwaters and SO_4 concentrations are low (typically 1 µg/l or less). Large concentrations of phosphate, bicarbonate, silicate and possibly organic matter can enhance the desorption of As competing for adsorption sites. A characteristic feature of areas with a groundwater arsenic anomaly is the large degree of spatial variability in As concentrations in groundwater. This means that it may be difficult, or impossible, to predict reliably the likely concentration of As in a particular well from the results of neighbouring wells and means that there is little alternative but to conduct intensive investigations of As occurrence. Arsenic-affected aquifers are restricted to certain environments and appear to be the exception rather than the rule.

Arsenic-rich groundwaters are also found in geothermal areas and hot springs, and are associated with decomposition and weathering of arsenopyrite. Arsenic concentrations in thermal waters of Iceland have been found to be in the range of 50–120 µg/l (White *et al.*, 1963).

Hernandez-Garcia and Custodio (2004) report high arsenic concentrations (up to 91 µg/l) in the Madrid Tertiary aquifer, one of the largest and most important aquifers of Spain for drinking water production. The aquifer is situated in the sedimentary basin of Madrid, consisting of cemented sands, silts, and clays of Tertiary age. The deeper regions of the aquifer are considered pristine and the groundwater chemistry has been derived from water-rock interaction processes. However, the water quality is affected by high arsenic concentrations, rising from <10 µg/l in the recharge areas to as high as 91 µg/l at discharge areas, thus showing an evolutionary trend from recharge to discharge areas. The natural arsenic contamination has its origin in pH-dependent anion exchange processes: with increasing residence time, the groundwater changes from Ca-HCO_3 to Na-HCO_3 type and pH increases to 9. The pH increase leads to desorption of anionic As-complexes from the fine sediment fraction which may contain as much as 20 ppm arsenic (Table 3.1.2).

Table 3.1.2   Sources and sinks processes of arsenic in groundwater.

	Subsurface conditions	Natural environment	Anthropogenic/polluted environment
Source process			
Anion competition and mineral weathering	High to near neutral pH under aerobic conditions	Arid regions with high evaporation rates	
Desorption due to reduction of As(V) to As(III)	Change to reducing conditions	River valleys and deltas with high burial rates and organic C	
Desorption due to reduction of surface area of oxide minerals by ageing	Near-neutral pH under aerobic conditions	?	
Desorption due to reduction binding strength by reductive dissolution of oxides	Strong reducing conditions	River valleys and deltas with high burial rates and organic C	
Mixing/dilution	Alkaline, high temperature	Geothermally influenced groundwater	
Dissolution of pyrite by oxidation	A change to aerobic conditions		Mining (Acid Mine Drainage), lowering of groundwater table
–	Contaminant source		Application of arsenic-based pesticides
Sink process			
Adsorption	Acidic to near neutral pH under aerobic conditions	Fluvial aquifers	Mining areas
Co-precipitation with metal sulphides	Sulphate reducing conditions	River valleys and deltas with high burial rates and organic C; hydrothermal systems	

Varsanyi *et al.* (1991) record high arsenic (up to 150 µg/l) in the Pleistocene groundwaters of the Great Hungarian Plain, as a part of the Pannonian Basin aquifer in Hungary. They found that arsenic concentrations were highly variable, even in the same well at different depths. They suggested that arsenic mobilisation may be controlled by complexation with organic molecules (humic substances).

3.1.3.3   Anthropogenic Sources of Arsenic in Groundwater

Anthropogenic arsenic contamination occurs with acid mine drainage (AMD) and where arsenic-based pesticides have been applied. In mining areas, extreme arsenic concentrations have been reported in highly acidic mine-effluent due to sulphide oxidation (up to 40 mg/l; e.g. Moncur *et al.*, 2005). Such extreme arsenic concentrations are often found with heavy metals, but arsenic is readily precipitated when AMD is neutralised to pH > 3. Arsenic mobilisation is also induced where dewatering of aquifers has resulted in a lowering of the groundwater table promoting pyrite oxidation. In addition, high arsenic concentrations have been found in the zone of fluctuation where water table oscillation occurs (Schreiber *et al.*, 2000).

The application of arsenic-based pesticides may be an accessory source for arsenic. Although the mobility of arsenic in top soils is low, leaching over long timescales may increase arsenic concentrations in groundwater under arable lands.

3.1.4 CHLORIDE

3.1.4.1 The Mobility and Natural Occurrence of Chloride in Groundwater

Chloride is extremely mobile and its mobility is not dependent on pH or redox conditions, neither does it form insoluble salts under environmental conditions. Chloride is an inert constituent and concentrations are derived from rainfall with a degree of evaporative concentration during recharge. The background Cl concentrations are also likely to be strongly linked to distance from coastlines in relation to the decreasing influence of the deposition of marine aerosols. Background levels of chloride concentrations in shallow unpolluted aquifers will be limited to a few tens of ppm. Baseline concentrations higher than this will be influenced either by non-marine formation evaporites, marine or non-marine formation waters or modern sea water. Groundwater transport of Cl in impermeable strata as clay may be affected by chemical osmosis (Neuzil, 2000; Garavito *et al.*, 2006), which is a very slow phenomenon.

Kunkel *et al.* (2004) estimate chloride baseline concentrations for a wide variety of aquifer lithologies in Germany, ranging from 1.0 mg/l in Alpine Limestones to 106 mg/l in unconsolidated Rhine Valley deposits. Unpolluted wells in a coastal alluvial aquifer in Mersin (Turkey) revealed Cl concentrations of about 40 mg/l (Demirel, 2004). Edmunds *et al.* (2002) report a median background concentration of 21 mg/l for the Chalk aquifer in Dorset (UK) with some seasonal variation for shallow groundwaters (Schürch *et al.*, 2004). Typical Cl concentrations in unpolluted groundwater sampled and analysed before 1945 vary from 5 to 55 mg/l in the Netherlands, where the lowest values refer to rain water origin and the highest to River Rhine infiltrate (Griffioen *et al.*, 2008).

High chloride concentrations in groundwater are due to additions from internal sources – either marine formation waters, recent seawater intrusion (which is often anthropogenically induced) or dissolution of marine evaporites; locally, thermal waters may be significant. Elevated chloride concentrations (e.g. $\sim$200–250 mg/l; Banaszuk *et al.*, 2005) may be found in shallow groundwaters directly under the plant root zone due to intensive evapotranspiration; this is likely in forested ecosystems and in semi-arid regions. Some other examples of brines, thermal circulation, and water-rock interactions (or combinations) are given below.

Conti *et al.* (2000) give an overview on the geochemistry of formation waters in the Po plain (northern Italy). The presence of brackish groundwaters results from mixing of meteoric water with deep-seated brines and subsequent upwelling to the surface due to a weak geothermal circulation system. Extreme chloride concentrations in the deep groundwaters influenced by brines reach up to 118,600 mg/l, whereas shallow groundwaters with a larger share of meteoric water may have up to $\sim$8000 mg/l.

Bein and Arad (1992) discuss saline groundwaters occurring in the Baltic region through freezing of seawater during glacial periods. The brines typically occur in crystalline and metamorphic rocks below fresh groundwater in various localities in Sweden and Finland. High total dissolved solids (up to 20 g/l in Sweden, and 120 g/l in Finland) are indicators of high chloride concentrations up to 16,800 mg/l. The brines are subsequently gradually diluted by fresh post-glacially infiltrated groundwater. Very high salinity groundwaters are also reported by Glasbergen (1985) and Grobe and Machel (2002). These authors observed chloride concentrations up to 62,000 mg/l resulting from evaporite (halite) dissolution, where the salts may occur as diapirs broken through younger geological formations.

In granitic rocks, Cl enrichment can result from water-rock interaction with rock-forming minerals rich in volatiles. Such cases are infrequent, but described for example from the Carnmenellis Granite (Cornwall, UK) by Edmunds *et al*. (1984a). Chloride concentrations up to 11,500 mg/l are found in thermal waters (around 54 °C) and the saline groundwaters are encountered in tin mines, and attributed to hydrolysis of Cl-rich biotites in the thermal aureole.

Hot thermal waters as a source of high Cl concentrations have been discussed by many workers in several countries. A good example comes from Valentino and Stanzione (2003), who describe three important Cl sources for groundwater associated with thermal activity: (1) inflow of magmatic HCl gas into deeply circulating groundwaters; (2) intensified rock leaching due to high temperatures; (3) uptake of seawater or marine components in powerful deep circulation systems. The combined effects of these processes led to very high Cl concentrations (up to 18,150 mg/l) in thermal groundwaters beneath the Phlegraean Fields (Naples, Italy).

3.1.4.2   Anthropogenic Sources of Chloride in Groundwater

There are many anthropogenic sources of chloride in groundwater, but in most cases Cl concentrations do not exceed 1000 mg/l (Hitchon, 1999). Ford and Tellam (1994) list industrial and domestic waste, sewage, road deicing salts, and processing chemicals as the main sources of Cl in shallow groundwater (up to 235 mg/l) of the Birmingham urban aquifer, UK. Edmunds *et al*. (1982) attributed chloride concentrations of 50–174 mg/l of the East Midlands Triassic sandstone aquifer either to induced recharge of nearby rivers or local pollution by agricultural and industrial activities.. Bank filtrated Rhine water in the Netherlands has led to Cl concentrations up to 160 mg/l (Stuyfzand, 1989). Stigter *et al*. (1998) suggest that high chloride concentrations (~350 mg/l) under irrigated fields in Campina de Faro (southern Portugal) are caused by the return flow of irrigation water into the aquifer. In this semi-arid climate, irrigation water is abstracted from local wells and Cl is subsequently concentrated due to evapotranspiration, which is at a rate of 86%. Application of fertilisers or manure is an important diffuse source of chloride beneath arable land. For example, in The Netherlands the chloride concentrations in modern groundwaters under arable land (~40–70 mg/l) are about a factor of 2 higher than in modern groundwater below nature areas (Van den Brink *et al*., 2007) (Table 3.1.3).

There are numerous studies on man-induced seawater intrusion due to overexploitation of coastal aquifers causing major problems in most, if not all Mediterranean countries

Table 3.1.3 Sources and sinks processes of chloride in groundwater.

	Natural environment	Anthropogenic/polluted environment
Source process		
Dissolution of evaporites	Evaporites	–
Degassing of magma	Deep thermal waters	
Seawater intrusion	Sea level changes coastal aquifers	Over exploited coastal aquifers
Thermal circulation	Thermal waters	
Leaching, rainwater infiltration, atmospheric deposition	Unconfined aquifers	Unconfined aquifers in urban and agricultural regions
Evapotranspiration		
Sink process		
There are no sinks for chloride in groundwater		

(Italy, Spain, Turkey, Greece, France, Cyprus), but to some extent in the coastal provinces in countries with temperate climate as well. The problem of overexploitation increases rapidly and is most profound in semi-arid regions due to ever-increasing urban developments and expanding resorts as built in the coastal plains. Chloride concentrations easily reach 1000 mg/l and this type of pollution can be easily recognised using major ion chemistry and stable isotopes of water.

3.1.5 MERCURY

3.1.5.1 The Mobility of Mercury in Water

The mobility of Hg is, like other heavy metals, mainly controlled by adsorption to solid particles and organic substances. The association of the Hg^{2+} ion with organic matter is very strong, and would be dominant when compared to adsorption to mineral surfaces (Schuster, 1991). Thus, the mobility of Hg in most soils is expected to be rather low. However, Hg^{2+} mobility is increased by ligands in solution (Cl^-, OH^-), and organic anions. The organic anions form strong complexes with mercury and have a weak positive or even negative net charge, thus diminishing the adsorptive tendency of Hg in water. Under reducing conditions and where sulphide is present, the mobility of Hg is likely to be determined by the formation of highly insoluble HgS, since the activity of free Hg^{2+} remains too low to exceed the solubility product of any other Hg solid. In thermal groundwaters, however, Hg would be more mobile due to increased solubility of HgS at high temperatures and salinity.

The mobility of Hg is also enhanced by methylation, a biochemical process leading to the formation of the very volatile monomethyl mercury ($HgCH_3^+$) and dimethyl mercury ($Hg(CH_3)_2$) compounds (MacLeod *et al.*, 1996). Methylation of Hg is enhanced in waters and sediments with low oxygen levels, low pH, and in the presence of sulphur-reducing bacteria. The methylated Hg species are entirely produced by microbial conversions and highly toxic for most life, including human.

3.1.5.2   Natural Occurrence of Mercury in Groundwater

In the survey of natural abundances in 25 European reference aquifers of sedimentary origins mercury had a median concentration of <0.05 µg/l (Shand and Edmunds, 2008). More specifically, Frengstad *et al.* (2000) reported a median Hg concentration of 0.018 µg/l and a maximum concentration of 0.13 µg/l for crystalline bedrock groundwaters in Norway, which was associated with Permian intrusive igneous rocks. In an extensive survey of German groundwaters, Kunkel *et al.* (2004) report background concentrations ranging from 0.03 µg/l in sandstone to 0.56 µg/l in unconsolidated sediments. Thus mercury has a very low natural aqueous abundance and anomalies in normal groundwaters are not to be expected.

Murphy *et al.* (1994) measured mercury species in 78 potable wells in southern New Jersey, and found that Hg^0 and $HgCl_2^0$ were the main species. Total mercury concentrations reported here as background values were in the range of 0.001–0.042 µg/l. A compilation of literature values by Allard (1995) revealed a wide range of background concentrations from a variety of geohydrological settings (0.0001–2.8 µg/l), but it is not clear whether or not these represent solely natural values.

Grassi and Netti (2000) found Hg concentrations up to 11.2 µg/l (drinking water limit is 1.0 µg/l) in some coastal alluvial aquifers in Tuscany (Italy), where increased groundwater abstraction has led to seawater intrusion. The elevated chloride concentrations enhance dissolution of mercury minerals originally present in the aquifer material by the formation of stable Hg-Cl complexes such as $HgCl_2^-$, $HgCl_3^-$, $HgCl_4^{2-}$, and $HgBrCl^-$. In a comparable study on Tuscany aquifers (Protano *et al.*, 2000) a maximum concentration as high as 39.7 µg/l was reported in unpolluted groundwater.

In some other studies, hydrothermal waters have been analysed for mercury. Thermal waters in the Phlegrean Fields (south-east of Rome, Italy) have up to 232 µg/l Hg (Valentino and Stanzione, 2003), while Aiuppa *et al.* (2000b) report incidental values slightly above 1.0 µg/l at Vulcano Island, Sicily. In these hot Na-Cl rich groundwaters, the main reason behind increased Hg concentrations is the formation of soluble chloride complexes such as $HgCl_2$ in combination with increased rock leaching.

3.1.5.3   Anthropogenic Sources of Mercury in Groundwater

The main anthropogenic sources for Hg are related to mining activities and metallurgical industrial activity. Passeriello *et al.* (2002) evaluated environmental contamination at an abandoned mining site located in the Tuscany region in Italy. The high mercury concentrations in local groundwaters (up to 5.2 mg/l) were attributed to a combination of mining activities and a geochemical aureole with naturally elevated background values. Cidu *et al.* (2001) report less extreme values (100 µg/l Hg) for an abandoned mine site at Monteponi, Sardinia. The reason for mercury mobilisation was increased weathering

Table 3.1.4 Sources and sinks processes of mercury in groundwater.

	Subsurface conditions	Natural environment	Anthropogenic/polluted environment
Source process			
Aqueous Cl complexation	Saline groundwaters	Coastal aquifers; brines; hydrothermal systems	Unconfined aquifers below fertilised soils with high salt input; seawater intrusion in coastal aquifers
Thermal sulphide dissolution	Thermal groundwaters	Hydrothermal systems	
Leaching from polluted soils			Unconfined aquifers below arable land treated with pesticides
Methylation	Sulphur reducing – methanogenic		Reduced aquifers below landfills pesticides
Sulphide oxidation	Transition from reduced to oxic environment		Mining areas and acid mine drainage
Sink process			
Adsorption	Reduced and oxic groundwaters	All aquifers with abundant organic matter	All aquifers with abundant organic matter
Sulphide formation	Sulphur reducing groundwaters	Sulphur reducing aquifers	Sulphur reducing aquifers

of primary sulphides present in the ore and mixing of formation water with shallow groundwater of higher salinity. The latter enhances both sulphide mineral dissolution and complex formation, increasing Hg mobility (Table 3.1.4).

Murphy *et al.* (1994) noticed that methyl-Hg originating from pesticides, comprised up to 8% of the total mercury concentration in groundwater, where the dominant aqueous forms were Hg^0 and $HgCl_2{}^0$.

3.1.6   NITROGEN (NITRATE AND AMMONIUM)

3.1.6.1   The Mobility of Nitrogen in Groundwater

Nitrogen is found in water systems in ionic form as NO_3 and NH_4 and to a minor extent as NO_2, and as dissolved gas. With the advent of NO_3 and NH_4 fertilisers in the twentieth century, regional contamination of groundwaters with nitrate has become a widespread problem both in Europe and outside (Griffioen *et al.*, 2005). Nitrogen is an essential nutrient for plant and animal growth. With the decay of dead organic matter under natural or agricultural conditions, NH_4 is released to the soil, where the nitrification process transfers ammonium into nitrite and nitrate. Nitrification within aquifers may also be recognised in groundwater plumes downstream from leaking organic-rich waste sites as landfills and septic tanks containing high ammonium concentrations. Nitrification is often observed along the flow path, especially across the fringe of the plume where oxic conditions prevail (Van Breukelen and Griffioen, 2004).

Ammonium competes with other cations for cation-exchange sites of the rock matrix. Sorption of NH_4 to clay minerals is stronger than to humic and fulvic acids. The clay mineral illite has a special preference for ammonium (and potassium), as it fits well in the clay mineral interlayer.

The oxidised nitrogen species – nitrate in particular – are highly mobile: if in excess for the needs of vegetation, nitrate will leach to the groundwater following nitrification under recharge conditions. Nitrate is chemically inert under persistent aerobic conditions. It may remain in solution for hundreds or thousands of years in non-reactive aquifers (Edmunds *et al.*, 1984b; Edmunds 1999). The reduction of nitrate to N_2 gas (denitrification) takes place in groundwaters when reductants are available. Feast *et al.* (1998) indicate that the optimal conditions for denitrification require availability of reductants, O_2 below 0.2 mg/l and pH in the range 7–8. Where denitrification is bacterially mediated, nutrients should also be present (Hiscock *et al.*, 1991). Microbial denitrification proceeds by heterotrophic and autotrophic bacterial pathways, which oxidise organic material (e.g. DOC, SOC). In environments poor in readily assimilable organic carbon, typical of many aquifers, denitrification may proceed autotrophically, mainly oxidising iron sulphide minerals (Ottley *et al.*, 1997). Instead of denitrification, dissimilatory nitrate reduction to ammonium may also happen: it is hypothesized that this happens when nitrate supplies are limiting instead of substrate limiting (Korom, 1992).

High denitrification rates and efficient nitrate removal from groundwater are typically found in topographically low areas with slow moving and shallow groundwater tables favouring anaerobic conditions. Typically, denitrification occurs in a narrow zone at or near the redox boundary, where the amount of electron donors in pore water and sediment increase over a short distance (e.g. Pedersen *et al.*, 1991). Rodvang and Simpkins (2001) conclude that unweathered Quaternary aquitards have the highest denitrification potentials with high amounts of labile organic matter and pyrite, protecting underlying aquifers for a long time from nitrate contamination. Denitrification is also intense in riparian zones, which have high ecological activity (e.g. Vidon and Hill, 2004), or during (natural or artificial) bank filtration due to active slurry within the river bed (Hiscock and Grischek, 2002). In sandy Quaternary aquifers denitrification is observed under anaerobic conditions, but capacity for denitrification may be much lower as reductants may be sparse and readily exhausted. In many aquifers a redox boundary can be identified (Edmunds *et al.*, 1984b), often – but not always – coinciding with a geological interface between oxidised and reduced sediments. In summary, in aerobic sections of the aquifer nitrate remains inert, but is becomes reduced upon depletion of oxygen.

3.1.6.2   Natural Occurrence of Nitrogen in Groundwater

Nitrate

Most groundwater studies on N focus on the fate of NO_3 and NH_4 in contaminated settings, not in pristine environments. Background values for nitrate in groundwater are difficult to establish, as many aerobic aquifers possibly containing background nitrate have been contaminated by anthropogenic nitrate sources. Studies of European aquifers from 25 countries confirm that a background concentration of around 1 mg/l NO_3-N is found quite widely in aerobic groundwaters where long-term records from the pre-war era exist in both carbonate and non-carbonate aquifers (Edmunds and Shand 2008a). Pore waters obtained from deep profiles in the Chalk of Dorset also illustrate well the presence of low nitrate (1–2 mg/l NO_3-N), beneath nitrate rich waters introduced by anthropogenic activitiy in recent groundwater (Edmunds and Shand, 2008b). This is also confirmed by Limbrick (2003) from the Chalk groundwater (Dorset, UK).

Elsewhere, however, high nitrate baselines have been reported. Rodvang and Simpkins (2001) reported extreme nitrate concentrations (100–400 mg/l) in groundwaters from organic-rich sedimentary rocks (Canada and several locations in USA). The occurrence of high nitrate groundwaters is associated with organic-rich shales, which bear NH_3 derived from mineralized organic nitrogen. There is also widely reported high nitrate from semi-arid regions where nitrogen accumulates in the unsaturated zone, percolating slowly to groundwater, associated with fixation by leguminous vegetation (Edmunds and Gaye, 1997). This is also a feature of deep aerobic palaeo-groundwaters in semi-arid areas.

Ammonium

Ammonium may occur naturally at low concentrations in many anaerobic groundwaters. Edmunds *et al*. (2002) showed for the Chalk aquifer in Berkshire (UK) that the natural baseline for ammonium in anaerobic groundwaters increased along the flowline due to leaching from minor amounts of marine clays, especially illite, reaching concentrations in excess of 0.5 mg/l. The median concentration for 25 European aquifers was 0.15 mg/l NH_4 (Shand and Edmunds, 2008)

Very high ammonium concentrations may be found in waters from ultrabasic rocks and thermal springs, such as Larderello-Traval geothermal waters, which have up to 162 mg/l NH_4 at 94 °C (Duchi *et al*., 1986). De Louw *et al*. (2000) found NH_4 concentrations around 35 mg/l in an aquifer in the coastal lowlands of the Western Netherlands, likely to have originated from anaerobic mineralisation of sedimentary organic matter as supported by extreme PO_4 concentrations and high CO_2 partial pressure (Griffioen, 2006). In a review of the mineralogical controls of ammonium Manning and Hutcheon (2004) reported anomalies in deep groundwaters and oilfield formation waters. Ammonium has been observed in concentrations ranging from 1 to 1000 mg/l and its abundance shows systematic variations with K. This suggests a mineralogical control, perhaps cation-exchange with illite or micas. All these studies indicate that natural concentrations of ammonium must be considered before citing pollution as the source (Table 3.1.5).

3.1.6.3 Anthropogenic Nitrogen Sources in Groundwater

The generally low natural background concentrations of both nitrate and ammonium in most groundwaters make these species good indicators of anthropogenic aquifer contamination. Several anthropogenic nitrogen sources are possible. The most important and widespread nitrogen sources are the application of inorganic fertilisers and manure from

Table 3.1.5   Sources and sinks processes of ammonium and nitrate in the subsoil.

	Subsurface conditions	Natural environment	Anthropogenic/polluted environment
Source process			
NH_4 release by organic matter decay	Reactive soil or sedimentary organic matter	In top soil and coastal lowlands	In agricultural soils
NH_4 release by cation exchange or desorption from clays or micas	Reduced and near neutral pH	Any natural aquifer containing clay mineral exchangers	Any aquifer containing clay mineral exchangers
Nitrification of NH_4	Oxic, unsaturated zone	In natural soils	Intense in fertilised soils
Nitrification of NH_4	Sub(oxic), fringe of contaminated groundwater plume		Below/downstream landfills, septic tanks, gasworks
Leaching of NO_3	Oxic to suboxic	Below natural soils, into phreatic aquifers	Below fertilised soils, into phreatic aquifers
NH_3 release by high temperature dissolution of igneous rocks	Reduced, high temperature	Geothermal areas	
Sink process			
Adsorption of NH_4 to clay minerals	Reduced, pH 5–8	Any natural aquifer containing clay mineral exchangers	Below/downstream landfills, septic tanks, gasworks
NO_3 conversion to N_2 by denitrification	Reduced, low O_2 (<0.2 mg/l)	Clayey or sandy aquifers/aquitards with shallow groundwater table, labile organic matter and/or inorganic reduced species; riparian buffer zones; river bank infiltrate	
NO_3 conversion to NH_4 by assimilatory reduction	Reduced	Reduced aquifers with NO_3 – poor groundwaters	

intensive livestock breeding on arable land and grass land. Nitrate is leached into the groundwater system via runoff or infiltration, especially in phreatic sandy aquifers and also in major carbonate aquifers such as the Chalk. In most EU countries a substantial part of the raw water supply from groundwater has nitrate concentrations above the EU drinking water limit of 50 mg/l nitrate (Strebel *et al.*, 1989).

Non-agricultural nitrogen sources may also lead to extreme nitrate and ammonium concentrations. Leakage from sewerage could lead to nitrate concentrations up to 150 mg/l in urban groundwater (Wakida and Lerner, 2005). The concentration of total nitrogen in effluents from septic tank systems ranges from 25 to 60 mg/l, with ammonia as the main component (20–55 mg/l). The ammonia in septic tank effluents may be oxidised to nitrate, which is easily leached to the groundwater. The anaerobic conditions under landfills may lead to groundwater plumes with NH_4 concentrations up to 1250 mg N/l, and also to high nitrate concentrations at the fringes of the plume where nitrification occurs (Christensen *et al.*, 2001). High N contamination also occurs below former coal gasification plant sites (50–1000 mg/l NH_4) and military facilities (20–200 mg N/l). Other N contamination sources include atmospheric deposition of NO_x, and mobilisation of N from the soil in large excavations at building sites promoting oxidation of organic N.

3.1.7 SULPHATE

3.1.7.1 The Mobility of Sulphate in Groundwater

The transport of sulphate in aquifers is mainly controlled by precipitation/dissolution of sulphate minerals and the redox transformations between sulphate and sulphide coupled with the precipitation/dissolution of metal sulphides. However, sulphate may generally be regarded as conservative in fresh groundwater as long as sulphate reduction is not taking place. Rates of sulphate reduction in aquifers are commonly low when compared to marine or limnic environments and controlled by fermentation rate of sedimentary organic matter (Jakobsen and Postma, 1999). In soils with abundant iron-oxyhydroxides and weak competition with other anions, sulphate retention by anion adsoption may occur.

Solubility limitations by gypsum, jarosite or other SO_4 minerals is relevant in volcanic and hydrothermal areas as well as in acid mine drainage areas and manipulated, aerated sulphide-rich clayey soils (so called acid-sulphate soils lacking carbonate buffering). The associated pore waters are likely to be brackish or saline, because of the high solubility of these SO_4 minerals.

3.1.7.2 Natural Occurrence of Sulphate in Groundwater

The natural (pre-industrial) background level sulphate concentration of the East Midlands Triassic sandstone aquifer (UK) was estimated to be well below 10 mg/l (Edmunds *et al.*, 1982; Edmunds *et al.*, 1996), while at greater depth in the same aquifer, anhydrite dissolution leads to a progressive sulphate increase (up to 365 mg/l, Edmunds *et al.*, 1982). It is interesting to note that sulphate reduction does not occur in the reducing section of this red-bed aquifer due to the lack of organic carbon and the slow rates of sulphur-reducing reactions.

Kunkel *et al.* (2004) estimated natural background levels for each of the main lithologies in Germany, and found the lowest concentrations in Alpine Limestone aquifers (13 mg/l), which was attributed to short residence times in these aquifers. In pristine parts of the unconsolidated aquifers in northern Germany, sulphate concentrations were in the range of 70 mg/l. For sandy aquifers in the Netherlands, sulphate concentrations of the older, pristine groundwaters have been estimated at 3.1 to 50 mg/l, where the lowest concentrations are associated with infiltrated rain water further away from the coast and the highest with infiltrated rain water along the coast or infiltrated River Rhine water. (Griffioen *et al.*, 2008).

Capaccioni *et al.* (2001) observed SO_4 concentrations up to 200 mg/l and incidentally 700 mg/l in carbonate formations with gypsiferous layers, yet the availability of gypsum was insufficient to reach gypsum saturation in the groundwater. In the Lincolnshire Limestone aquifer (UK), Bottrell *et al.* (2000) presented sulphur isotopic data for fissure-waters and pore-waters in the deep confined zone of this carbonate aquifer. They found good evidence of sulphate reduction of formation water taking place, removing sulphate from the fissure waters to a level below 25 mg/l, whereas the pore waters contained up to 230 mg/l SO_4. It was concluded that sulphate-reducing bacteria were unable to enter micro-pores due to small-sized pore throats, and sulphate reduction only could start from the moment that sulphate had diffused out from the pores into the fissures.

Grassi and Cortecci (2005) detected high sulphate concentrations (up to 1680 mg/l) in the Pisa alluvial plain (Tuscany, Italy), which were linked to thermal groundwater circulation within Mesozoic carbonate formations. The deep thermal sulphate-rich groundwaters move upwards through a fluvial gravel aquifer used for drinking water production. It was found that within the gravel aquifer sulphate was partially converted to sulphide minerals (pyrite) before it reached production wells. Cortecci *et al.* (2001) studied cold thermal waters from the Volcano Porto area, Vulcano Island, Italy. Isotopic signatures suggested that fumarolic SO_2 was the main source of sulphate in local groundwaters (up to 3190 mg/l). The contribution of seawater to the recharge of the groundwater system on the island was excluded. Another interesting example of sulphate mobilisation comes from the Krafla and Námafjall geothermal areas, northern Iceland. In a study by Gumundsson and Arnórsson (2002), high sulphate groundwaters (up to 525 mg/l) were found to be controlled by anhydrite dissolution. The increase of sulphate in thermal groundwater was found to be a reflection of cooling, as anhydrite solubility increases with decreasing temperature.

Mallén *et al.* (2005) showed that up-welling of sulphate-rich formation waters (~2500 mg/l) from Zechstein evaporite deposits below the Elbe Basin (Germany) have led to enrichment of sulphide and locally elevated sulphate concentrations in a Quaternary aquifer. Although most sulphate was reduced at the base of the Quaternary aquifer, still elevated levels (up to ~430 mg/l) were found near fractures, where up-welling is more intense and faster than sulphate reduction. This study is an interesting example of sulphate mobilisation/immobilisation processes occurring in natural groundwaters.

3.1.7.3   Anthropogenic Sources of Sulphate in Groundwater

The main anthropogenic sources of sulphate in groundwater are associated with mining of sulphide minerals, application of agrochemicals, coal processing activities, and contemporary acid rain or aerosol input. High sulphate concentrations could also be caused by the oxidation of sulphide-bearing aquifer material where anthropogenic intervention in the natural hydrogeological regime leads to aeration of the subsurface or increased oxidation of pyrite upon nitrate leaching from agricultural soils. Seawater intrusion is another process that increases sulphate concentrations together with Cl and other major seawater constituents, as found in many over-pumped coastal aquifers. Worth mentioning is that the introduction of increased SO_4 concentrations in wetlands may give rise to eutrophication leading to increased SO_4 reduction rates and extra mineralisation of organic matter, as well as accumulation of toxic sulphide, which reduces uptake of NH_4, PO_4 and K by living plants (Lamers *et al.*, 1998).

Polluted groundwaters in mining areas contain considerable amounts of sulphate. Sulphate and protons are principally released by sulphide oxidation and leached from ore-bearing rocks into the groundwater. The acidification may be buffered by carbonate dissolution or weathering. In Monteponi (Italy), groundwater contains up to 2620 mg/l sulphate (Cidu *et al.*, 2001), and near the Aznalcollar mining area in Spain 1362 mg/l sulphate was found in shallow phreatic aquifers (Santos *et al.*, 2002). Other sources of heavy sulphate pollution are coal processing plants, where sulphur is produced as a waste

Table 3.1.6 Sources and sinks processes of sulphate in groundwater.

	Subsurface conditions	Natural environment	Anthropogenic/polluted environment
Source process			
Dissolution of sulphate minerals		Limestone aquifers; evaporite deposits	
Oxidative dissolution of sulphide minerals	Aerobic	Young sandy aquifers; mineralised areas	Mining areas; coal processing plants; pyrite-bearing aquifers below fertilised soils
Sulphate leaching from surface	No specific environment		Unconfined aquifers below fertilised soils
SO_2 uptake	No specific environment	Uptake of fumarole gases	
Acid rain	No specific environment		Industrialised areas
Sink process			
Sulphide precipitation	Sulphate reducing	Highly reducing aquifers rich in organic matter	
Precipitation of gypsum, jarosite and other SO_4 minerals	Groundwater that is brackish or saline	Evaporite deposits, volcanic and hydrothermal areas	Mining areas, aerated sulphide-rich clay soils ('acid sulphate soils')

product. For example, Binotto *et al*. (2000) reported a maximum sulphate concentration of 6700 mg/l in a phreatic aquifer below a coal processing site (Rio Grande do Sul State, Brasil) (Table 3.1.6).

Minor sulphate contamination is found under arable lands, where the main source is excess fertilisers infiltrating in the aquifer. This type of contamination is generally not as high as industrial or mining inputs. For instance Moncaster *et al*. (2000) reported maximum sulphate concentrations of 105 mg/l under fertilised fields in the Lincolnshire Limestone aquifer (UK). Reduction of agricultural nitrate in association with pyrite oxidation will increase the groundwater sulphate concentration as described by Postma *et al*. (1991). They observed sulphate concentration up to 70 mg/l in a Danish unconfined sandy aquifer. Historic atmospheric deposition alone leads to mean SO_4 concentrations of 70 mg/l in shallow, oxic groundwater below natural areas in the Netherlands (Van den Brink *et al*., 2007), which is about ten times higher than the natural background concentration (Griffioen *et al*., 2008).

Semi-natural causes for elevated sulphate concentrations in groundwater were reported by Massmann *et al*. (2003). In their study, describing the Oderburch polder system in northeastern Germany, isotope data showed that the main sulphate sources were bank infiltration of polluted river water and the oxidation of sulphides occurring naturally in the aquifer underneath. Sulphate concentrations in groundwater ranged from ∼600 mg/l – which was close to the local river water composition – to up to 3500 mg/l in the centre of the polder. The latter represented oxidative pyrite dissolution in the subsurface, which was triggered by a hydraulic gradient reversal, as a result of the artificially maintained hydrological situation in that polder.

The effects of historical groundwater abstraction on groundwater quality from the Chalk of the London Basin (UK) have been described by Kinniburgh *et al*. (1994). They reported sulphate concentrations as high as 33,000 mg/l in the overlying Basal Sands. The extreme sulphate concentrations resulted from pyrite oxidation, which is present in only small quantities (<0.1–1.2%), but it is still forming (Kimblin and Johnson, 1992). These extreme sulphate concentrations indicate that the extent of oxidation is controlled by the entry of air on dewatering the aquifer rather than by normal concentrations of dissolved oxygen or nitrate in groundwater or recharge water. Due to over-extraction of the aquifer following dewatering, the groundwater table has now risen again, and the solutes were leached into the groundwater.

Recent seawater intrusion has been found to be the reason for increased sulphate concentrations (up to about 2000 mg/l) in several coastal aquifers in southern Spain (Martos *et al*., 1999; Martos *et al*., 2002), Italy (Grassi and Netti, 2000), and Cyprus (Milnes and Renard, 2004).

3.1.8 TRACE METALS: ZINC, CADMIUM, COPPER, LEAD AND NICKEL

3.1.8.1 The Mobility of Zn, Cd, Cu, Ni, and Pb in Groundwater

The mobility of Zn, Cd, Cu, Ni and Pb in water is largely dictated by pH-dependent sorption on clay minerals, iron oxyhydroxides or organic matter. The affinity for a sorbent is expressed as the distribution coefficient (K_d), expressing the partitioning of the metal between the solid and the aqueous phase. The adsorbed metal fraction increases abruptly from a critical pH value onward (sorption edge). The sorption edges of these metals are between pH 4 to 7 for Pb, Cu, Zn, Ni, and Cd in increasing order. Generally, the sorption edges coincide with increasing hydrolysis. However, the position of sorption edges, and thus distribution coefficients, may vary with type of sorbent and the presence of complexing agents.

Surface complexation of Cd, Ni and Zn by dissolved organic carbon has been demonstrated (e.g. Christensen and Christensen, 2000), and depends on metal concentration, metal type, pH, and type of DOC. The tendency for complexing increases with decreasing metal concentration, increasing pH and number of functional groups (carboxylic and phenolic groups) of the humic or fulvic acids. Although aqueous complexation increases Cd, Ni and Zn mobility, these workers suggested that the overall effect would be small compared to retardation by adsorption to solid sorbents.

Both Cu and Pb form even stronger complexes with organic matter. Dissolved organic matter (DOC) could thus be the dominant carrier of these metals in groundwater with high DOC at a neutral pH (Weng *et al*., 2002). Often, free Cu^{2+} concentrations in natural waters are far below the solubility of solid phases, suggesting that Cu is readily adsorbed to aquifer materials, or complexed by DOC (Jensen *et al*., 1999). Lead concentrations are generally low because of the extreme low solubility of Pb solids. Therefore, the transport of Pb may largely take place in adsorbed form with DOC.

Zinc, Cd, Cu, Ni and Pb do not participate in redox processes directly (valence is stable at +2), but they are affected by redox-dependent mineral precipitation/dissolution. For example, these metals will co-precipitate with sulphide under strongly reducing conditions, but will be mobilised when redox conditions change to oxic and sulphides get oxidized and dissolve.

3.1.8.2 Natural Occurrence of Zn, Cd, Cu, Ni, and Pb in Groundwater

The median baseline concentrations of these five metals in representative European reference aquifers (Shand and Edmunds, 2008) was found to be 11.5, <0.05, 1.2, 1.0 and 0.39 μg/l, respectively and their occurence is discussed further for each aquifer in Edmunds and Shand (2008a). Table 3.1.7 gives a brief overview of heavy metal concentrations in groundwaters. A study by Ledin *et al*. (1989) on background concentrations in Swedish groundwaters from crystalline rocks reports 2σ ranges for Zn (2–40 μg/l), Cd (0.006–0.1 μg/l), Cu (0.3–4 μg/l), Pb (0.02–0.3 μg/l), but no data for Ni. Although metal concentrations were low in these groundwaters, the effect of pH was proposed as the main factor controlling variations. The highest concentrations were found with lowest pH's, which were related to infiltrated acidic surface waters. In an extensive survey of bedrock groundwaters in Norway, Frengstad *et al*. (2000) report median concentrations of 14 μg Zn/l, 0.017 μg Cd/l, 16 μg Cu/l, 0.53 μg Ni/l and 0.36 μg Pb/l, all except for Cu within or close to the ranges found by Ledin *et al*. (1989). The study by Frengstad confirms low background values in crystalline bedrock and the median values are comparable to the European reference aquifers cited above.

In an extensive survey on German groundwater composition Kunkel *et al*. (2004) estimate background values for as many as fifteen different hydrogeological environments. The background value ranges are: 9.9–196 μg Zn/l, 0.02–0.54 μg Cd/l, 0.89–10 μg Cu/l, 0.13–16 μg Ni/l, and 0.07–8.5 μg Pb/l, with systematically highest background concentrations in unconsolidated aquifers, and the lowest in basalt and/or carbonate aquifers. The high background values in unconsolidated aquifers may be anthropogenically influenced, but naturally elevated metal contents would be related to abundant clay minerals and other secondary minerals in these young sediments (Table 3.1.7).

High natural concentrations are often found in thermal waters and formation waters. For example, Valentino and Stanzione (2002) found Pb concentrations of 1.3 to 29 μg/l

Table 3.1.7 Representative heavy metal concentrations in groundwater (after Hitchon (1999) and Allard (1995), and completed with Christensen *et al*. (2001), Aiuppa *et al*. (2000a,b; 2005)).

Water type	Zn (mg/l)	Cd (mg/l)	Cu (mg/l)	Ni (mg/l)	Pb (mg/l)
groundwater	0.001–1.0	<0.0001–0.005	<0.0001–0.1	0.0001–0.03	<0.001–0.05
groundwater (polluted)	<30	3	>150	13 ?	up to 0.5
groundwater (maximum)	~30	>3	>150	23.4 ?	>3
formation water	575	1.02	up to 2.1	?	360
thermal water	>0.1	?	~0.05	0.0008–0.6	~0.001

in the thermal waters of the Phlegraean Fields (Naples, Italy). Metal concentrations in thermal groundwaters of Vulcano Island (Sicily) ranged from $10-100\,\mu g/l$ for Cu and Zn, $1-10\,\mu g/l$ for Ni and Pb, and $0.1-1\,\mu g/l$ for Cd (Aiuppa *et al.*, 2000a). Moderate to high concentrations for zinc ($88\,\mu g/l$), cadmium ($1.0\,\mu g/l$), copper ($10.8\,\mu g/l$), Ni ($1.6\,\mu g/l$) and Pb ($0.6\,\mu g/l$) in Mt Etna groundwaters have been reported by Aiuppa *et al.* (2000b). In the thermal waters described above, the elevated concentrations of trace metals – and also many other elements – mainly depend on (1) the chemical composition of the host rock; (2) the physico-chemical conditions of the weathering solution (temperature, acidity, redox conditions); (3) input of magmatic gases; (4) the formation of soluble complexes with increasing salinity. In formation waters, the usually extremely long residence time is another factor to increase metal concentrations due to prolonged water-rock interactions.

Pauwels *et al.* (2002) demonstrated that increased metal concentrations also occur in groundwaters in contact with ore deposits as found in the Iberian Pyrite Belt deposits, southern Spain. Pristine groundwaters in unmined parts of this sulphide mineralisation contain up to $2.0\,\mu g/l$ lead, and also elevated concentrations for other heavy metals, indicating the interaction of sulphide minerals with groundwater/formation water.

Saline groundwaters in Münsterland (Germany) have as high as $3.6\,mg$ Zn/l, $0.25\,mg$ Cu/l and $0.10\,mg$ Ni/l, but no detectable cadmium and lead (Grobe and Machel, 2002). Most examples found in the literature on elevated background concentrations for Zn, Cd, Cu, Ni, and Pb are associated with saline groundwaters in coastal aquifers, hypersaline groundwaters (formation waters) or mineralized areas.

Schürch *et al.* (2004) reported several natural processes that could enrich groundwater in Zn, Cd and other trace metals in chalk aquifers. Marking dissolution of impure chalks as the principal source, trace metals are secondarily released from iron hydroxides where oscillations in redox conditions occur, leading to an anomalously high Cd baseline concentration ($146\,\mu g/l$), and elevated Zn ($214\,\mu g/l$) concentration. Although very locally observed, this demonstrates that metals commonly diagnostic of anthropogenic activity could have a natural origin as well.

3.1.8.3 Anthropogenic Sources of Trace Metals in Groundwater

The natural cycling of heavy metals has been perturbed by human activities since early industrial times. Ever since, large quantities of metals have been released into the environment by ore mining and smelting, fossil fuel combustion, the metal industry, fertilisers (mainly Cd, Cu, Zn), sewage sludge, landfills, and countless other activities and sources. Most studies on trace metal contamination focus on their mobility in the atmosphere, surface waters and soils, but much less on groundwater. The reason may be that groundwater pollution by metals is rather limited due to significant retardation in soils and aquifer materials. In addition, trace metals are readily scavenged by colloidal particles, which are not likely to infiltrate to deeper groundwater. Nevertheless, high heavy metal input occurs via infiltration of acidic tailings in mining areas, leachates from landfills and contaminated areas where permeable soils, low sorption capacities and high recharge rates exist (Tanji and Valoppi, 1989). Some examples of elevated heavy metal concentrations and responsible processes are given below.

Santos *et al*. (2002) report high heavy metal concentrations in the alluvial aquifers along the Guadiamar River, just north of the Doñana estuary (southern Spain). Infiltration of acid mine spills produced by metal sulphide oxidation lead to extreme concentrations for Zn (238 mg/l), Cd (0.57 mg/l), Cu (3 mg/l), Ni (0.32 mg/l) and Pb (0.24 mg/l) in groundwater. Other mining activities in southern Spain led to heavy metal concentrations in the range: 2.3 mg Zn/l, 0.01 mg Cd/l, 1.7 mg Cu/l, 0.023 mg Ni/l, and 0.45 mg Pb/l (Pauwels *et al*., 2002). Cidu *et al*. (2001) also measured extreme concentrations in shallow groundwaters at an abandoned mining site in southwestern Sardinia (Italy), with 57.4 mg Zn/l, 0.11 mg Cd/l, and 0.47 mg Pb/l (no data for Ni). In these studies, the highest concentrations are found closest to the mining sites, and often in acidified shallow aquifers. Heavy metal concentrations tend to drop drastically further away from the source where contaminated waters had time to mix with unpolluted groundwater. These unpolluted groundwaters have higher pH enhancing metal sorption to the aquifer material, thus decreasing metal concentrations. To this end Van der Grift and Griffioen (2008) reported regional groundwater contamination with Cd and Zn by historic smelter emissions and soil leaching in acid, sandy soils in the Kempen region, the Netherlands. Zinc concentrations around 1000 µg/l and Cd concentrations around 10–20 µg/l were observed, where a clear pH-dependency was observed.

The concentrations of heavy metals in landfills show major variations but are generally low. Ranges and average concentrations were reported in a review on landfill leachate chemistry by Christensen *et al*. (2001), as summarised in Table 3.1.8. However, only a small part of the metals would occur as free metal ions, as is shown by numbers published by Jensen *et al*. (1999) indicating that a substantial part of heavy metals would be associated with colloidal fractions of both organic and inorganic origin (last column in Table 3.1.8. Taking into account colloidal matter, the total metal content in groundwater may be substantially higher than the ranges and average values listed in Table 3.1.7, but Christensen *et al*. (1996) already found that the overall metal mobility would still be confined to about 1–2% of the ambient groundwater velocity. The overall low mobility of heavy metals is attributed to dominant sorption processes, and co-precipitation with sulphide and carbonate commonly formed in landfill leachates (Kjeldsen *et al*., 2002) (Table 3.1.8).

Kjøller *et al*. (2004) described the effect of groundwater acidification on the mobility of trace metals in a sandy non-calcareous aquifer (Grinsted, Denmark). Acidification was caused by the infiltration of acid rain, and heavy metals like Cd and Ni were subsequently

Table 3.1.8 Heavy metals in landfill leachates (after Christensen *et al*., 2000).

Metal	range (mg/l)	average (mg/l)	colloidal fraction (%)
Zn	0.03–1000	0.6	24–45
Cd	0.0001–0.4	0.005	38–45
Cu	0.005–10	0.065	86–95
Ni	0.015–13	0.17	27–56
Pb	0.001–5	0.09	96–99

Table 3.1.9 Sources and sinks processes of trace metals in groundwater.

	Subsurface conditions	Natural environment	Anthropogenic/polluted environment
Source process			
Carbonate dissolution	pH neutral	Chalk aquifers	
Reductive dissolution of Fe-oxyhydroxides or Mn-oxides	oscillating redox environment	Shallow groundwater near redox cline	Polluted hyporheic zones, overexploited sulphide-bearing aquifers
Desorption	Acidic	High P_{CO2} waters	Mining areas, acidified aquifers
Sulphide oxidation	Change from anoxic to oxic	Sulphide-rich aquifers	Mining areas (ore and coal), sulphide-rich aquifers below agricultural fields
Complexation by dissolved organic matter	Reduced	High DOC waters	Landfill leachates, river bank filtrate
Chloride complexation	High salinity	Natural brines, formation waters, mineralized areas, saline groundwaters in arid sedimentary basins	Brines, waste disposals, coal burning sites
Thermal rock leaching	High temperature	Thermal groundwaters	
Sink process			
Adsorption	Near-neutral pH	Most natural aquifers	Down-gradient of polluted sites
Carbonate coprecipitation	Alkaline pH	Calcareous aquifers	No specific environment
Sulphide coprecipitation	Anoxic, sulphate reducing	Strongly reducing aquifers	No specific environment

released from the sediment and accumulated in a narrow zone around the acidification front. The maximum observed metal concentrations were 21.1 µg/l Ni and 0.56 µg/l Cd, being a sharp increase with respect to the background concentrations in the Grinsted aquifer, which are at the ng/l level (Table 3.1.9).

Lowering of the water table by groundwater abstraction may not only cause increased sulphate concentration in the presence of pyrite, it may also trigger Ni mobilisation by desorption from Mn-oxides in the re-submerged zone in examples where abstraction stops (Larsen and Postma, 1997). The Ni concentrations found were up to 230 µg/l. Seasonal mobilisation of trace elements by reductive dissolution of Mn-oxides and associated desorption was described by Von Gunten *et al.* (1991) for groundwater recharged from a polluted alpine river. The maximum mean concentrations observed were 5.8 Cu µg/l, 10 µg Zn/l and 0.28 µg Cd/l.

3.1.9 CONCLUDING REMARKS

The above presented survey of the hydrogeochemistry of common, inorganic groundwater pollutants indicates that a wide variety of conditions is encountered in Europe with respect to both natural systems and kinds of anthropogenic contamination. Some considerations arising from studies on individual substances are as follows.

- The problems arising from natural arsenic contamination are well recognised, though the processes at work not yet entirely understood,

- The problems with nitrate arising from anthropogenic activity are widely described for many different aquifer settings, but more attention needs to be paid to natural background values which may vary.

- Aluminium does not present a problem in most groundwaters where near-neutral pH conditions prevail. Its mobility is mainly restricted to acidified soils and shallow groundwaters in poorly buffered lithologies.

- For trace metals (Zn, Cd, Ni, Cu, Pb) the main conclusion is that anomalies in groundwater problems have usually anthropogenic causes rather than natural ones, because in pristine groundwaters at near-neutral pH, concentrations are very low. In addition, a large part of these metals would not occur as the free metal species, but as complexes or adsorbed on particulate matter or associated with dissolved organic molecules. The main exceptions are hydrothermal waters, which could be rich in heavy metals.

- Sulphate has received little attention as a possible groundwater pollutant. However, high sulphate concentration may happen due to both natural and anthropogenic causes: dissolution of gypsum or anhydrite, presence of thermal water or sea water, mine drainage, application of agrochemicals, coal processing activities, and contemporary acid rain or aerosol input.

- The natural occurrence of Hg is very low and may be restricted to surface waters rather than groundwater, because of its strong affinity for soil organic matter.

REFERENCES

Aiuppa A., Allard P., D'Allessandro W., *et al.* (2000a) Mobility and fluxes of major and trace metals during basalt weathering and groundwater transport at Mt. Etna volcano (Sicily). *Geochim Cosmochim. Acta*, 64: 1827–41.

Aiuppa A., Dongarra G., Capasso G. and Allard P. (2000b) Trace elements in the thermal groundwaters of Vulcano Island (Sicily). *J. Volcanology and Geothermal Res.* 98: 189–207.

Aiuppa A., Federico C., Allard P., Gurrieri S. and Velenz, M. (2005) Trace metal modeling of groundwater-gas-rock interactions in a volcanic aquifer: Mount Vesuvius, *Southern Italy. Chem. Geol.* 216: 289–311.

Allard B. (1995) Groundwater. In: Salbu, B., Steinnes, E. (eds), *Trace Elements in Natural Waters*. CRC Press, Boca Raton, U.S.. pp. 151–76.

Banaszuk P., Wysocka-Czubaszek A. and Kondratiuk, P. (2005) Spatial and temporal patterns of groundwater chemistry in the river riparian zone. *Agriculture Ecosystems & Environ.* 107: 167–79.

Bauer M. and Blodau C. (2006) Mobilization of arsenic by dissolved organic matter from iron oxides, soils and sediments. *Sci. Total Environ.* 354: 179–90.

Bein A. and Arad, A. (1992) Formation of saline groundwaters in the Baltic region through freezing of seawater during glacial periods. *J. Hydrol.* 140: 75–87.

Binotto R. B., Sachez J. C. D., Migliavacca D. and Nanni A. S. (2000) Environmental assessment: contamination of phreatic aquifer in areas impacted by waste from coal processing activities. *Fuel* 79: 1547–60.

Bottrell S. H., Moncaster S. J., Tellam J. H., Lloyd J. W., Fisher Q. J. and Newton R. J. (2000) Controls on bacterial sulphate reduction in a dual porosity aquifer system: the Lincolnshire Limestone aquifer, England. *Chem. Geol.* 169: 461–70.

Capaccioni B., Didero M., Paletta C. and Salvadori P. (2001) Hydrogeochemistry of groundwater from carbonate formations with basal gypsiferous layers: an example from the Mt Catria-Mt Nerone ridge (Northern Appennines, Italy). *J. Hydrol* 253: 14–26.

Chen J. H., Lion L. W., Ghiorse W. C. and Shuler, M. L. (1995) Mobilization of adsorbed cadmium and lead in aquifer material by bacterial extracellular polymers. *Water Res.* 29: 421–30.

Christensen J. B. and Christensen T. H. (2000) The effect of pH on the complexation of Cd, Ni and Zn by dissolved organic carbon from leachate-polluted groundwater. *Water Res.* 34: 3743–54.

Christensen J. B., Jensen D. L. and Christensen T. H. (1996) Effect of dissolved organic carbon on the mobility of cadmium, nickel and zinc in leachate polluted groundwater. *Water Res.* 30: 3037–49.

Christensen T. H., Kjeldsen P., Bjerg P. L., *et al.*, (2001) Biogeochemistry of landfill leachate plumes. *Appl. Geochem.* 16: 659–718.

Cidu R., Biagini C., Fanfani L., La R. G. and Marras I. (2001) Mine closure at Monteponi (Italy): effect of the cessation of dewatering on the quality of shallow groundwater. *Appl. Geochem.* 16: 489–502.

Coetsiers M. and Walraevens K. (2008) The Neogene aquifer, Flanders, Belgium. In: Edmunds, W. M. and Shand, P. (eds). *Natural Groundwater Quality*. Blackwell, Oxford, pp. 263–86.

Conti A., Sacchi E., Chiarle M., Martinelli G. and Zuppi G. M. (2000) Geochemistry of the formation waters in the Po plain (Northern Italy): an overview. *Appl. Geochem.* 15: 51–65.

Cortecci G., Dinelli E., Bolognesi L., Boschetti T. and Ferrara G. (2001) Chemical and isotopic compositions of water and dissolved sulfate from shallow wells on Vulcano Island, Aeolian Archipelago, Italy. *Geothermics* 30: 69–91.

Cox P. A. (1995) *The Elements on Earth: Inorganic Chemistry in the Environment*. Oxford University Press, Oxford.

De Louw P. G. B., Griffioen J., Van Eertwegh G. A. P. H. and Calf B. (2002) High nutrient and chloride loads in polder areas due to groundwater exfiltration. *2nd Int. Conf. New Trends in Water and Environ. Eng. for Safety and Life: Eco-compatible Solutions for Aquatic Environments. Capri, Italy, June 2002.*

Demirel Z. (2004) The history and evaluation of saltwater intrusion into a coastal aquifer in Mersin, Turkey. *J. Environ. Management* 70: 275–82.

Duchi, V., Minissale, A. A. and Rossi, R. (1986) Chemistry of thermal springs in the Larderello-Travale geothermal region, southern Tuscany, Italy. *Appl. Geochem.*, 1: 659–67.

Edmunds W. M. (1999) Groundwater nitrate as a palaeoenvironmental indicator. In: *5th International Symposium on the Geochemistry of the Earth's Surface. Balkema, Rotterdam, Reykjavik, Iceland*, pp. 35–8.

Edmunds W. M. and Gaye C. B. (1997) High nitrate baseline concentrations in groundwaters from the Sahel. *J. Environ. Qual.* 26: 1231–9.

Edmunds W. M. and Kinniburgh D. G. (1986) The susceptibility of UK groundwaters to acid deposition. *J. Geol. Soc. of London* 143: 707–20.

Edmunds W. M. and Shand P. (2008a) *Natural Groundwater Quality*. Blackwell, Oxford.

Edmunds W. M. and Shand P. (2008b) The Chalk aquifer of Dorset, UK. In: Edmunds, W. M. and Shand, P. (eds), *Natural Groundwater Quality*. Blackwell, Oxford. pp. 195–215.

Edmunds W. M., Bath A. H. and Miles D. L. (1982) Hydrochemical evolution of the East Midlands Triassic sandstone aquifer, England. *Geochim Cosmochim. Acta* 46: 2069–81.

Edmunds W. M., Andrews J. N., Burgess W. G., Kay R. L. F. and Lee D. J. (1984a) The evolution of saline and thermal groundwaters in the Carnmenellis granite. *Mineralogical Magazine* 48: 407–24.

Edmunds W. M., Miles D. L. and Cook J. M. (1984b) A comparative study of sequential redox processes in three British aquifers. IAHS Publication No. 150, Ed. E Eriksson. In: *Hydrochemical Balances of Freshwater Systems*. Eriksson, E., pp. 55–70.

Edmunds W. M., Smedley P. L. and Spiro B. (1996) Controls on the geochemistry of sulphur in the East Midlands Triassic aquifer, UK. In: *Isotopes in Water Resources Management*. IAEA, Vienna, pp. 107–22.

Edmunds W. M., Doherty P., Griffiths K. J. and Peach D. (2002) The Chalk of Dorset. In: Baseline Report Series 4. British Geological Survey Report CR/02/268N.

Edmunds, W. M., Shand, P., Hart, P. and Ward, R. S. (2003) The Natural (baseline) quality of groundwater in England and Wales: A UK pilot study. *The Science of the Total Environment*, 310: 25–35.

Feast N. A., Hiscock K. M., Dennis P. F. and Andrews J. N. (1998) Nitrogen isotope hydrochemistry and denitrification within the Chalk aquifer system of north Norfolk, UK. *J. Hydrol.* 211, 233–52.

Fest E. P. M. J., Temminghoff E. J. M., Griffioen J., Van der Grift B. and Van Riemsdijk W. H. (2007) Groundwater chemistry of Al under Dutch acid sandy soils: effects of land use and depth. *Appl. Geochem* 22, 1427–38.

Ford M. and Tellam J. H. (1994) Source, type and extent of inorganic contamination within the Birmingham urban aquifer system, UK. *J. Hydrol.* 156: 101–35.

Frengstad B., Midtgard S. A. K., Banks D., Reidar K. J. and Siewers U. (2000) The chemistry of Norwegian groundwaters: III. The distribution of trace elements in 476 crystalline bedrock groundwaters, as analysed by ICP-MS techniques. *Sci. Total Environ.* 246: 21–40.

Garavito A. M., Kooi H. and Neuzil C. E. (2006) Numerical modeling of a long-term in situ chemical osmosis experiment in the Pierre Shale, South Dakota. *Adv. Water Resour.* 29: 481–92.

Glasbergen P. (1985) The origin of groundwater in Carboniferous and Devonian aquifers at Maastricht. *Geol. Mijnbouw* 64: 123–9.

Grassi S., and Cortecci G. (2005) Hydrogeology and geochemistry of the multilayered confined aquifer of the Pisa plain (Tuscany, central Italy). *Appl. Geochem.* 20: 41–54.

Grassi S. and Netti R. (2000) Sea water intrusion and mercury pollution of some coastal aquifers in the province of Grosseto (Southern Tuscany, Italy). *J. Hydrol.* 237: 198–211.

Griffioen J. (2006) Extent of immobilization of phosphate during aeration of nutrient-rich, anoxic groundwater. *J. Hydrol.* 320: 359–69.

Griffioen J., Brunt R., Vasak S. and Van der Gun J. (2005) A global inventory of groundwater quality: First results. In 'Bringing groundwater quality research to the watershed scale', ed. N. R. Thomson. *Int. Assoc. Hydrol. Sci.*, public. no. 297, pp. 3–10

Griffioen J., Passier H. and Klein J. (2008) Comparison of selection methods to deduce natural background levels for groundwater units. *Env. Sci. Technol.* 42: 4863–9.

Grobe M. and Machel H. G. (2002) Saline groundwater in the Munsterland Cretaceous Basin, Germany: clues to its origin and evolution. *Marine and Petroleum Geology* 19: 307–22.

Gudmundsson B. T. and Arnorsson S. (2002) Geochemical monitoring of the Krafla and Namafjall geothermal areas, N-Iceland. *Geothermics* 31: 195–243.

Harvey F., Swartz H., Badruzzaman A., *et al.* (2002) Arsenic mobility and groundwater extraction in Banladesh. *Science* 298: 1602–6.

Hernandez-Garcia E. and Custodio E. (2004) Natural baseline quality of Madrid Tertiary Detrital Aquifer groundwater (Spain): a basis for aquifer management. *Environ. Geol.* 46: 173–88.

Hiscock, K. M. and Grischek, T. (2002) Attenuation of groundwater pollution by bank filtration, *Journal of Hydrology* 266, 139–44.

Hiscock K. M., Lloyd, J. W., Lerner, D. N. (1991) Review of natural and artificial denitrification of groundwater. *Water Res.* 25: 1099–1111.

Hitchon B., Perkins E. H. and Gunter W. D. (1999) *Introduction to Ground Water Geochemistry*. Geosciene Publishing Ltd., Sherwood Park, Alberta, Canada.

Jakobsen R. and Postma D. (1999) Redox zoning, rates of sulfate reduction and interactions with Fe-reduction and methanogenesis in a shallow sandy aquifer, Romo, Denmark. *Geochim. Cosmochim. Acta* 63: 137–51.

Jensen D. L., Ledin A. and Christensen T. H. (1999) Speciation of heavy metals in landfill-leachate polluted groundwater. *Water Res.* 33: 2642–50.

Kimblin R. T. and Johnson A. C. (1992) Recent localised sulphate reduction and pyrite formation in a fissured Chalk aquifer. *Chem. Geol.* 100: 119–27.

Kinniburgh D. G., Gale I. N., Smedley P. L., *et al.* (1994) The effects of historic abstraction of groundwater from the London Basin aquifers on groundwater quality. *Appl. Geochem.* 9: 175–95.

Kjeldsen, P., Barlaz, M. A., Rooker, A. P., Baun, A., Ledin, A., Christensen, T. H. (2002) Present and long-term composition of MSW landfill leachate: a review. *Crit.. Rev. Environ. Sci. Technol.* 32(4): 297–336.

Kjøller C., Postma D. and Larsen F. (2004) Groundwater acidification and the mobilization of trace metals in a sandy aquifer. *Environ. Sci. Technol.* 38: 2829–35.

Korom S. F. (1992). Natural denitrification in the saturated zone: A review. *Water Resour. Res.* 28: 1657–68.

Kunkel R., Voigt H, J., Wendland F. and Hannappel S. (2004) Die natürliche, ubiquitär überprägte Grundwasserbeschaffenheit in Deutschland. In: *Schriften des Forschungszentrums Jülich Reihe Umwelt Environment*. Forschungszentrum Jülich GmbH, Jülich, Germany.

Lamers L. P. M., Tomassen H. B. M. and Roelofs J. G. M. (1998) Sulfate-induced eutrophication and phytotoxicity in freshwater wetlands. *Environ. Sci. Technol.* 32: 199–205.

Larsen F. and Postma, D. (1997) Nickel mobilization in a groundwater well field: Release by pyrite oxidation and desorption from manganese oxides. *Env. Sci. Technol.* 31: 2589–95.

Ledin A., Petterson C., Allard B. and Aastrup M. (1989) Background concentration ranges of heavy metals in Swedish groundwaters from crystalline rocks: a review. *Water, Air and Soil Pollut.* 47: 419–26.

Limbrick K. J. (2003) Baseline nitrate concentration in groundwater of the Chalk in south Dorset, U.K. *Sci. Total Environ.* 2003: 314–16.

MacLeod C. L., Borcsik M. P. and Jaffé, P. R. (1996) Effect of infiltrating solutions on the desorption of mercury from aquifer sediments. *Environ. Technol.* 17: 465–75.

Mallen G., Trettin R., Gehre M., Flynn R., Grischek T. and Nestler W. (2005) Influence of upwelling Zechstein sulphate on the concentration and isotope signature of sedimentary sulphides in a fluvioglacial sand aquifer. *Appl. Geochem.* 20: 261–74.

Manning D. A. C. and Hutcheon I. E. (2004) Distribution and mineralogical controls on ammonium in deep groundwaters. *Appl. Geochem.* 19: 1495–1503.

Massmann G., Tichomirowa M., Merz C. and Pekdeger A. (2003) Sulfide oxidation and sulfate reduction in a shallow groundwater system (Oderbruch Aquifer, Germany). *J. Hydrol.* 278: 231–43.

Milnes E. and Renard P. (2004) The problem of salt recycling and seawater intrusion in coastal irrigated plains: an example from the Kriti aquifer (Southern Cyprus). *J. Hydrol.* 288: 327–43.

Moncaster S. J., Bottrell S. H., Tellam J. H., Lloyd J. W. and Konhauser K. O. (2000) Migration and attenuation of agrochemical pollutants: insights from isotopic analysis of groundwater sulphate. *J. Cont. Hydrol.* 43: 147–63.

Moncur M. C., Ptacek C. J., Blowes D. W. and Jambor J. L. (2005) Release, transport and attenuation of metals from an old tailings impoundment. *Appl. Geochem.* 20: 639–59.

Murphy E. A., Dooley J. and Windom H. L. (1994) Mercury species in potable ground water in southern New Jersey. *Water, Air Soil Pollut.* 78: 61–72.

Neuzil C. E. (2000) Osmotic generation of 'anomalous' fluid pressures in geological environments. *Nature* 403: 182–4.

Nordstrom (1982) The effect of sulfate on aluminium concentrations in natural waters: some stability relations in the system Al_2O_3-SO_3-H_2O at 298 K. *Geochim. Cosmochim.* Acta 46: 681–92.

Ottley C. J., Davidson W. and Edmunds W. M. (1997) Chemical catalysis of nitrate reduction by Fe (II). *Geochim Cosmochim. Acta* 61: 1819–28.

Passariello B., Giuliano V., Quaresima S., *et al.* (2002) Evaluation of the enviromental contamination at an abondoned mining site. *Microchemical J.* 73: 245–50.

Pauwels H., Tercier-Waeber M., Arenas M., *et al.* (2002) Chemical charecteristics of groundwater around two massive sulphide deposits in an area of previous mining contamination, Iberian Pyrite Belt, Spain. *J. Geochemical Expl.* 75: 17–41.

Pedersen J. K., Bjerg P. L. and Christensen T. H. (1991) Correlation of nitrate profiles with groundwater and sediment characteristics in a shallow sandy aquifer. *J. Hydrol.* 124: 263–77.

Postma D., Boesen C., Kristiansen H. and Larsen F. (1991) Nitrate reduction in an unconfined sandy aquifer: water chemistry, reduction processes, and geochemical modelling. *Water Resour. Res.* 27: 2027–45.

Protano G., Riccobono F. and Sabatini G. (2000) Does salt water intrusion constitute a mercury contamination risk for coastal fresh water aquifers? *Environ. Poll.* 110: 451–8.

Rodvang, S. J. and Simpkins, W. W. (2001) Agricultural contaminants in Quaternary aquitards: A review of occurrence and fate in North America. *Hydrogeology J.*, 9: 44–59.

Sanchez-Martos, F., Pulido-Bosch, A. and Calaforra, J. M., 1999. Hydrogeochemical processes in an arid region of Europe (Almeria, SE Spain). *Appl. Geochem.* 14: 735–45.

Sánchez-Martos, F., Pulido-Bosch, A., Molina-Sánchez, L. and Vallejos-Izquierdo, A. (2002) Identification of the origin of salinization in groundwater using minor ions (Lower Andarax, Southeast Spain). *Science of the Total Environment* 297: 43–58.

Santos A., Alonso E., Callejon M. and Jimenez J. C. (2002) Heavy metal content and speciation in groundwater of the Guadiamar river basin. *Chemosphere* 48: 279–85.

Schreiber M. J., Simo J. and Freiberg P. (2000) Stratigraphic and geochemical controls on naturally occurring arsenic in groundwater, eastern Wisconsin, USA. *Hydrogeol. J.* 8: 161–76.

Schürch M., Edmunds W. M. and Buckley D. (2004) Three-dimensional flow and trace metal mobility in shallow Chalk groundwater, Dorset, United Kingdom. *J. Hydrol.* 292: 229–48.

Schuster E. (1991) The behavior of mercury in the soil with special emphasis on complexation and adsorption processes - a review of the literature. *Water, Air Soil Pollut.* 56: 667–80.

Shand P. and Edmunds W. M. (2008) The baseline inorganic chemistry of European groundwaters. In: Edmunds, W. M. and Shand, P. (eds). *Natural Groundwater Quality*. Blackwell, Oxford, pp. 22–58.

Smedley, P. L. and Edmunds, W. M. (2002) Redox patterns and trace-element behavior in the East Midlands Triassic Sanstone aquifer, U.K. *Groundwater* 40: 44–58.

Smedley P. L. and Kinniburgh D. G. (2002) A review of the source, behaviour and distribution of arsenic in natural waters. *Appl. Geochem.* 17: 517–68.

Stigter T. Y., Van Ooijen S. P. J., Post V. E. A., Appelo C. A. J. and Carvalho Dill A. M. M. (1998) A hydrogeological and hydrochemical explanation of the groundwater composition under irrigated land in a Mediterranean environment, Algarve, Portugal. *J. Hydrol.* 208: 262–97.

Strebel O., Duynisveld W. H. M. and Bottcher J. (1989) Nitrate pollution of groundwater in western Europe. *Agriculture Ecosystems & Environ.* 26: 189–214.

Stuyfzand P. J. (1989) Hydrology and water quality aspects of Rhine bank groundwater in The Netherlands. *J. Hydrol.* 106: 341–63.

Tanji K. and Valoppi L. (1989) Groundwater contamination by trace elements. *Agriculture Ecosystems & Environ.* 26: 229–74.

Valentino G. M. and Stanzione D. (2003) Source processes of the thermal waters from the Phlegraean Fields (Naples, Italy) by means of the study of selected minor and trace elements distribution. *Chem. Geol.* 194: 245–74.

Van Breukelen B. M. and Griffioen J. (2004) Biogeochemical processes at the fringe of a landfill leachate pollution plume: potential for dissolved organic carbon, Fe(II), Mn(II), NH$_4$ and CH$_4$. oxidation. *J. Cont. Hydrol.* 73: 181–205.

Van den Brink C., Frapporti G., Griffioen J. and Zaadnoordijk W. J. (2007) Statistical analysis of anthropogenic versus geochemical-controlled differences in groundwater composition in the Netherlands. *J. Hydrol.* 336: 470–80.

Van der Grift B. and Griffioen J. (2008) Modelling assessment of regional groundwater contamination due to historic smelter emissions of heavy metals. *J. Cont. Hydrol.* 96: 48–68.

Varsanyi I., Fodre Z. and Batha A. (1991) Arsenic in drinking water and mortality in the southern Great Plains, Hungary. *Environ. Geochem. and Health* 13: 14–22.

Vidon P. and Hill A. R. (2004) Denitrification and patterns of electron donors and acceptors in eight riparian zones with contrasting hydrogeology. *Biogeochemistry* 71: 259–83.

Von Gunten H. R., Karametaxas G., Kraehenbuehl U., Kuslys M., Giovanoli R., Hoehn E. and Keil R. (1991) Seasonal biogeochemical cycles in riverborne groundwater. *Geochim. Cosmochim. Acta* 55: 3597–3609.

Wakida, F. T. and Lerner, D. N. (2005) Non agricultural sources of groundwater nitrate: a review and a case study. *Water Research* 39: 3–16.

Weng L., Temminghoff E. J. M., Lofts S., Tipping E. and Van Riemsdijk W. H. (2002). Complexation with dissolved organic matter and solubility control of heavy metals in a sandy soil. *Environ. Sci. Technol.* 36: 4804–10.

White, D. E., Hem, J. D., and Waring, G. A. (1963). Chemical composition of sub-surface waters. In M. Fleischer (ed.), *Data of Geochemistry* 6th edn. US Geol. Survey Prof. paper 440F, U.S. Geol. Survey.

3.2

Contaminant Behaviour of Micro-Organics in Groundwater

Jan Gerritse, Bas van der Grift and Alette Langenhoff

Deltares, Subsurface and Groundwater Systems, Geosciences laboratories, Utrecht, The Netherlands

Groundwater Monitoring Edited by Philippe Quevauviller, Anne-Marie Fouillac, Johannes Grath and Rob Ward
© 2009 John Wiley & Sons, Ltd

3.2.1 INTRODUCTION

This chapter deals with the behaviour of xenobiotic micro-organic compounds (MOCs) in groundwater. Xenobiotic organic compounds can broadly be defined as 'all organic compounds that are released in any compartment of the environment by the action of man and thereby occur in a concentration that is higher than natural' (Leisinger, 1983). To date, more than 1000 organic chemicals have been identified in groundwater (Christensen *et al*., 2002). Even though MOCs typically constitute less than a few percent of the total dissolved C in groundwater, their behaviour in the affected aquifer is of major concern because they pose a potential health risk (Brown and Donnelly, 1988). Strict drinking water standards are enforced in many countries, with acceptable concentrations often as low as 0.1 µg/l for individual MOCs. This review focuses on the attenuation of organic compounds of the Water Framework Directive (WFD) pollutants and EC Priority Substances in groundwater in different redox environments (Table 3.2.1). Attenuation of MOCs is (besides dilution) due to sorption, (bio)degradation and volatilization. Volatilization is only important for volatile compounds at shallow depths and in unsaturated zones, and as such will not be discussed here.

3.2.2 SORPTION OF MICRO-ORGANICS IN GROUNDWATER

Sorption of organic compounds is one of the key processes governing their environmental behaviour (Schmidt *et al*., 2005). Sorption influences transport, in particular in porous media but also across compartment interfaces, bioavailability and, consequently, biotransformation of organic compounds. The phase transfer process to a solid can involve *ad*sorption occurring on surfaces, and solution into a bulk phase appropriately termed *ab*sorption. Depending on the type of soil organic matter (SOM), adsorption, and/or absorption into the three-dimensional SOM matrix are important. Typically, adsorption dominates at low aqueous concentrations, and partitioning becomes more important at higher concentrations, and nonlinear isotherms are found (Kleineidam *et al*., 2002). Adsorption to the particles is an equilibrium process while partitioning into the three-dimensional matrix is a much more slower process. This causes 'aging' or 'hystereses' which means that the sorption relation between the concentration in aqueous phase and in the solid during the time of net adsorption is different than during the period of net desorption. Many micro-organics are hydrophobic, which indicates that these substances have a low affinity for solution in (polar) water, and prefer solution in a-polar liquids. These pollutants are readily taken up in organic matter of sediment and are seldom found in groundwater. When looking at groundwater systems, focus should be given to the more hydrophilic micro-organics. The tendency to become sorbed is related to the distribution coefficient of the chemical and an a-polar liquid like octanol. The distribution coefficient between octanol and water (K_{ow}) can easily be measured and is highly reproducible. Therefore it is a good parameter to distinguish the micro-organics that can potentially be a threat to groundwater systems from the hydrophobics that will not spread in groundwater. Table 3.2.1 gives the log K_{ow} values of the WFD pollutants and the EC Priority Substances. There is a range in values from 1.25 for the highly hydrophilic

Table 3.2.1 Categories and substances of Water Framework Directive priority micro-organic contaminants and their potential fate in groundwater.

Category	Substances	Typical half-life ranges under "field" conditions (days)[1]		Major transformation products	Log K_{ow}	Potential threat to groundwater systems
		Oxic	Anoxic			
Aromatic hydrocarbons	Benzene	$1.4-\infty$[2]	$10-\infty$	Phenol, catechol	2.13	High
	Anthracene	i.d.[3]	i.d.	Hydroxylated polyaromatcs	4.54	Moderate
	Fluoranthene	i.d.	i.d.	Hydroxylated polyaromatcs	5.16	Moderate
	Naphthalene	11-169	$84-\infty$	Hydroxylated polyaromatcs, naphtoic acids	3.36	Moderate-High
Fuel oxygenates	Methyl *tert*-butyl ether	$574-\infty$	$617-\infty$	*Tert*-butyl alcohol	1.06	High
Chlorinated aliphatic hydrocarbons	Dichloromethane	143	91-108	Chloromethane	1.25	High
	Trichloromethane	i.d.	22	Dichloromethane, CS_2, CO	1.97	High
	1,2-Dichloroethane	32-42	$63-\infty$	Chloroethane, vinyl chloride, ethene	1.48	High
	Trichloroethylene	$25-\infty$	$30-\infty$	Chloroethene-epoxides, CO, vinyl chloride, di- and trichloroethylenes	2.29	High
	Tetrachloroethylene	∞	69-693	Tetra- and trichlorobutadienes	2.88	High
	Hexachlorobutadiene	$>1000-\infty$	>120	unknown	4.8	Low
	C10-13-chloroalkanes	i.d.	i.d		6^5	Low
Chlorinated aromatic hydrocarbons	Hexachlorobenzene	1000-2000	2.000-4.000	Tri- and dichlorobenzenes,	5.5	Low
	Pentachlorobenzene	i.d.	194-345	chlorobenzene	5.2	Low
	Trichlorobenzenes	i.d.	i.d.		4.05	Low
Substituted phenols	Pentachlorophenol	>14	$23-\infty$	tri- and dichlorophenols	3.32	Moderate
	Nonylphenols	7-21	$46-\infty$	unknown	4.48	Low
	Octylphenols	7-21	i.d.	unknown	3.7-5.3	Low

(continued overleaf)

Table 3.2.1 *(continued)*.

Category	Substances	Typical half-life ranges under "field" conditions (days)[1]		Major transformation products	Log K_{ow}	Potential threat to groundwater systems
		Oxic	Anoxic			
Chlorinated pesticides	Hexachlorocyclohexane	980–∞	187–∞	Benzene, chlorobenzene, chlorophenol	3.5-3.9	Moderate
	Endosulfan	32	148	Endosulphan sulphate, chlorendic acid	2.23–3.62	Low
	Alachlor	20	5	2-Hydroxy alachlor, alachlor-ethanesulfonic acid, -mercapturate, -oxanilic acid, -sulphonic acid	2.97	Moderate
	Atrazine	>60	159–∞	2-Hydroxyatrazine, deisopropylatrazine, deethylatrazine, deethylisopropylatrazine	2.5	High
	Simazine	110	71	Deethylsimazine, deethylhydroxysimazine, hydroxysimazine	2.2	Moderate-High
	Chlorfenvinphos	i.d.	i.d.	Diethylphosphate	3.84-4.22	Low
	Chlorpyrifos	113	136	3,5,6-trichloro-2-pyridinol, chloropyrivos oxygen analog, diethylphosphate, diethylthiophosphate	5.3	Low
Polar pesticides	Diuron	90–372	995	3,4-dichloroanilin, 3,4-dichloromethylphenyl urea, 3,4-dichlorophenyl urea	2.8	Moderate

Compound			Degradation products		Sorption
Isoproturon	12–40	>119	4,4'-diisopropylazobenzene, desmethylisoproturon, 4-(2-hydroxyisopropyl)aniline	2.5	Moderate
Trifluralin	1–169	21–126	3,4,5-benzotriamine	5.34	Low
Other organic compounds					
Brominated diphenyl ether	i.d.	700–1400	Brominated hydroxylated diphenyl ethers, -phenols, -mucoic acids, lower or non-brominated diphenyl ethers	4.28–9.9	Low
Di(2-ethylhexyl)phthalate	1–150	∞	Unknown	7.5	Low
Tributylin compounds	30–90	500–∞	Dibutylin, butylin, tin	3.2–4.5	Low

[1] Values reported from field studies or microcosms and columns under *in situ* groundwater conditions.
[2] ∞ Compound reported to be persistent.
[3] i.d. Insufficient data were found to report half-life.

dichloromethane to around 6 for the strongly hydrophobic C10-13 chloroalkanes. For substances with log K_{ow} values higher than 3.0–3.5 there is no real threat for groundwater contamination at a regional scale. However, local groundwater contamination with these substances possibly occurs through leakage from point sources like contaminated sites, landfills or obsolete pesticides stockpiles.

3.2.3   BIODEGRADATION OF MICRO-ORGANICS IN GROUNDWATER

Although abiotic chemical reactions sometimes result in partial transformation, biodegradation is the most important process for the complete removal of MOCs from groundwater systems. The term biodegradation is used liberally in the literature. It has been defined as compound disappearance (compared with no disappearance in sterile control experiments), by identification of end products (using e.g. radio-labelled molecules or isotopes), or by revealing degradation pathways, intermediate compounds, and active bacterial species or enzymes. In the groundwater and attached to the soil matrix a wide variety of microorganisms (mainly bacteria and fungi) performs biochemical reactions that decompose complex organic molecules into inorganic minerals. By using specific enzymes the microbes transfer electrons from electron donor substrates to electron acceptors (Figure 3.2.1). Through this mineralization process the microbes obtain their energy (e.g. ATP, NADH) and elements (e.g. C, N, P, S) for life and reproduction (new cells).

In many cases MOCs can be used as a sole source of carbon and electrons for growth. Aerobic bacteria and fungi use O_2 as terminal electron acceptor for the oxidation of

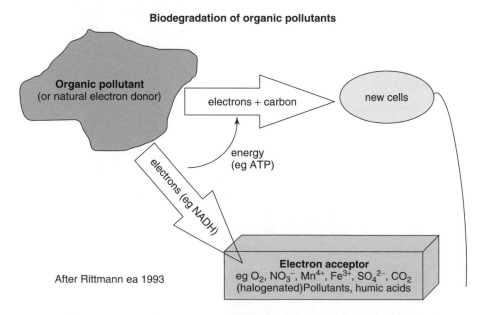

Figure 3.2.1   Principle scheme of the microbial biodegradation process (Rittman, 1993).

organic compounds. Important alternative electron acceptors that can be available in anoxic groundwater systems are nitrate, manganese(IV)oxides, iron(III)oxides, sulphate, carbon dioxide and humic acids. Many (poly)halogenated MOCs can also serve as electron acceptor and are reduced by 'halorespiring' or 'dehalorespiring' bacteria. Some MOCs can act as both electron donor and acceptor, yielding a mixture of reduced and oxidised products, in a process called fermentation. Sometimes microbes transform an organic compound without using energy or carbon from it. This 'cometabolism' can occur under oxic and anoxic conditions. It is mediated by enzymes that the microorganisms use to degrade their natural growth substrates, but have a broad substrate range and also transform the xenobiotic compound. The cometabolic transformation products are often used by other bacteria as growth substrates, resulting in complete degradation.

Degradation of MOCs in the environment depends on many factors. Obviously, microorganisms with proper biodegradation capabilities (enzymes) should be present or evolve in the polluted environment. When the properties of a xenobiotic MOC deviates much from natural substrates the microorganisms may not be able to readily decompose it. Hydrophobic compounds may be recalcitrant because they are not 'bioavailable' due to low solubility in the groundwater and sorption to the soil matrix. MOCs can also be toxic for the biodegrading bacteria. Finally, the environmental conditions must be suited for microbial activity. Favourable conditions for pollutant degradation include the availability of nutrients, pH in the range from 5 to 9, temperature between 5 and 30 °C and the presence of sufficient water. The availability of electron acceptors is crucial for the biodegradation of specific MOCs (Figure 3.2.2). Many organic compounds are best degraded in an oxidising environment. One reason is that the amount of energy that bacteria can obtain from the transfer of electrons from a reduced compound to an electron acceptor generally decreases in the order: $O_2 > NO_3^- > Mn^{4+} > Fe^{3+} > SO_4^{2-} > CO_2$. A second reason is that O_2 is not only used as terminal electron acceptor, but also as a direct

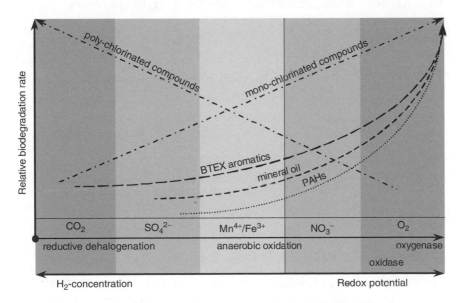

Figure 3.2.2 Trends in biodegradation of groundwater pollutants under various redox conditions.

reactant for the initial conversion of aromatics, such as PAHs, BTEX, and aliphatics such as mineral oil. In contrast, oxidised polychlorinated compounds are best dechlorinated in an anaerobic reducing environment. Under these conditions specific anaerobic bacteria can use organic substrates or H_2 to fuel reductive dechlorination reactions. The (partially) dechlorinated products often accumulate in groundwater systems, but can further be degraded when alternative electron acceptors become available.

3.2.4   AROMATIC HYDROCARBONS

Contamination of groundwater with benzene occurs at many sites, especially those associated with petrochemical industry like gasoline stations, former gaswork sites and landfills. Benzene contaminations often occur in combinations with other fuel components such as toluene, ethylbenzene, xylenes (TEX), oil alkanes and methyl *tert*-butyl ether (MTBE). In these situations benzene is usually the primary risk-determining contaminant. Polycyclic aromatic hydrocarbons (PAHs) consist of two or more fused aromatic rings. Their wide distribution and toxic, mutagenic and carcinogenic properties have aroused global concern. Combustion sources are thought to account for over 90% of the environmental burden of PAHs (UNEP, 2002; Howsam and Jones, 1998). Production and use of creosote and coal-tar are other sources. The contamination of groundwater with PAHs is seen at many sites, especially those associated with petrochemical industry like gasoline stations. Low aqueous solubilities of PAHs and high octanol-water partition coefficients (K_{OW}) often result in their accumulation in soils and sediments to levels several orders of magnitude above aqueous concentrations. In groundwater systems naphthalene, anthracene and fluoranthene are among the most common PAH contaminants.

3.2.4.1   Aerobic Biodegradation of Aromatic Hydrocarbons

Aerobic biodegradation of aromatic hydrocarbons starts via oxidation(s) with molecular oxygen (O_2) as a reactant (van Agteren *et al.*, 1998). These reactions transform the xenobiotics into compounds which are structurally more similar to natural substrates and can be funnelled into in the common metabolic pathways, resulting in full mineralization to CO_2 and water. Aerobic biodegradation of benzene starts with the oxidation to catechol by a dioxygenase (Figure 3.2.3).

The aromatic ring of catechol is subsequently cleaved by a catechol-dioxygenase, between the two hydroxyl substituents (ortho cleavage) or adjacent to them (meta cleavage) leaving non-aromatic organic acid(s), which are readily mineralized via the tricarboxylic acid cycle. The availability of oxygen is also important for biodegradation of PAHs. Aerobically, PAHs can be degraded by bacteria, fungi, algae and yeasts. Many aerobic bacteria have been isolated that can use naphthalene, anthracene or fluoranthene as the sole source(s) for growth and energy. Their metabolism involves dioxygenases which incorporate molecular oxygen in one of the aromatic rings (van Agteren *et al.*, 1998). The resulting dihydroxynapthalene, dihydroxyanthracene and dihydroxyfluoranthene are ring-cleaved by a second dioxygenase and further transformed to mono-aromatic acids or catechol, after which complete mineralisation can occur. Fungi cannot use PAHs as a sole source of carbon and energy for growth but transform these compounds cometabolically with cytochrome P450-like monooxygenases or extracellular ligninases and peroxidises.

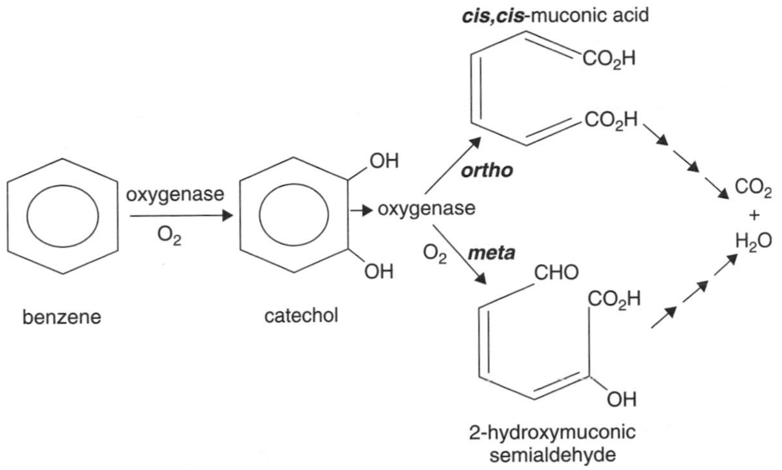

cis,cis-muconic acid

Figure 3.2.3   Pathway for aerobic degradation of benzene (Van Agteren *et al.*, 1998).

These enzymes partially oxidise the aromatic ring, resulting in the release of epoxides and/or hydroxylated aromatic compounds in the environment, which may serve as growth substrates of aerobic bacteria. The specific contributions of bacterial and fungal (co)metabolism to PAHs degradation in groundwater systems is currently unknown.

3.2.4.2   Anaerobic Biodegradation of Aromatic Hydrocarbons

Anaerobic biodegradation of aromatic hydrocarbons is important because these compounds frequently occur in contaminated groundwater where the potential consumption of oxygen exceeds the supply. The lack of molecular oxygen as reactant for ring activation and cleavage makes anaerobic degradation of aromatic hydrocarbons more difficult than aerobic conversion. In many cases, degradation of benzene and PAHs in anoxic groundwater systems appears to occur at very low or even undetectable rates. Anaerobic biodegradation of aromatic hydrocarbons can be coupled to reduction of nitrate, iron(III), manganese(IV) and sulphate, or to methanogenesis as terminal electron accepting processes (Bianchin *et al.*, 2006; Grbić-Galić and Vogel, 1987; Lovley, 2000; Meckenstock *et al.*, 2004). Recently, several bacteria which can grow anaerobically on benzene or PAHs have been isolated and their degradation mechanisms have been partially resolved. Anaerobic pathways for breakdown of aromatic hydrocarbons are different and quite distinct from the aerobic pathways (Karthikeyan and Bhandari, 2001). The aromatic ring is destabilised through activation reactions, which may involve hydroxylation, carboxylation, methylation, fumarate addition or acetyl-CoA addition. Activation is followed by reduction of the aromatic ring and hydrolytic cleavage. The resulting acids are further mineralised. The bacteria involved in benzene degradation under methanogenic or sulphate-reducing conditions are unknown. Molecular analysis of dominant microbial populations in Fe(III) reducing sediments demonstrated that *Geobacter* species may be involved in the anaerobic oxidation of benzene to CO_2 (Lovley, 2000). However, the known *Geobacter* species that are available in pure culture have not been shown to oxidise benzene. The only four bacteria known today, that are capable of anaerobic

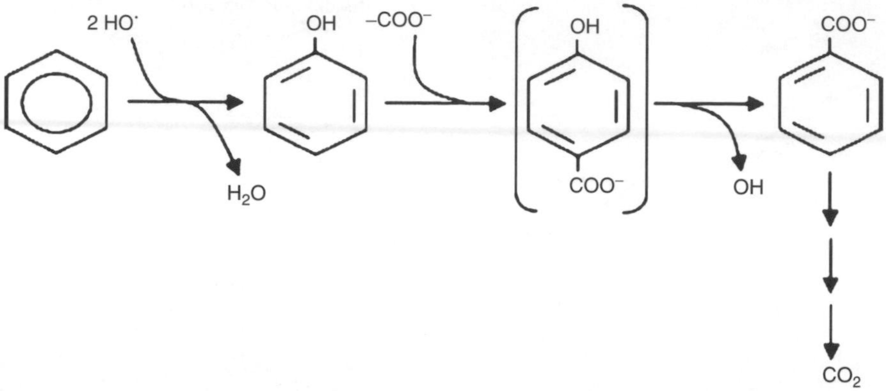

Figure 3.2.4 Pathway for anaerobic degradation of benzene (Chakraborty and Caotes, 2005). Reproduced from Applied & Environmental Microbiology, Chakraborty and Caotes, 2005 with permission from American Society of Microbiology.

growth on benzene with nitrate as electron acceptor are *Dechloromonas* strains RCB and JJ, and *Azoarcus* strains DN11 and AN9 (Coates *et al.*, 2002; Kasai *et al.*, 2006). *Dechloromonas* sp. strain RCB degrades benzene, toluene, ethylbenzene and xylene compounds aerobically or anaerobically with (per)chlorate or nitrate as electron acceptor. Benzene is completely oxidised to CO_2 (Figure 3.2.4).

The crucial steps of anaerobic benzene degradation by this bacterium are a hydroxylation and carboxylation reaction (Chakraborty and Coates, 2005). The authors suggest that all anaerobic benzene-degrading bacteria may use this pathway, regardless of their terminal electron acceptor. This is in agreement with the detection of phenol and benzoate as intermediates of anaerobic benzene degradation under various terminal electron accepting conditions (Caldwell and Suflita, 2000; Grbić-Galić and Vogel, 1987).

The first indications on anaerobic degradation of PAHs are from Mihelcic and Luthy (1988) who reported degradation of naphthalene, naphtol and acenaphthene in microcosms under denitrification conditions. Napthol was also degraded in the absence of nitrate but naphthalene and acenaphthene were not. Subsequent studies revealed that the anaerobic oxidation of naphthalene, fluoranthene and anthracene to CO_2 can be coupled to the reduction of nitrate, sulphate, iron(III) or manganese(IV) (Chang *et al.*, 2002; Coates *et al.*, 1997; Langenhoff *et al.*, 1996; Meckenstock *et al.*, 2004). Early studies on bacterial enrichment cultures suggested that the initial reaction of the sulphate-coupled naphthalene oxidation was a hydroxylation to naphtol (Bedessem *et al.*, 1997) or a carboxylation to 2-naphthoic acid (Zhang and Young, 1997). A pure culture of the sulphate-reducing bacterium strain NaphS2 was able to grow on naphthalene and 2-naphthoic acid, but not on naphtol, which suggested a carboxylation pathway (Galushko *et al.*, 1999). Recently however, Safinowski and Meckenstock (2006) found that, in a sulphate-reducing enrichment culture, degradation of naphthalene was via methylation to 2-methylnaphthalene, with subsequent fumarate addition, oxidation to 2-naphthoic acid, followed by ring reduction and cleavage. The latter authors propose that methylation is a general mechanism of activation reactions in anaerobic degradation of unsubstituted aromatic hydrocarbons.

3.2.4.3 Biodegradation Rates of Aromatic Hydrocarbons in Groundwater

Demonstrating the *in situ* biodegradation of specific contaminants is critical in the assessment of their natural attenuation in groundwater. Proving the *in situ* biodegradation of aromatic hydrocarbons under strictly anoxic conditions is difficult because: (i) concentrations of aromatic hydrocarbons are often low but still of concern, (ii) degradation rates are low (Rügge *et al.*, 1999) and (iii) patterns of electron acceptors supporting evidence for biodegradation are biased because 'natural' dissolved organic carbon acts as the primary electron donor (Christensen *et al.*, 2000). Natural attenuation studies suggest that anaerobic biodegradation of aromatic hydrocarbons occurs with first-order degradation rates typically 1 to 2 orders of magnitude lower than aerobic biodegradation (Christensen, 2002; Nielsen *et al.*, 1996; Rifai *et al.*, 1995; Suarez and Rifai, 1999). The rate of biodegradation of a particular contaminant can be very different in diverse groundwater systems (Table 3.2.1). In a number of field studies and laboratory studies under '*in situ*' groundwater conditions first order degradation rates for benzene were in the range from 0 to 0.45 day^{-1} under oxic conditions, whereas under anoxic conditions the rates ranged from 0 to 0.071 day^{-1} (US-EPA, 1999; Suarez and Rifai, 1999). First order decay rates of benzene obtained under *in situ* conditions indicated mean half-lives of 2.1 days and 231 days under oxic and anoxic conditions, respectively (Suarez and Rifai, 1999). In many anoxic groundwater systems biodegradation of benzene appears to be essentially absent. In eleven anoxic landfill leachate studies benzene was reported as not degraded while it was demonstrated to be degraded in only one study (Christensen, 2002). Baun *et al.* (2003) also showed that benzene was not removed in an anaerobic part of a landfill leachate plume influenced aquifer. Under oxic conditions in acclimated sediments half-lives of naphthalene and anthracene ranged from 0.2 to 0.4 and 1.8 to 11.7 days, respectively. In non-acclimated sediments the half-lives were 10 to 400 times longer. In oxic groundwater microcosms at 22 °C naphthalene was depleted with half-lives between 11 and 169 days (Durant *et al.*, 1995). PAHs were often found to be recalcitrant in anoxic groundwater systems. In eight landfill leachate studies naphthalene was reported as not degraded (Christensen, 2002), but Bianchin *et al.* (2006) reported a zero-order degradation rate of 5 µg/L-day in an anoxic creosote-contaminated aquifer. In a series of 11 anoxic groundwater microcosms at 22 °C, naphthalene degradation occurred in only two microcosms in the presence of nitrate with half-lives of 84–144 days (Durant *et al.*, 1995). US-EPA (1999) reported a most likely first order biodegradation rate for naphthalene of 0 day^{-1} under anaerobic conditions, together with a maximum of 0.03 (half-life 23 days) and a standard deviation of 0.008. This estimate is based on 18 field and laboratory studies with different temperature, pH and redox conditions.

3.2.5 METHYL *TERT*-BUTYL ETHER (MTBE)

Methyl tertiary-butyl ether (MTBE) is the most commonly used fuel oxygenate added to petrol to improve fuel combustion and reduce the resulting concentrations of carbon monoxide and unburned hydrocarbons. Since the 1970s, MTBE has been used in the United States and since 1988 in Europe. Its oxygenating effects resulted in more extensive use in recent years as a result of legislation on air quality (e.g. 1990 Clean Air Act Amendments in the USA). Other fuel oxygenates include ethanol, methanol, ethyl tertiary-butyl ether (ETBE), and tertiary-butyl ether (TBA). In the United States, MTBE concentrations in petrol vary from 15 to 30%, and in Europe from 1.5% to 15% (Squillace *et al.*, 1997). The massive production and use of MTBE, combined with its high mobility due to its solubility, poor sorption properties and low rates of intrinsic degradation, make MTBE one of the most frequently detected contaminants in groundwater. Of thousand sites tested, more than 80% were contaminated with MTBE, with concentrations as high as 23 mg/l (US-EPA, 1998b). As a result, MTBE was placed on the 1998 Contaminant Candidate List (CCL), published by the EPA (US-EPA, 1998a). Compared to the benzene, toluene, ethylbenzene and xylene (BTEX) components of petrol, MTBE is much more soluble and is less likely to adsorb to soil organic matter (Table 3.2.1). MTBE moves through aquifers as fast as a conservative tracer and travels at nearly the same velocity as the groundwater (Barker *et al.*, 1990; Borden *et al.*, 1997). Consequently, MTBE is present at the leading edge of contaminant plumes. Initially, MTBE was thought to be recalcitrant towards degradation. In many petroleum hydrocarbon plumes in the USA, MTBE replaced benzene as the risk determining contaminant and similar findings emerged in Europe (Schmidt *et al.*, 2002). The degradation of MTBE has only been studied in the past years (Deeb *et al.*, 2000; Fayolle *et al.*, 2001; Fayolle and Monot, 2005; Prince, 2000). So far, only a few aerobic bacteria have been isolated that can grow on MTBE, e.g. *Methylibium petroleiphilim* PM1, *Hydrogenophaga* ENV 735, *Mycobacterium austroafricanum* and *Aquincola tertiacarbonis* (Fayolle and Monot, 2005; Lechner *et al.*, 2007; Rohweder *et al.*, 2006). Cometabolic degradation of MTBE has also been described for a variety of oxygenase-expressing bacteria. Aerobically, MTBE is degraded via an initial hydroxylation, resulting in cleavage of the ether bond, followed by the formation of TBF (*tert*-butyl formate) and TBA. Other degradation intermediates are formate, acetate, 2-hydroxybutyrate, formaldehyde and acetone. Anaerobic MTBE or TBA degradation has been studied in sediment microcosms, but no degrading organisms have been identified. Degradation has been demonstrated in the presence of nitrate, sulphate, Fe(III) or Mn(IV) as electron acceptor and under methanogenic conditions. Under anoxic conditions, MTBE was in most cases not completely degraded, resulting in accumulation of TBA (Kuder and Philp, 2008). Recalcitrance of TBA under anoxic conditions is of concern since it is more toxic than MTBE.

3.2.6   CHLORINATED ALIPHATICS

The chlorinated aliphatic compounds are a diverse group of chemicals. They are produced on a large scale for various industrial activities, but may also have a natural origin. The chlorinated aliphatics are often relatively mobile in groundwater and belong, together with the low molecular weight hydrocarbons (BTEX, alkanes) and MTBE to the most frequently detected pollutants in the subsurface (National Research Council Committee on Intrinsic Remediation, 2000). The chlorinated aliphatics include many toxic and carcinogenic compounds and are considered to belong to the most serious groundwater pollutants

Table 3.2.2 Biological transformation and degradation mechanisms of chlorinated aliphatic compounds under various conditions (Bradley, 2003).

Chlorinated aliphatic	Degradation mechanism			
	Aerobic	Anaerobic		
	Oxidation	Oxidation	Fermentation	Reduction
Trichloromethane	+ (Co)	+ (Co)	−	+ (Co)
Dichloromethane	+ (Me)	+ (N, Me)	+ (Me)	+ (Co)
1,2-Dichloroethane	+ (Me)	+ (N, Fe)	+ (Me)	+ (Me)
Trichloroethylene	+ (Co)	−	−	+ (Me)
Tetrachloroethylene	+ (Co)	−	−	+ (Me)
Hexachlorobutadiene	−	−	−	+
C10-13-chloroalkanes	+(<50%)	−	−	−

Abbreviations: + biological transformation reported, − no biological transformation reported, (Co) only cometabolic transformation reported, (Me) compound can be used for metabolic energy generation, (N) with nitrate as electron acceptor, (Fe) with ferric iron as electron acceptor, (<50%) only compounds with chlorine content less than 50% (w/w) degraded.

worldwide. Different mechanisms are involved in the microbial degradation of chlorinated aliphatic compounds in oxic or anoxic environments (Table 3.2.2; Bradley, 2003).

Oxidative degradation pathways dominate in oxygen and nitrate containing groundwater. Anaerobic oxidation of chlorinated alphatics with iron(III) or manganese(IV) as electron acceptor can also be an important natural attenuation process. The end products of oxidation of chlorinated aliphatics are CO_2, H_2O and chloride. Under more reducing sulfidogenic and methanogenic conditions, reductive dechlorination pathways are dominant. Reductive dechlorination usually involves substitution of a chloride (Cl^-) for a proton (H^+) on the aliphatic compound. Sequential dechlorinations yields partially or completely dechlorinated aliphatic compounds. Chlorinated alkanes can also be dechlorinated through removal of two chlorides in one step (dihaloelimination) from two adjacent carbon atoms. This results in the formation of a double $C=C$ bond. During reductive dechlorination the chlorinated aliphatics act as an electron acceptor. This implies that this process must be fuelled by electron donor substrates such as H_2, naturally occurring carbon compounds or petroleum hydrocarbons (Christensen *et al.*, 2002). Fermentation is a third mechanism for anaerobic degradation of chlorinated aliphatic pollutants. Although the information on fermentative degradation is still limited, this mechanism may play a role in groundwater with limiting amounts of electron acceptors and electron donor substrates.

Trichloromethane, or chloroform (TCM), is mainly used as feedstock (>90%) for the manufacture of other chemicals or as a solvent, for example in extraction of antibiotics, pesticides, fats, oils, rubbers, alkaloids and waxes. TCM is also a degradation product from carbon tetrachloride. TCM is observed in landfill leachates in a concentration range of 1–70 µg/l (Kjeldsen *et al.*, 2002). There are currently no microorganisms known capable of using TCM as substrate for growth or energy production. Yet, TCM can be transformed cometabolically by various aerobic and anaerobic bacteria. Under oxic conditions TCM is effectively oxidised by monooxygenase enzymes of bacteria which use substrates as methane, butane or ammonia as electron donor (Arp *et al.*, 2001). The soluble methane monooxygenases of methanotrophic or butane-grown bacteria oxidises TCM to metabolites, which are further degraded to CO_2 and HCl. The relative importance of aerobic cometabolic degradation of TCM in natural groundwater systems is largely unknown. Under anoxic conditions TCM is cometabolically trans- formed by methanogenic, sulphate reducing, nitrate reducing, fermentative and specific dechlorinating bacteria. They form a variety of products, including reductive dechlorina- tion products dichloromethane and chloromethane, CS_2, CO, CO_2, and other unidentified products (Mohn and Tiedje, 1992; Yu and Smith, 2000). In landfill leachate plumes under nitrate reducing or oxic conditions TCM was reported as not degraded (Chris- tensen, 2002). On a site contaminated with TCM and 1,2-dichloroethane, TCM rapidly degraded to dichloromethane near the source area where the groundwater was anoxic (Cox *et al.*, 2000). Further down-gradient from the source area the groundwater became oxic and the remaining dichloromethane was cometabolised in the presence of methane with a half-life of 143 days. US-EPA (1999) reported as most likely first order biodegra- dation rate for TCM of 0.0315 day^{-1} (half-life 22 days) under anoxic conditions, with a maximum of 0.25 and standard deviation of 0.0884. This estimate is based on 11 field and laboratory studies with temperature $>15\,^{\circ}$C, pH <6 to 8 and undefined redox conditions.

Dichloromethane (DCM) is used as solvent, degreasing agent, paint remover and pressure mediator in aerosols. In anoxic systems it can be a degradation product of tetra- and trichloromethane. DCM can serve as growth substrate for a number of aerobic and anaerobic bacteria. In addition, cometabolic transformations have been documented under oxic and anoxic conditions. Aerobic cometabolism involves oxidation of DCM by monooxygenases. Reduced coenzymes of anaerobic methanogenic or acetogenic bac- teria cometabolise DCM to chloromethane. Aerobic facultative methylotrophic bacteria that are able to grow on DCM can easily be isolated from groundwater systems. These bacteria transform DCM to HCl and S-chloromethylglutathione (Fetzner, 1998b). The lat- ter compound hydrolyzes to glutathione, chloride and formaldehyde, which is a central metabolite for methylotrophic growth. Growth of methylotrophs on DCM with nitrate as electron acceptor has also been described. No scientific literature on iron(III)- or manganese(IV)-coupled anaerobic oxidation of DCM has been found. In methanogenic microcosms which used DCM as the sole growth substrate, it was degraded to mainly CO_2 and CH_4 (Freedman and Gossett, 1991). CO_2 and acetate were the major degradation products when methanogenesis was inhibited. The acetogenic bacterium *Dehalobacterium formicoaceticum* is able to grow on DCM, forming formate, acetate and chloride (Mägli *et al.*, 1998). DCM was observed in landfill leachates in a concentrations range of $1.0-827\,\mu$g/l (Kjeldsen *et al.*, 2002). US-EPA (1999) reported a first order biodegra- dation rate for DCM of 0.0064 day^{-1} (half-life 108 days) under anoxic conditions, based on a field study at $15\,^{\circ}$C, pH $6-8$ and methanogenic redox conditions.

1,2-Dichloroethane, or ethylene dichloride (DCA), is mainly used as intermediate for production of vinyl chloride, PVC and other chlorinated chemicals ($>95\%$). Less than 5% is used as raw material for ethyleneamines, trichloroethylene, perchloroethy- lene, extraction and cleaning solvent, or as lead scavenger for gasoline. Many different mechanisms can result in the biological transformation of DCA (Figure 3.2.5). It serves as a growth substrate for a number of aerobic bacteria. In most of the known species, such as *Xanthobacter, Ancylobacter, Rhodococcus* and *Mycobacterium* strains the initial transformation step is a hydrolytic dehalogenation yielding chloroethanol and HCl as products (Janssen *et al.*, 2005).

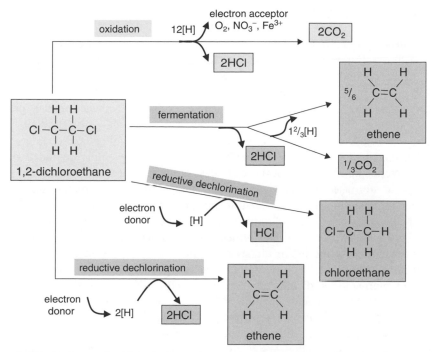

Figure 3.2.5 Pathways for biological transformations of 1,2-dichloroethane in microcosms with contaminated aquifer material incubated under various redox conditions (Gerritse *et al*., 1999).

In contrast, *Pseudomonas* sp. strain DCA1 uses a monooxygenase-mediated transformation yielding the unstable 1,2-dichloroethanol, which spontaneously decomposes to chloroacetaldehyde (Hage and Hartmans, 1999). Both aerobic pathways completely mineralize chloroacetaldehyde via chloroacetic acid and glycolic acid to CO_2. The aerobic oxidation of DCA is also mediated by monooxygenase enzymes of methanotrophic bacteria, in a cometabolic process. Under anoxic conditions cometabolic dechlorination to mainly ethene and/or chloroethane is carried out by various methanogenic, acetogenic or sulphate reducing bacteria. Vinyl chloride and ethane may be formed as additional products. Anaerobic bacteria capable of metabolic respiration with DCA as an electron acceptor have also been described (Grostern and Edwards, 2006). The nitrate and iron(III)-coupled oxidation of DCA, and the fermentation of DCA to ethene, CO_2 and HCl are additional biotransformation pathways under anaerobic conditions (Gerritse *et al*., 1999; Dijk *et al*., 2005; Dinglisan-Panlilio *et al*., 2006). Dichloroethane is observed in landfill leachates in concentrations <6 μg/l (Kjeldsen *et al*., 2002). US-EPA (1999) reported a first order biodegradation rate for DCA of 0.0076 day^{-1} under anaerobic conditions. This estimate, corresponding to a half-life of 91 days, is based on one field study under pH neutral, methanogenic conditions. Bosma *et al*. (1998) reported half-lives over a wide range from less than 1 to over 30 years in an anaerobic aquifer, heavily contaminated with DCA. Abiotic transformation of DCA, including alkaline hydrolysis to vinyl chloride and hydrolytic substitution to ethylene glycol, also occurs under anaerobic conditions. The half-lives of these processes are generally, in the order of 10 to 70 years (Barbash and Reinhard,

1989; Jeffers *et al.*, 1989). This indicates that microbial transformation is usually the dominating degradation process in groundwater systems.

Trichloroethylene (TCE) is used as a solvent and in manufacturing to clean grease from machinery (>80%), in adhesives and for synthesis in the chemical industry. It is one of the most abundant groundwater pollutants. Under oxic conditions TCE is degraded cometabolically by a wide variety of bacteria expressing non-specific monooxygenase or dioxygenase enzymes. Examples are bacteria oxidising methane, ethene, propane, propene, toluene, phenol, ammonium, isoprene and vinyl chloride (Bradley, 2003). Oxidation of TCE can yield many different compounds, including TCE-epoxide, which may be detected in groundwater during active aerobic cometabolic degradation. This unstable intermediate spontaneously degrades in dichloroacetate, carbon monoxide, glyoxylate, or formate. In methanogenic and sulphate-reducing groundwater systems anaerobic halorespiring bacteria catalyse the reductive dechlorination of TCE to cis-1,2-dichloroethylene, vinyl chloride, and ethylene. Trans-1,2-dichloroethylene and 1,1-dichloroethylene may also be formed in low quantities. During the anaerobic dechlorination process the intermediates accumulate sequentially in groundwater. This is significant, because especially vinyl chloride is a potent human carcinogen. The less chlorinated intermediates, dichloroethylene and vinyl chloride are mainly dechlorinated reductively under methanogenic conditions. They can however also be degraded through anaerobic oxidative or fermentative pathways. Although the bacteria which are responsible for these processes are unknown, they may play an important role in the natural attenuation of chlorinated ethylenes in nitrate and iron-reducing groundwater systems. US-EPA (1999) reported as most likely first order biodegradation rate for trichloroethylene of 0.0016 day^{-1} under anaerobic conditions, together with a maximum of 0.04 and a standard deviation of 0.00889. This estimate is based on a large number of field studies under a variety of conditions. Higher rates with a mean of 0.003 day^{-1}, with a maximum of 0.023 and standard deviation of 0.005, were reported for reductive dechlorination of trichloroethylene under anaerobic field *in situ* conditions (Suarez and Rifai, 1999). This corresponds to a half-life of about 230 days. Under conditions stimulating aerobic cometabolism, *in situ* decay rates of TCE may correspond to a half-life of less than 1 day.

Tetrachloroethylene, or perchloroethylene (PCE) is a good solvent used to clean machinery, electronic parts, and clothing. PCE is a suspected carcinogen and one of the most abundant environmental pollutants of groundwater. It was believed that PCE could not be degraded aerobically, until recently it was discovered that it can be cometabilically oxidised by the toluene-*o*-xylene monooxygenase of a *Pseudomonas stutzeri* strain (Ryoo *et al.*, 2000). Nevertheless, PCE is considered very resistant to oxidative degradation and often found recalcitrant in oxic environments. The anaerobic oxidation of PCE, for example with nitrate, manganese or iron(III) as electron acceptor, has never been reported. In contrast, the reductive dechlorination of PCE to TCE readily occurs in anaerobic reducing groundwater. Methanogenic and sulphate-reducing conditions favour this dechlorination, but in a predominantly iron(III) reducing environment the reaction also occurs. Certain methanogenic, acetogens and sulphate reducing bacteria can slowly cometabolise PCE and TCE reductively, but specific dehalorespiring bacteria, such *Sulfurospirillum, Dehalobacter, Desulfitobacterium, Desulfuromonas* and *Dehalococcoides*, are considered to be mainly responsible for chloroethylene dechlorination in contaminated groundwater systems. Most organisms studied convert PCE to

TCE or *cis*-1,2-dichloroethylene. *Dehalococcoides ethenogenes* species are the only reported bacteria that can completely dechlorinate PCE to ethylene. US-EPA (1999) reported as most likely first order dechlorination rate for PCE to TCE of 0.00186 day^{-1} under anoxic conditions, together with a maximum of 0.071 and a standard deviation of 0.0223. This estimate is based on a number of field studies under a variety of conditions. A higher rate, 0.010 day^{-1} with a maximum of 0.080 and a standard deviation of 0.022 was reported for reductive dechlorination of PCE in field and *in situ* studies (Suarez and Rifai, 1999). This corresponds to a half-life of about 70 days. Under oxic *in situ* conditions the degradation rate constant of 0.000 confirmed the persistence of PCE under these conditions.

Hexachlorobutadiene (HCBD) is an intermediate in the manufacture of rubber compounds and it was used in the production of lubricants, as a fluid for gyroscopes, as a heat transfer liquid, as solvent, fungicide and in hydraulic fluids. HCBD is mainly formed as a by-product during the manufacture of chlorinated hydrocarbons such as TCE, PCE and tetrachloromethane. The world annual production was estimated to be 10,000 tonnes in 1982, but there is probably no commercial production of HCBD any more (van de Plassche and Schwegler, 2005). Once HCBD is released into the environment, intercompartmental transport will occur chiefly by volatilization from water and soil, adsorption to particulate matter in water and air, and subsequent sedimentation or deposition (Vermeire, 1994). HCBD does not migrate rapidly in soil and accumulates in sediment. Therefore, it is not a major thread for contamination of groundwater. Chemical hydrolysis of HCBD does not occur. There are no bacteria known that can mineralize HCBD. This compound has been considered recalcitrant for a period of three years under conditions with oxygen or nitrate as the major electron acceptor (Bosma *et al.*, 1994; van Agteren *et al.*, 1998). Tabak *et al.* (1981) observed complete loss of HCBD (5 and 10 mg/l) in 7 days. The flasks were incubated under an air atmosphere, but the addition of yeast extract and the fact that they were not shaken may have created anaerobic conditions in the water phase. In methanogenic soil columns with Rhine sediment HCBD was reductively dechlorinated via pentachlorobutadiene to mainly 1,2,3,4-tetrachloro-1,3-butadiene (>90%) and traces (<5%) of a trichlorobutadiene (Bosma *et al.*, 1994). This process occurred after an acclimation period of 4 months. This substance – which is known as an antifungal agent according to the authors – may be degraded further aerobically. No half-lives are presented.

Chlorinated paraffins are complex mixtures of straight chain chlorinated hydrocarbon molecules with a range of chain lengths (short C10-13, intermediate C14-17 and long C18-30) and degrees of chlorination (between 30 and 70% weight basis). Over 200 commercial formulations with a range of physical and chemical properties exist which make them useful in a wide range of applications, such as secondary plasticizers in PVC and other plastics (C14–17), extreme pressure additives, flame retardants, sealants and paints. The widespread uses of chlorinated paraffins result in contamination, particularly to the aquatic environment. Only the short chain chlorinated paraffins (SCCPs) are considered as priority substances. Risk assessment and risk management showed a need for regulating the SCCPs and restrict their use in main fields of current applications. Chlorinated paraffins are not readily biodegradable. A *Rhodococcus* strain was able to use various chlorinated paraffins with chlorine content <50% (w/w) as the sole carbon and energy source (Allpress and Gowland, 1999). Therefore biodegradation of these compounds may also be expected to occur in an oxic environment. Chlorinated par-rafins with chlorine content >58% were not used as a growth substrate by this bacterium, but certain bacteria have also been shown to dechlorinate SCCPs with high chlorine contents in a cometabolic process. No information on the anaerobic biodegradation of SCCPs is available (European Commission, 2005). SCCP residues have been detected in sediment cores dating back to the 1920s and 1930s. From observations like this, it can be concluded that those components degraded very slowly.

3.2.7   CHLORINATED AROMATICS

Hexachlorobenzene (HCB) is a fungicide that was first introduced in 1945 for seed treatments of grain crops (UNEP, 2002). It is an impurity in several pesticide formulations, including pentachlorophenol and dicloram and may be present as an impurity in others. HCB may still be found in the food chain from its former use as a pesticide; however, today the main release into the environment is as by-product of industrial processes such as aluminium smelting, the manufacture of industrial chemicals including carbon tetrachloride, PCE, TCE, vinyl chloride and pentachlorbenzene and inadequate incineration of chlorine-containing wastes. The use of HCB was discontinued in many countries in the 1970s owing to concerns about adverse effects on the environment and human health. HCB is still observed in landfill leachates in a concentrations range of 0.025–10 μg/l (Kjeldsen *et al.*, 2002). HCB belongs to the most persistent environmental pollutants because of its chemical stability and resistance to degradation. The half-life of HCB is estimated to range from 2.7 to 5.7 years in surface water and from 5.3 to 11.4 years in groundwater.

   Pentachlorobenzene was used to make the fungicide pentachloronitrobenzene (quintozene) and as a fire retardant. Nowadays, quintozene is manufactured via a production process without pentachlorobenzene. Pentachlorobenzene entered the environment when quintozene was used, resulting in agriculture related diffuse exposure. Pentachlorobenzene absorbs strongly to soil and is expected not leach to the groundwater (van de Plassche *et al.*, 2005).

   The three **trichlorobenzene** isomers (TCBs) – 1,2,3-trichlorobenzene (1,2,3-TCB), 1,2,4-trichlorobenzene (1,2,4-TCB) and 1,3,5-trichlorobenzene (1,3,5-TCB) – are used for the production of herbicides, pesticides, pigments and dyes or as heat transfer medium. TCBs can be released to the environment directly from production, from their uses, final treatment and waste disposal (e.g. leakage from landfills) and through other sources such as combustion of plastics or degradation of higher chlorinated benzenes. TCBs have been detected in fresh water in concentrations up to 0.4 μg/l, whereas concentrations from 0.002 to 0.007 μg/l were detected in marine waters in open sea areas and concentrations from 0.02 to 0.03 μg/l in dispersion zones of rivers or important waste water treatment plants. High TCB-concentrations have been detected occasionally in river sediment on specific locations. OSPAR (2005a) reported that there are no recent data for TCB in groundwater, but they also reported that two decades ago, concentrations were found at or below the detection limit (<0.001 μg/l). 1,2,3-TCB was observed in landfill leachate plumes but the concentration range was not specified, 1,2,4-TCB is observed in a concentration of 4.3 μg/l (Kjeldsen, 2002). Soils that are rich in organic matter and aquatic sediments are probably the major environmental sinks for these compounds. Trichlorobenzenes are compounds which are not readily biodegradable (WHO, 1991). The most likely degradation mechanisms involve photochemical reactions and microbial action. Higher chlorinated chlorobenzenes are much more resistant to aerobic degradation than mono- and dichlorobenzenes. Under anaerobic conditions the opposite has been found. Microorganisms able to degrade HCB or pentachlorobenzene under aerobic conditions are not known (van Agteren *et al.*, 1998). Some *Dehalococcoides* strains can use chlorinated benzenes as electron acceptor for anaerobic respiration. Under anoxic conditions HCB can be reductively dechlorinated via penta-, tertra-, tri- and dichlorobenzenes to chlorobenzene. In most cases 1,3,5-trichlorobenzene and 1,4- and 1,3-dichlorobenzenes accumulate. Chlorobenzene is generally thought to resist reductive dechlorination, but some studies suggest that it may be biodegraded through anaerobic oxidative mechanisms (Kaschl *et al.*, 2005). Some aerobic bacteria are able to use mono-, di-, and some tri- and tetrachlorobenzenes as a growth substrate. The degradation pathways involve the oxygenation to chlorocatechols, *ortho* cleavage to chloromucoic acid, which is subsequently dechlorinated and further metabolised. Complete aerobic mineralization of chlorinated benzenes is possible, but toxic *meta* cleavage intermediates often accumulate in un acclimated microbial populations, resulting in cell death. Although anaerobic degradation occurs, persistence of chlorobenzenes in the environment is high. In sediment cores pentachlorobenzene was removed with half-lives of several years, but an anaerobic enrichment culture showed a half-life of several days (Beurskens *et al.*, 1994). Beck and Hansen (1974) observed a half-life for pentachlorobenzene of 194–345 days in soils.

3.2.8   SUBSTITUTED PHENOLS

Pentachlorophenol (PCP) and its salts have been widely used as pesticides, fungicides and herbicides with a variety of applications in industrial, agricultural, and domestic fields. Its salt, sodium pentachlorophenate (C_6Cl_5NaO – NaPCP), is used for similar purposes and readily degrades to PCP. World production of PCP is estimated to be of the order of 30,000 tonnes per year. Since 1991, EU legislation has limited the use and application of PCP-based products in order to minimise human exposure (EC 91/179/EEC). The relatively high volatility of PCP and the water solubility of its ionised form and release into the environment from a number of diffuse sources have led to widespread contamination with this compound. Because of its high vapour pressure, PCP easily evaporates from treated wood surfaces, and the loss may be as high as 30–80% a year (UNEP, 2002). Volatilization from water to air is dependent on pH (only the non-ionized form is volatile) and temperature. Emissions, discharges and losses of PCP and NaPCP are likely from wood in use – most of which will have been treated with PCP in the past. Volatilization can be an important source of loss of PCP from water and soil surfaces as well as from PCP-treated materials. A variety of aerobic bacteria is able to grow on PCP (van Agteren *et al.*, 1998). The aerobic degradation mechanism involves the partial dechlorination to 2,3,5,6-tetrachlorohydroquinone before further dechlorination and ring cleavage occurs. For most degrading bacteria PCP is toxic at relatively low concentrations. Some PCP-degrading bacteria also O-methylate chlorinated phenolic compounds (Häggblom *et al.*, 1988). This results in formation of chlorinated anisoles which are less toxic than the corresponding chlorophenols. Under anoxic conditions PCP is degraded by reductively dechlorinating bacteria. Reductive dechlorination of chlorophenols has been observed under denitrifying, iron-reducing, sulphate-reducing and methanogenic conditions. Certain anaerobic bacteria are able to use chlorinated phenols as electron acceptors for anaerobic respiration (Dennie *et al.*, 1998; Gerritse *et al.*, 1996; Utkin *et al.*, 1995). Different patterns of reductive chlorine removal occur. Lesser chlorinated phenols often accumulate as end products, but complete dechlorination to phenol has also been demonstrated. Phenol can further be mineralized under anaerobic conditions. The less chlorinated phenols can also be degraded via anaerobic oxidative pathways, coupled to the reduction of sulphate, nitrate or iron(III) (Bae *et al.*, 2002; Häggblom and Young, 1990). The rate of photodecomposition increases with pH (half-life 100 hr at pH 3.3 and 3.5 hr at pH 7.3). Although enriched through the food chain, it is rapidly eliminated after discontinuing the exposure (half-life = 10–24 h for fish). Not much is known on the degradation of PCP in the environment. It can be degraded in with half-lives in the range of several weeks (van Agteren *et al.*, 1998). Complete decomposition in soil suspensions takes >72 days, other authors reports half-life in soils of 23–178 days. In one landfill leachate plume study under anoxic conditions PCP was reported as not degraded (Christensen, 2002). Higher chlorinated phenols are easily absorbed on soil particles and can be incorporated in the humic fraction of sediments and aquifers. Thus they can become very recalcitrant to biodegradation.

Nonylphenols (NP) and **octylphenols** (OP) are the starting material in the synthesis of alkylphenol ethoxylates (APEs), first used in the 1960s. These compounds are highly effective cleaning agents or surfactants that have been widely used in a number of industrial sectors including textiles, pulp and paper, paints, adhesives, resins and protective coatings. Alkylphenols can also be used as plasticizers, stabilizers for rubbers, lube oil additives, and the alkylphenol phosphite derivatives can be used as UV stabilizers in plastics. NP and OP mainly reach the aquatic environment via industrial and municipal wastewaters and sewage, landfills and wastewater treatment plants (OSPAR, 2001; OSPAR, 2003). This can be directly as NP and OP or as breakdown products of NP- and OP-ethoxylates. As a result of the industry led voluntary agreement, the use of NP-ethoxylates in domestic detergents in most European countries has reduced in recent years. Groundwater monitoring results in the river Glatt area in Switzerland reported decreasing levels of NP from about 1 µg/l to 0,20 µg/l, from a depth of 2.5 to 13 m, during infiltration of river water into groundwater (OSPAR, 2001). However, this older study may not reflect the current levels of NP, particularly where the major source was thought to be from NP-ethoxylate use in detergents. The OSPAR background document on OP does not contain groundwater monitoring data (OSPAR, 2003). NP was observed in landfill leachates in a concentration range of 6–7 µg/l (Kjeldsen *et al.*, 2002). NP and OP are degradation products of APEs under both aerobic and anaerobic conditions. Therefore, the major part is released to water and concentrated in sewage sludges. NP and OP are persistent in the environment with half-lives of 30–60 years in marine sediments, 1–3 weeks in estuarine waters and 10–48 hours in the atmosphere. Due to their persistence they can bioaccumulate to a significant extent in aquatic species. However, excretion and metabolism is rapid. In spite of their persistence, the biodegradation of NP has been demonstrated mainly under aerobic and once under anaerobic conditions (Chang *et al.*, 2004; Ushiba *et al.*, 2003). Anaerobic degradation was observed under sulphate-reducing, methanogenic and nitrate-reducing conditions. In anoxic sediment samples half-lives were reported from 46.2 to 69.3 days. The NP and OP degradation pathways remain largely unresolved.

3.2.9 PESTICIDES

The production, distribution and use of pesticides provide many possible sources and opportunities for groundwater contamination. The ideal outcome of pesticide use occurs when a pesticide accomplishes the purposes for which it was applied and then rapidly breaks down into harmless components such as carbon dioxide and water. This happens in most cases, but the process and time vary among pesticide chemicals, and the application is affected by the physical, chemical, and biological characteristics of the pesticide and the associated environment. Pesticide chemicals that dissolve readily in water are highly soluble and, thus, are generally carried with the water flow. Such pesticides have a tendency to leach from the soil to groundwater. However, many pesticides do not leach because they are adsorbed on the soil particles or organic matter even though they may have a relatively high solubility. Highly volatile pesticides are easily lost to the atmosphere and less likely to leach to groundwater. Degradation affects the potential for a pesticide to reach groundwater. The persistence of the pesticide influences the potential for contamination. The longer the compound lasts before it is broken down, the longer it is subject to the forces of leaching. However, many highly persistent pesticides have not been found in groundwater because of their low solubility and strong adsorption to soil particles. On the other hand, some relatively soluble pesticides of low persistence have been found in groundwater. The risk of leaching to the groundwater is also influenced by soil factors including texture and organic matter. These factors have an effect on pesticide adsorption and degradation. Soils with a high organic and clay content have a high adsorption capacity. Degradation of many pesticides is dependent on the redox status of the soil. Generally, soils with a high organic matter content have more reduced anoxic conditions. Besides these chemical controls, soil permeability, or how readily water moves through the soil, is important. The more permeable a soil, the greater potential for pesticide leaching: a sandy soil is much more permeable than a clay soil. Degradation reactions (microbial-, chemical- and photodegradation) change pesticides often to residues which are inactive, less toxic, and less harmless in the environment, but in some cases these residues may be toxic and have significant half-lives as well. The rate of pesticide breakdown depends on a variety of factors including temperature, soil pH, soil microbe content and exposure to light, water and oxygen. Therefore different types of half-lives are used to describe degradation of pesticides:

- Soil half-life: The amount of time required for half of the pesticide to degrade in soil. This half-life is governed by the types of soil organisms that are present that can break down the pesticide, the soil type (e.g. sand, loam, clay), pH, and temperature.

- Photolysis half-life: The amount of time required for half of the pesticide to degrade from exposure to light.

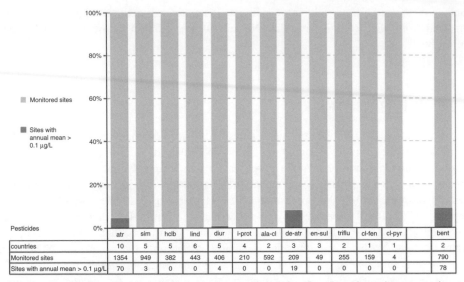

Pesticides	atr	sim	hclb	lind	diur	i-prot	ala-cl	de-atr	en-sul	triflu	cl-fen	cl-pyr		bent
countries	10	5	5	6	5	4	2	3	3	2	1	1		2
Monitored sites	1354	949	382	443	406	210	592	209	49	255	159	4		790
Sites with annual mean > 0.1 µg/L	70	3	0	0	4	0	0	19	0	0	0	0		78

Note: atr...Atrazine; sim... Simazin; hclb...Hexachlorobenzene; lind...lindan; diur...diuron; i-prot...Isoproturon; ala-cl...Alachlor; de-atr...Desethylatrazine; en-sul...Endosulfan; triflu... Trifluralin; cl-fen...chlorfenvinphos; cl-pyr...Chlorpyriphos; bent...Bentazon.

Source: WATERBASE data collected through EUROWATERNET.

Figure 3.2.6 Proportion of groundwater wells in Europe with pesticide concentrations greater than 0.1 µg/l (Lindinger and Scheidleder, 2004). Reproduced by permission of the European Environment Agency.

- Hydrolysis half-life: The amount of time required for half of the pesticide to degrade from reaction with water.

The European Environmental Agency has made an inventory of observed pesticide concentrations in groundwater in Europe (Lindinger and Scheidleder, 2004). The main conclusion is that for Europe there is limited information available and a lack of reliable data on pesticides in groundwater. However, from national State of the Environment Reports written for the European Environmental Agency (EEA, 2000) it appears that there is a danger of pesticide pollution. Although typically less than 10% of the groundwater wells has pesticide concentrations greater than 0.1 µg/l, monitoring activities should be further developed to improve knowledge on the occurrence and behaviour of pesticides in groundwater (Figure 3.2.6).

3.2.9.1 Chlorinated Pesticides

Hexachlorocyclohexane (HCH) is a mixture of isomers, and a technical HCH mixture typically contains α-HCH (55–80%), β-HCH (5–14%), γ-HCH (8–15%), δ-HCH (10–12%), and ε-HCH (3–5%). Lindane (γ-HCH) is the best known and effective insecticide component of HCH and its effective use was discovered in the early 1940s.

The importance of technical HCH or lindane as insecticide is declining in Europe (UNEP, 2002). While in a period from 1970 to 1979 HCH made up around 13% of the insecticides used in Europe, it was reduced to less than 5% in the period of 1991 to 1996 (Breivik *et al.*, 1999). However, there is a large regional variance in the use of insecticides depending on the latitude (less in the north and more in southern parts of Europe) and the crops grown (cereals and soya requiring relatively less insecticides than, for example, maize, rape and further rice and cotton). Estimates of γ-HCH use and emissions in European countries for every year from 1970 to 1996 were made by Breivik *et al.* (1999). According to these estimates total γ-HCH emission in Europe decreased 3.5 fold between 1970 and 1996. Its physicochemical properties suggest that it is predominantly gaseous at moderate temperatures and volatilisation will be an important route of dissipation to the atmosphere, particularly in high temperature conditions. Lindane enters water from direct application from use in agriculture and forestry, from precipitation and, to a lesser extent, from occasional contamination of wastewater from manufacturing plants. A survey of pesticides in groundwater for 1990–1995 was carried out in Germany (UBA, 2001), when lindane was still in agricultural use. Lindane was ranked number 20 of the most common active substances found in groundwater. In the following years lindane was still found regularly at 5–7% of the observation points, and in few cases with more than 0.1 µg/l. The EEA (2000) reports lindane concentrations in groundwater in different European countries. Slovakia indicates lindane concentrations > 0.1 µg/l in 2 of 8 observed groundwater wells. The Czech Republic reports 0% of 215 wells. Lindane has been found in a significant number of groundwater samples in the USA and in Italy at concentrations <1 µg/l (Howard, 1991), and in landfill leachates in a concentration range of 0.025–0.95 µg/l (Kjeldsen *et al.*, 2002). Lindane and other HCH isomers are relatively persistent in soils and water, with half lives generally greater than 1 and 2 years, respectively. HCH isomers are much less bioaccumulative than other organochlorines because of their relatively low liphophilicity. HCH isomers are readily dechlorinated in anaerobic methanogenic and sulphate-reducing ecosystems (Langenhoff *et al.*, 2002; van Agteren *et al.*, 1998). Both microbiological and abiotic processes are involved. The major products of anaerobic HCH transformation are chlorobenzene, benzene and chlorophenol. Most HCH isomers can also be degraded under oxic conditions. Various aerobic bacteria that can grow on HCH isomers have been isolated. Their HCH degradation pathways involve hydrolytic and reductive dechlorination reactions leading to the formation of chlorophenols which are further metabolized. The aerobic biodegradation rates generally decrease in the order γ-HCH$> \alpha - \text{HCH} > \beta - \text{HCH} > \delta$-HCH.

Endosulfan is used as an insecticide and acaricide in a great number of food and nonfood crops (e.g. tea, vegetables, fruits, tobacco, cotton) and it controls over 100 different insect pests. It was first introduced in 1954. Endosulfan does not easily dissolve in water. It has a moderate capacity to adsorb to soils and it is not likely to leach to groundwater. Transport of this pesticide most likely occurs if endosulfan is attached to soil particles in surface runoff. Large amounts of endosulfan can be found in surface water near areas of application (Farm Chemicals Handbook, 1992). It has also been found in surface water at very low concentrations and has been detected in the air at minute levels. It is has been found, but not quantified, in well water in California (Smith, 1991). It is not expected to pose a threat to groundwater (Extoxnet, 1996). Endosulfan is moderately persistent in the soil environment with a reported average field half-life of about 50 days. The two isomers have different degradation times in soil (half-lives of 35 and 150 days for α- and β-isomers, respectively, in neutral conditions). Breakdown products of endosulfan are endosulfan sulphate and chlorendic acid. The sulfate is more persistent than the parent compound.

Alachlor is an aniline herbicide used to control annual grasses and certain broadleaf weeds in field corn, soybeans and peanuts. It is a selective systemic herbicide, absorbed by germinating shoots and by roots. Alachlor has moderate mobility in sandy and silt soils and thus can migrate to groundwater (Extoxnet, 1996). A large groundwater testing program for pesticides in the USA (National Alachlor Well Water Survey) was conducted for alachlor throughout the last half of the 1980s. Over six million private and domestic wells were tested for the presence of the compound (Holden and Graham, 1992). Less than 1% of all of the wells had detectable levels of alachlor. Detection of the chemical in well water was more common where the herbicide was used more intensively. Concentrations of the pesticide in the 1% of the wells where it was detected ranged from 0.1 μg/l to 1.0 μg/l. The majority of these wells had concentrations around 0.2 μg/l. Authorisations for plant protection products containing alachlor are withdrawn by the European Commission at 18 December 2006. This decision was made due to expected concentration of Alachlor metabolites in groundwater above the maximum acceptable limit of 0.1 μg/l and the fact that it could not be precluded that alachlor has a carcinogenic potential.

Atrazine is a selective triazine herbicide used to control broadleaf and grassy weeds in corn, sorghum, sugarcane, pineapple, christmas trees, and other crops, and in conifer reforestation plantings. It was discovered and introduced in the late 50's. In many countries, atrazine has been found in groundwater at levels of 0.01–6 μg/litre. In Germany atrazine concentrations >0.1 μg/l were found in 4% of 11 690 observed groundwater wells and in Denmark in 0% of 625 wells (EEA, 2000). Hungary reports 16% of 174 wells and Slovenia 32% of 84 wells. With a total of 8% contaminated wells, atrazine is the most important pesticide for groundwater in Europe. Atrazine has been banned in the European Union since 2004. It is still widely used in the USA today because it is economical and effectively reduces crop losses due to weed interference. It is probably the most commonly used herbicide in the world. Atrazine is moderately to highly mobile in soils, especially those with low clay or organic matter content. Because it does not absorb strongly to soil particles (K_{oc} 100 g/ml) and it has a lengthy soil half-life (60–100+ days), it has high potential for groundwater contamination, even though it is only moderately soluble in water (28 mg/l) (Extoxnet, 1996). Chemical hydrolysis, followed by degradation by soil microorganisms, probably accounts for most of the breakdown of atrazine in soil. Hydrolysis is rapid in acidic or basic environments, but is slower at neutral pHs. Addition of humic material increased the rate of hydrolysis. Atrazine can persist for longer than one year under conditions which are not conducive to chemical or biological activity, such as dry or cold climates (Howard, 1991). Rügge *et al*. (1999) found in an injection experiment in a landfill plume in Denmark that atrazine was recalcitrant under strongly anoxic conditions.

Simazine is a selective triazine herbicide. It is used to control broad-leaved weeds and annual grasses in field, berry fruit, nuts, vegetable and ornamental crops, turfgrass, orchards, and vineyards. At higher rates, it is used for non-selective weed control in industrial areas. In the past, simazine was used to control submerged weeds and algae in large aquaria, farm ponds, fish hatcheries, swimming pools, ornamental ponds, and cooling towers (Extoxnet, 1996). The EEA (2000) gives simazine concentrations in groundwater in different European countries. In Germany simazine concentrations >0.1 μg/l were found in 0.9% of 11,630 observed groundwater wells. Denmark reports 0.3% of 625 wells, Austria 0.2% of 1248 wells, Hungary 6.8% of 174 wells and Slovenia 8.4% of 84 wells. With a total of 7% contaminated wells simazine is after atrazine the second most detected pesticide for groundwater in Europe (EEA, 2000). Since 2004 it is banned in European Union states. Simazine binds moderately to poorly to soils (Wauchope *et al.*, 1992). It does, however, adsorb to clays and mucks. Its low water solubility makes it less mobile, limiting its leaching potential. Simazine is moderately persistent with an average field half-life of 60 days. Soil half-lives of 28–149 days have been reported (Wauchope *et al.*, 1992). Residual activity may remain for a year after application (2–4 kg/ha) in high pH soils.

Chlorfenvinphos is an insecticide. It was widely used to control household pests such as flies, fleas, and mice. There is insufficient data to classify Chlorfenvinphos as a groundwater contaminant (http://www.pesticideinfo.org). The European Union cancelled use of chlorfenvinphos in 1999.

Chlorpyrifos is a broad-spectrum organophosphate insecticide. Originally it was used primarily to kill mosquitoes, but it is also effective in controlling cutworms, corn rootworms, cockroaches, grubs, flea beetles, flies, termites, fire ants, and lice. It is used on grain, cotton, field, fruit, nut and vegetable crops, as well as on lawns and ornamental plants. It is directly used on sheep, for horse site treatment, dog kennels, domestic dwellings, farm buildings, storage bins, and commercial establishments. Chlorpyrifos adsorbs strongly to soil particles and it is not readily soluble in water. It is therefore immobile in soils and unlikely to leach or to contaminate groundwater. US EPA (1998) has reported chlorpyrifos in groundwater in less than 1% of the wells tested, with the majority of measurements below 0.01 μg/litre (TCP 3,5,6-Trichloro-2-pyridinol), the principal metabolite of chlorpyrifos, adsorbs weakly to soil particles and appears to be moderately mobile and persistent in soils (Extoxnet, 1996). Chlorpyrifos is moderately persistent in soils. The half-life of chlorpyrifos in soil is usually between 60 and 120 days, but can range from 2 weeks to over 1 year, depending on the soil type, climate, and other conditions (Extoxnet, 1996).

3.2.9.2 Polar Pesticides

Diuron is a general use herbicide (phenylurea herbicide). Diuron is a substituted urea herbicide used to control a wide variety of annual and perennial broadleaf and grassy weeds, to control weeds and mosses on non-crop areas and among many agricultural crops such as fruit, cotton, sugar cane and legumes. Diuron works by inhibiting photosynthesis. Diuron is moderately to highly persistent in soils. It was commonly found in ground- and surface waters in Europe. Residue half-lives are from one month to a few years. Mobility in the soil is related to organic matter and to the type of the residue. The metabolites are less mobile than the parent compound (Howard, 1991). The European Union banned the use of Diuron in 1999.

Isoproturon is a selective, systemic herbicide used in the control of annual grasses and broadleaved weeds in cereals. Isoproturon is mobile in soil and has been detected in both surface water and groundwater. In groundwater, it has been detected at concentrations between 0.05 and 0.1 µg/litre (WHO, 1996). Levels above 0.1 µg/litre have occasionally been detected in drinking-water (WHO, 1996). In water it hydrolyses with a half-life of about 30 days. It is quite persistent under anoxic conditions (92% remained in the system after 119 days). In soil, isoproturon undergoes enzymatic and microbial demethylation at the urea nitrogen and hydrolysis of the phenylurea to form 4-(2-hydroxyisopropyl)aniline. Under field conditions, its half-life is about 40 days in temperate climates and 15 days in tropical climates. Isoproturon is allowed in the European Union.

Trifluralin is a dinitroaniline herbicide used to control a wide spectrum of annual grasses and broadleaf weeds in agriculture, horticulture, viticulture, amenity and home garden. The major crops it is used on are oilseed rape and sunflowers and, to a lesser extent, cotton and cereals (OSPAR, 2005b). Trifluralin strongly adsorbs on soils and is nearly insoluble in water. Because adsorption is highest in soils high in organic matter or clay content and adsorbed herbicide is inactive, higher application rates may be required for effective weed control on such soils. Occurrence of trifluralin in groundwater is rare. Of a considerable number of sites (almost 3500) and groundwater samples analyzed (over 12,000), only a small number of positive findings were reported (OSPAR, 2005b). These positive findings come from Austria (1 of 7000 samples, 1992–97), France (1 of 336 samples, 1997) and the UK (6 of 517 samples, 1999). The highest concentration reported was 0,1 µg/l. Hydrolysis is not expected to be a significant route of dissipation of trifluralin in the environment. A study by OSPAR (2005b) determined a half-life in water for trifluralin in the water/sediment system to be 1–2 days. Half-life in sediment was calculated to be 7–15 days. Trifluralin was more persistent in soils with a half-life is 3–18 weeks. Hence sediments and soils are major reservoirs for trifluralin and assessment of persistence should be based on these compartments.

3.2.10   OTHER ORGANIC COMPOUNDS

Polybrominated diphenyl ethers (PBDEs) represent important additive flame retardants with numerous uses in industrial and domestic electronic equipment and textiles. PBDEs are similar in behaviour (hydrophobic, lipophilic, thermally stable) to the well-studied contaminants PCBs. More than 67,400 tons of PBDEs were produced annually worldwide, including penta-BDE, octa-BDE and deca-BDE (Kim *et al.*, 2007). Growing evidence suggests that PBDEs are widespread global environmental pollutants and that they are capable of bioaccumulation in food chains. They have already been found in high concentrations in marine birds and mammals from remote areas. PBDEs enter air, water, and soil during their manufacture and use in consumer products. When PBDEs are suspended in air, they can be present as particles. They eventually return to land or water as the dust settles and are washed out by snow and rainwater. PBDEs do not dissolve easily in water, and therefore, high levels of PBDEs are not found in water. The very small amounts of PBDEs that do occur in water stick to particles and eventually settle to the bottom. Rainwater is not expected to spread them much below the soil surface; thus, it is unlikely that PBDEs will enter groundwater. Data on environmental fate, although limited, suggest that PBDEs are recalcitrant. Aerobically, mono, di-, and tribrominated diphenylethers were transformed cometabolically to corresponding hydroxylated brominated hydroxydiphenylethers, phenols and mucoic acids by *Sphingomonas* species or fungi (Kim *et al*., 2007). Mixed bacterial cultures derived from soil used a commercial PBDE mixture as sole carbon source, and almost completely removed parent compounds within minutes (Vonderheide *et al*., 2006). Under anoxic conditions reductive debromination of PBDEs may occur resulting in production of lower or non-brominated diphenyl ethers (Rayne *et al*., 2003; Gerecke *et al*., 2005). Recent studies demonstrated abiotic debromination of PBDEs using zero valent iron or UV light (Rayne *et al*., 2003, Kim *et al*., 2007).

Di(2-ethylhexyl)phthalates (DEHP) are a family of industrial chemicals used as softeners, adhesives or solvents by a variety of industries. They are mainly used in the polymer industry as plasticizer in PVC and to a lesser extent in the non-polymer industry for different consumer products such as sealants, paints, printing inks, cosmetics, coatings of different products such as cars, coils, cables or fabrics (EU Risk Assessment DEHP, 2001). Accidental releases of DEHP have caused these compounds to become ubiquitous pollutants, in marine, estuarine and freshwater sediments, sewage sludges, soils and food. DEHP will adhere to soil, and so will neither evaporate nor leach into groundwater. Abiotic degradation in aquatic environments is reported to be very slow with a hydrolysis half-life of approx. 2000 years and a very slow photooxidation as well (OSPAR 2005c). In most reported tests for ready biodegradability, the mineralization is lower than the pass level for judging DEHP as readily biodegradable. In one test, however a degradation of 82% was determined (OSPAR, 2005c). Based on simulation test results, a half-life of 50 days has been proposed for the freshwater environment (EU Risk Assessment DEHP, 2001) EUSES. UNEP (2002) reports that degradation half-life values for DEHP are generally in a range from 1 to 30 days in soils and freshwaters. A half-life of 150 days is indicated by the EU Technical Guidance Document on risk assessment (European Commission, 2003).

Tributylin compounds are mainly used as antifouling paints (tributyl and triphenyl tin) for underwater structures and ships. The main primary source of tributyl tin is leaching from ship hulls. Minor identified applications are as antiseptic or disinfecting agents in textiles and industrial water systems, such as cooling tower and refrigeration water systems, wood pulp and paper mill systems, and breweries. They are also used as stabilizers in plastics and as catalytic agents in soft foam production. It is also used to control the shistosomiasis in various parts of the world. Because of the low water solubility Tributyltin compounds binds strongly to suspended material and sediments. Therefore, they have not been found in groundwater (Extoxnet, 1996). Tributyltin compounds may be moderately to highly persistent. Degradation depends on temperature and the presence of microorganisms. Under oxic conditions, TBT takes 1–3 months to degrade, but in anaerobic soils may persist for more than 2 years. The breakdown of TBT leads eventually to the tin ion. All of the breakdown products are less toxic than TBT itself. The European Union has forbidden its Member States to bring TBT on the market and is developing a legislation to prohibit the presence of TBT on ships.

3.2.11 CONCLUSIONS AND OUTLOOK

In the 20th and 21st centuries a wide variety of organic chemicals have been synthesized and introduced into the environment. In the European Water Framework Directive, from the more than thousand known organic pollutants, a selection of priority compounds has been defined as an instrument to detect, monitor and, if possible, diminish environmental contamination. The potential threat of these WFD priority pollutants to our groundwater resources depends on the specific properties of the particular chemicals and the exposed environment. The MOCs that can be a threat for the good status of a groundwater body have in common that they, and/or their degradation products, are toxic, poorly or non-degradable in subsurface environments, hydrophilic, relatively soluble in water and are frequently used. Typical examples of such chemicals are benzene, chlorinated aliphatic hydrocarbons, atrazine and simazine. Substances with log K_{ow} values higher than 3.0–3.5 do not pose a real treat for groundwater contamination at a regional scale, because they will strongly adsorb to the soil matrix and organic matter. However, local groundwater contamination with these substances is possible by leakage of point sources like contaminated sites, landfills or obsolete pesticides stockpiles.

Outside the group of WFD pollutants and priority substances more MOCs can be important for groundwater. Among these chemicals the fuel additive MTBE is most frequently detected. In an intensive screening survey in groundwater protection zones around drinking water pumping stations contaminants such as 1,2-dichlorpropane (DCP), various anilines, (cloro-)phenols, sulfone compounds and organophosphate compounds like triethyl phosphate and tributyl phosphate were detected (Wuijts *et al.*, 2007). Another groundwater screening survey revealed that pharmaceuticals can be present in drinking water and drinking water resources in very low concentrations. The most frequently detected medicines in drinking water are almost non-degradable in the environment and/or frequently used. Medicines frequently found in groundwater at drinking water production sites were analgesics, salicylic acid (mainly from aspirin), phenazon and the anti-epileptic carbamazepin. Traces of the tranquilizer, prozac, were found in a few samples (Versteegh *et al.*, 2007).

In groundwater, chemical transformation of some WFD priority chemicals may occur, but biodegradation by indigenous microorganisms is the most important mechanism for their complete removal. Thus microorganisms play a crucial role in the eventual cleanup of our groundwater systems. Different biodegradation mechanisms may occur, largely depending of the presence of microorganisms with specific metabolic capacities and their activity under the residing geochemical conditions. Half-lives of MOCs in the environment can range from days to years, and at some locations potentially biodegradable compounds appear to be essentially persistent. Therefore it is unclear, even when introduction of new pollution could be stopped today, how long MOCs can remain a hazard to groundwater resources. An important challenge is to identify limitations of biodegradation in contaminated aquifers. The question why some contaminants appear persistent although microorganisms that could degrade them can be isolated from the aquifer is largely unsolved. Novel (molecular) monitoring techniques, targeting biodegrading microorganisms and their enzymes, compound-specific metabolites, and stable isotope fractionation studies can help to monitor, predict and enhance *in situ* degradation rates of MOCs in contaminated aquifers.

REFERENCES

Accardi-Dey A. and Gschwend P.M., 2002. Assessing the combined roles of natural organic matter and black carbon as sorbents in sediments, *Environ. Sci. Technol.*, **36**, 21–9.

van Agteren M., Keuning S. and Jansen D.B., 1998. *Handbook on Biodegradation and Biological Treatment of Hazardous Organic Compounds*, Kluwer Academic Publishers, Dordrecht.

Allpress J.D. and Gowland P.C., 1999. Biodegradation of chlorinated paraffins and long-chain chloroalkanes by *Rhodococcus* sp. S45-1, *International Biodeterioration and Biodegradation*, **43**, 173–9.

Arp D.J., Yeager C.M. and Hyman M.R., 2001. Molecular and cellular fundamentals of aerobic cometabolism of trichloroethylene, *Biodegradation*, **12**, 81–103.

Bae H.-S., Yamagishi T. and Suwa Y., 2002. Evidence for degradation of 2-chlorophenol by enrichment cultures under denitrifying conditions, *Microbiology*, **148**, 221–7.

Barbash J.E. and Reinhard M., 1989. Abiotic dehalogenation of 1,2-dichloroethane and 1,2-dibromoethane in aqueous solution containing hydrogen sulfide, *Environ. Sci. Technol.*, **23**, 1349–57.

Barker J.F., Hubbard E. and Lemon L.A., 1990. Presented at the Petroleum Hydrocarbons and Organic Chemicals in Groundwater Prevention, Detection, and Restoration, Houston, Texas.

Baun A., Reitzel L.A., Ledin A., Christensen T.H. and Bjerg P.L., 2003. Natural attenuation of xenobiotic organic compounds in a landfill leachate plume (Vejen, Denmark), *J. Cont. Hydrol.*, **65**, 269–91.

Beck J. and Hansen K.E., 1974. The degradation of quintozene, pentachlorobenzene, hexachlorobenzene and pentachloroaniline in soil, *Pestic. Sci.*, **5**, 41–8.

Bedessem M.E., Swoboda-Colberg N.G. and Colberg P.J.S., 1997. Naphthalene mineralization coupled to sulfate reduction in aquifer- derived enrichments, *FEMS Microbiology Letters*, **152**, 213–18.

Beurskens J.E.M., Dekker C.G.C., van den Heuvel H., Swart M. and de Wolf J., 1994. Dechlorination of chlorinated benzenes by an anaerobic microbial consortium that selectively mediates the thermodynamic most favorable reactions, *Environ. Sci. Technol.*, **28**, 701–6.

Bianchin M., Smith L., Barker J.F. and Beckie R., 2006. Anaerobic degradation of naphthalene in a fluvial aquifer: A radiotracer study, *J. Cont. Hydrol.*, **84**, 178–96.

Borden R.C., Daniel R.A., Lebrun L.E. and Davis C.W., 1997. Intrinsic biodegradation of MTBE and BTEX in a gasoline-contaminated aquifer, *Water Resources Research* **33**, 1105–15.

Bosma T.N.P., Cottaar F.H.M., Posthumus M.A., *et al.*, 1994. Comparison of reductive dechlorination of hexachloro- 1,3-butadiene in Rhine sediment and model systems with hydroxocobalamin, *Environ. Sci. Technol.*, **28**, 1124–8.

Bosma T.N.P., van Aalst-van Leeuwen M., Gerritse J., van Heiningen E., Taat J. and Pruijn M., 1998. Intrinsic dechlorination of 1,2-dichloroethane at an industrial site. *Contaminated Soil '98*, Thomas Telford Publishing, London.

Bradley P.M., 2003. History and ecology of chloroethene biodegradation: a review, *Biorem. J.* **7**(2), 81–109.

Breivik K., Pacyna J.M. and Munch J., 1999. Use of a-, β- and γ-hexachlorocyclohexane in Europe 1970–1996, *Sci. Tot. Environ.*, **239**, 151–63.

Brown K.W. and Donnelly K.C., 1988. Estimation of the Risk Associated with the Organic Constituents of Hazardous and Municipal Waste Landfill Leachates, Hazardous Waste and Hazardous Materials HWHME2 5, 1-30.

Caldwell M.E. and Suflita J.M., 2000. Detection of phenol and benzoate as intermediates of anaerobic benzene biodegradation under different terminal electron-accepting conditions, *Environ. Sci. Technol.*, **34**, 1216–20.

Chakraborty R. and Coates J.D., 2005. Hydroxylation and carboxylation – Two crucial steps of anaerobic benzene degradation by *Dechloromonas* strain RCB. *Appl. Environ. Microbiol.*, **71**, 5427–32.

Chang B.V., Shiung L.C. and Yuan S.Y., 2002. Anaerobic biodegradation of polycyclic aromatic hydrocarbon in soil, *Chemosphere*, **48**, 717–24.

Chang B.V., Yu C.H. and Yuan S.Y., 2004. Degradation of nonylphenol by anaerobic microorganisms from river sediment, *Chemosphere*, **55**, 493–500.

Christensen T.H., Bjerg P.L. and Kjeldsen P., 2000. Natural attenuation: a feasible approach to remediation of groundwater pollution at landfills? *Ground Water Monit. Remediat.*, **20**, 69–77.

Christensen T.H., Kjeldsen P., Bjerg P.L., *et al.*, 2002. Biogeochemistry of landfill leachate plumes, *Appl. Geoch.*, **16**, 659–718.

Coates J.D., Chakraborty R. and McInerney M.J., 2002. Anaerobic benzene biodegradation – A new era, *Research in Microbiology*, **153**, 621–8.

Coates J.D., Woodward J., Allen J., Philp P. and Lovley D.R., 1997. Anaerobic degradation of polycyclic aromatic hydrocarbons and alkanes in petroleum-contaminated marine harbor sediments. *Appl. Environ. Microbiol.*, **63**, 3589–93.

Cox E.E., Major D. and Edwards E., 2000. Natural attenuation of 1,2-dichloroethane in groundwater at a chemical manufacturing facility. In *Natural Attenuation Considerations and Case Studies: Remidiation of Chlorinated and Recalcitrant Compounds*, G.B. Wickramanayke, A.R. Gavasker, M.E. Kelley (eds), Battelle Press, Monterey.

Deeb R.A., Scow K.M. and Alvarez-Cohen L., 2000. Aerobic MTBE biodegradation: an examination of past studies, current challenges and future research directions, *Biodegradation* **11**, 171–85.

Dennie D., Gladu I., Lépine F., Villemur R., Bisaillon J.-G. and Beaudet R., 1998. Spectrum of the reductive dehalogenation activity of *Desulfitobacterium frappieri* PCP-1, *Appl. Environ. Microbiol.*, **64**, 4603–6.

Dijk J.A., de Bont J.A.M., Lu X., *et al.*, 2005. Anaerobic oxidation of (chlorinated) hydrocarbons. In *Bioremediation and Phytoremediation of Chlorinated and Recalicitrant Compounds*, G.B. Wickramanayake, A.R. Gavaskar, B.C. Alleman and V.S. Magar (eds), Batelle Press, Monterey, California.

Dinglasan-Panlilio M.J., Dworatzek S., Mabury S., Edwards E., 2006. Microbial oxidation of 1,2-dichloroethane under anoxic conditions with nitrate as electron acceptor in mixed and pure cultures, *FEMS Microbiology Ecology*, **56**(3), 355–64.

Durant N.D., Wilson L.P. and Bouwer E.J., 1995. Microcosm studies of subsurface PAH-degrading bacteria from a former manufactured gas plant, *J. Cont. Hydrol.*, **17**, 213–37.

Duyzer J.H. and Vonk A.W., 2003. *Atmosferic deposition of pesticides, PAH en PCB's in the Netherlands (in Dutch)*, STOWA report 2003-01.

European Commission, 2002. *Review Report for the Active Substance Isoproturon SANCO/3045/99-final*.

European Commission, 2003. *Technical Guidance Document in Support of Commission Directive 93/67/EEC on Risk Assessment for New Notified Substances and Commission Regulation (EC) No 1488/94 on Risk Assessment for Existing Substances and Commission Directive (EC) 98/8 on Biocides*.

European Commission, 2005. *Risk Profile and Summary Report for Short-chained Chlorinated Paraffins (SCCPs)*, Dossier prepared for the UNECE Convention on Longrange Transboundary Air Pollution, Protocol on Persistent Organic Pollutants.

EEA, 2000. Groundwater quality and quantity in Europe. *Environmental Assessment Report No 3*, Copenhagen.

EU Risk Assessment, 2001. *Bis (2-ethylhexyl) phthalate (DEHP)*, CAS-No. 117-81-7. Consolidated Final Report.

Extoxnet, 1996. Extension Toxicology Network, (http://extoxnet.orst.edu/pips/ghindex.html)

Farm Chemicals Handbook, 1992. Meister Publishing Company, Willoughby, OH.

Fayolle F., Vandecasteele J.P. and Monot F., 2001. Microbial degradation and fate in the environment of methyl tert-butyl ether and related fuel oxygenates, *Appl. Microbiol. Biotechnol.*, **56**, 339–49.

Fayolle F. and Monot F., 2005. Biodegradation of fuel ethers, In Magot, M. Ollivier, B. (eds), *Petroleum Microbiology*, ASM. Washington DC, 301–16.

Fetzner S., 1998a. Bacterial degradation of pyridine, indole, quinoline, and their derivatives under different redox conditions, *Appl. Microbiol. Biotechnol.*, **49**, 237–50.

Fetzner S., 1998b. Bacterial dehalogenation, *Appl. Microbiol. Biotechnol.*, **50**, 633–57.

Freedman D.L. and Gossett J.M., 1991. Biodegradation of dichloromethane and its utilization as a growth substrate under methanogenic conditions, *Appl. Environ. Microbiol.*, **57**, 2847–57.

Galushko A., Minz D., Schink B. and Widdel F., 1999. Anaerobic degradation of naphthalene by a pure culture of a novel type of marine sulphate-reducing bacterium, *Environ. Microbiol.*, **1**, 415–20.

Gerecke A.C., Hartmann P.C., Heeb N.V., *et al.*, 2005. Anaerobic degradation of decabromodiphenyl ether, *Environ. Sci, Technol.*, **39**, 1078–83.

Gerritse J., Drzyzga O., Kloetstra G., *et al.*, 1999. Influence of different electron donors and acceptors on dehalorespiration of tetrachloroethene by *Desulfitobacterium frappieri TCE1*, *Appl. Environ. Microbiol.*, **65**, 5212–21.

Gerritse J., Renard V., Pedro Gomes T.M., Lawson P.A., Collins M.D. and Gottschal J.C., 1996. *Desulfitobacterium* sp. strain PCE1, an anaerobic bacterium that can grow by reductive dechlorination of tetrachloroethene or ortho-chlorinated phenols, *Arch. Microbiol.*, **165**, 132–40.

Grbić-Galić D. and Vogel T.M., 1987. Transformation of toluene and benzene by mixed methanogenic cultures, *Appl. Environ. Microbiol.*, **53**, 254–60.

Grostern A. and Edwards E.A., 2006. Growth of Dehalobacter and Dehalococcoides spp. during Degradation of Chlorinated Ethanes, *Appl. Environ. Microbiol* **72** 428–36.

Hage J.C. and Hartmans S., 1999. Monooxygenase-mediated 1,2-dichloroethane degradation by *Pseudomonas* sp. strain DCA1, *Appl. Environ. Microbiol.*, **65**, 2466–70.

Haggblom M.M. and Young L.Y., 1990. Chlorophenol degradation coupled to sulfate reduction, *Appl. Environ. Microbiol.*, **56**, 3255–60.

Haggblom M.M., Nohynek L.J. and Salkinoja-Salonen M.S., 1988. Degradation and O-methylation of chlorinated phenolic compounds by *Rhodococcus and Mycobacterium strains*, *Appl. Environ. Microbiol.*, **54**, 3043–3052.

Harwood C.S. and Gibson J., 1997. Shedding light on anaerobic benzene ring degradation: a process unique to prokaryotes?, *J Bacteriol.*, **179**, 301–9.

Holden L. and Graham J.A., 1992. Results of the national alachlor well water survey, *Environ. Sci. Technol.*, **26**, 935–43.

Howard P.H., 1991. *Handbook of Environmental Fate and Exposure Data for Organic Chemicals*, Lewis Publishers, Chelsea, MI.

Howsam M. and Jones K.C., 1998. Sources of PAHs in the Environment. *Handbook of Environmental Chemistry*, Vol. **3**, Springer-Verlag.

Janssen D.B., Dinkla I.J.T., Poelarends G.J. and Terpstra P., 2005. Bacterial degradation of xenobiotic compounds: Evolution and distribution of novel enzyme activities, *Environ. Microbiol.*, **7**, 1868–82.

Jeffers P.M., Ward L.M., Woytowitch L.M. and Wolfe N.L., 1989. Homogeneous hydrolysis rate constants for selected chlorinated methanes, ethanes, ethenes, and propanes, *Environ. Sci. Technol.*, **23**, 965–9.

Karthikeyan R. and Bhandari A., 2001. Anaerobic biotransformation of aromatic and polycyclic aromatic hydrocarbons in soil microcosms: a review, *J. Hazardous Substance Res.*, **3**, 1–19.

Kasai Y., Takahata Y., Manefield M. and Watanabe K., 2006. RNA-Based stable isotope probing and isolation of anaerobic benzene-degrading bacteria from gasoline-contaminated groundwater, *Appl. Environ. Microbiol*, **72**, 3586–92.

Kaschl A., Vogt C., Uhlig S., *et al.*, 2005. Isotopic fractionation indicates anaerobic monochlorobenzene biodegradation, *Environ. Toxicol. Chem.*, **24**, 1315–24.

Kim Y.-M., Nam I.-H., Murugesan K., Schmidt S., Crowley D.E. and Chang Y.-S., 2007. Biodegradation of diphenyl ether and transformation of selected brominated congeners by *Sphingomonas* sp. PH-07. *Appl. Microbiol. Biotechnol.*, **77**, 187–94.

Kjeldsen P., Barlaz M.A., Rooker A.P., Baun A., Ledin A. and Christensen H., 2002. Present and long-term composition of MSW landfill leachate: a review, *Environ. Sci. Technol.*, **32**, 297–336.

Kleineidam S., Rugner H., Ligouis B. and Grathwohl P., 1999. Organic matter facies and equilibrium sorption of phenanthrene, *Environ. Sci. Technol.*, **33**, 1637–44.

Kleineidam S., Schuth C. and Grathwohl P., 2002. Solubility-normalized combined adsorption-partitioning sorption isotherms for organic pollutants, *Environ. Sci. Technol.*, **36**, 4689–97.

Kuder T. and Philp P., 2008. Modern geochemical and molecular tools for monitoring in-situ biodegradation of MTBE and TBA, *Rev. Environ. Sci. Biotechnol.*, **7**, 79–91.

Langenhoff A.A.M., Zehnder A.J.B. and Schraa G., 1996. Behaviour of toluene, benzene and naphthalene under anaerobic conditions in sediment columns, *Biodegradation*, **7**, 267–74.

Langenhoff A.A.M, Staps J.J.M., Pijls C.G.J.M., Alphenaar A., Zwiep G. and Rijnaarts H.H.M., 2002. Intrinsic and stimulated *in situ* biodegradation of hexachlorocyclohexane (HCH), *Water, Air and Soil Pollution*, **2**, 171–81.

Lechner U., Brodkorp D., Geyer R., *et al.*, 2007. *Int. J. Syst. Evol. Microbiol.*, **57**, 1295–1303.

Leisinger T., 1983. Microorganisms and xenobiotic compounds, *Experientia*, **39**, 1183–91.

Lindinger H. and Scheidleder A., 2004. Indicator Fact Sheet (WHS1a) Pesticides in Groundwater, EEA.

Lovley D.R., 2000. Anaerobic benzene degradation, *Biodegradation*, **11**, 107–16.

Magli A., Messmer M. and Leisinger T., 1998. Metabolism of dichloromethane by the strict anaerobe Dehalobacterium formicoaceticum, *Appl. Environ. Microbiol.*, **64**, 646–50.

Meckenstock R.U., Safinowski M. and Griebler C., 2004. Anaerobic degradation of polocyclic aromatic hydro-carbons, *FEMS Microbiol. Ecol.*, **49**, 27–36.

Mihelcic J.R. and Luthy R.G., 1988. Degradation of polyclinic aromatic hydrocarbons compounds under various redox conditions in soil–water systems, *Appl. Environ. Microbiol.*, **54**, 1182–7.

Mohn W.W. and Tiedje J.M., 1992. Microbial reductive dehalogenation, *Microbiol. Mol. Biol. Rev.*, **56**, 482–507.

National Research Council Committee on Intrinsic Remediation, 2000. *Natural Attenuation for Groundwater Remediation*. National Academy Press, Washington, D.C.

Nielsen P.H., Bjerg P.L., Nielsen P., Smith P. and Christensen T.H., 1996. *In Situ* and laboratory determined first-order degradation rate constants of specific organic compounds in an aerobic aquifer, *Environ. Sci. Technol.*, **30**, 31–7.

OSPAR Commission, 2001. *OSPAR Background Document on NP/NPethoxylates*, Publication Number 136.

OSPAR Commission, 2003. *OSPAR Background Document on Octylphenol*, Publication Number 173.

OSPAR Commission, 2005a. *OSPAR Background Document on Trichlorobenzenes*, Publication Number 170.

OSPAR Commission, 2005b. *OSPAR Background Document on Trifluralin*, Publication Number 203.

OSPAR Commission, 2005c. *OSPAR Background Document on Phthalates*, Publication Number 226.

Prince R.C., 2000. Biodegradation of methyl tertiary-butyl ether (MTBE) and other fuel oxygenates, *Crit. Rev. Microbiol.*, **26**, 163–78.

Rayne S., Ikonomou M.G., Whale M.D., 2003. Anaerobic microbial and photochemical degradation of 4,4'-dibromodiphenyl ether, *Water Research*, **37**, 551–60.

Rifai H.S., Borden R.C., Wilson J.T. and Ward C.H., 1995. Intrinsic bioattenuation for subsurface restoration. In *Intrinsic Bioremediation*, R.E. Hinchee, J.T. Wilson and D.C. Downey (eds), vol. **1**, Battelle Press, Columbus, OH, 1–29.

Rohwerder T., Breuer U., Benndorf D., Lechner U. and Muller R.H., 2006. The alkyl tert-butyl ether interme-diate 2-hydroxyisobutyrate is degraded via a novel cobalamin-dependent mutase pathway, *Appl. Environ. Microbiol.*, **72**, 4128–35.

Rügge K., Bjerg P.L., Pedersen J.K., Mosbæk H. and Christensen T.H., 1999. An anaerobic field injection experiment in a landfill leachate plume (Grindsted, Denmark): 1. Experimental set up, tracer movement and fate of aromatic and chlorinated compounds, *Water Resour. Res.*, **35**, 1231–46.

Ryoo D., Shim H., Canada K., Barbieri P. and Wood T.K., 2000. Aerobic degradation of tetrachloroethylene by toluene-o-xylene monooxygenase of *Pseudomonas stutzeri* OX1, *Nature Biotechnology*, **18**, 775––8.

Safinowski M. and Meckenstock R.U., 2006. Methylation is the initial reaction in anaerobic naphthalene degradation by a sulfate-reducing enrichment culture, *Environmental Microbiology* **8**, 347–52.

Schmidt T.C., Morgenroth E., Schirmer M., Effenberg M. and Haderlein S.B., 2002. Use and occurrence of fuel oxygenates in Europe, 58-79. In A.F. Diaz and D.L. Drogos (ed.), *Oxygenates in Gasoline: Environmental Aspects*, ACS, Washington.

Schmidt T.C., Grathwohl P., Gocht T., Bi E. and Werth C.J., 2005. Sorption of organic compounds in the subsurface: experimental approaches, data evaluation and predictive methods. *Proc. ConSoil 2009, 9 Int. Conf. on Contaminated Soil, Bordeaux, France*.

Smith A.G., 1991. Chlorinated Hydrocarbon Insecticides. In *Handbook of Pesticide Toxicology*, Volume 3, Classes of Pesticides, W.J. Hayes Jr. and E.R. Laws Jr. (eds), Academic Press, Inc., NY.

Squillace P.J., Pankow J.F., Korte N.E. and Zogorski J.S., 1997. Review of the environmental behavior and fate of methyl tert-butyl ether, *Env. Toxicol Chem.*, **16**, 1836–44.

Suarez M.P. and Rifai H.S., 1999. Biodegradation rates for fuel hydrocarbons and chlorinated solvents in groundwater, *Biorem. J.*, **3**, 337–62.

Tabak H.H., Quave S.A., Mashni C.I. and Barth E.F., 1981. Biodegradability studies with organic priority pollutant compounds, *J. Water Pollut. Control Fed.*, **53**, 1503–18.

UBA, 2000. *Daten zur Umwelt – Der Zustand der Umwelt in Deutschland*, Hrsg. Umweltbundesamt, E. Schmidt (ed.), Springer Verlag, Berlin.

UNEP, 2002. *Regionally Based Assessment of Persistent Toxic Substances*, Europe regional report. United Nations Environment Programme Chemicals.

US-EPA, 1990. *National Survey of Pesticides in Drinking Water Wells*, Phase I Report. Washington, DC.

US-EPA, 1998. *Drinking Water Assessment of Chlorpyrifos*. Washington, DC.

US-EPA, 1998a. *Contaminant Candidate List EPA/600/R-98/048*, U.S. EPA.

US-EPA, 1998b. Oxygenates in water: Critical information and research needs EPA/600/R-98/048, U.S. EPA.

US-EPA, 1999. *Anaerobic Biodegradation Rates of Organic Chemicals in Groundwater, a Summary of Field and Laboratory Studies*, U.S. Environmental Protection Agency Office of Solid Waste Washington, DC 20460.

Ushiba Y., Takahara Y. and Ohta Y., 2003. *Sphingobium amiense* sp. Nov., a novel nonylphenol-degrading bacterium isolated from a river sediment, *Int. J. of Syst. and Evol.Microbiol.*, **53**, 2045–8.

Utkin I., Dalton D.D. and Wiegel J., 1995. Specificity of reductive dehalogenation of substituted ortho-chlorophenols by *Desulfitobacterium dehalogenans* JW/IU-DC1, *Appl. Environ. Microbiol.*, **61**, 346–51.

van de Plassche E. and Schwegler A., 2005. *POP Hexachlorobutadiene*, European Commission Task Force on Persistent Organic Pollutants.

van de Plassche E., Schwegler A., Rasenberg M. and Schouten G., 2005. *POP Pentachlorobenzene*, European Commission Task Force on Persistent Organic Pollutants.

Vermeire T., 1994. *Environmental health criteria for Hexachlorobutadien (nr 156)*, International Programme on Chemical Safety (IPCS), World Health Organization.

Versteegh J.F.M., van der Aa N.G.F.M. and Dijkman E., 2007. Pharmaceuticals in drinking water and drinking water resources. RIVM report 703719016, Bilthoven, the Netherlands.

Vogel T.M., Criddle C.S. and McCarty P.L., 1987. Transformations of halogenated aliphatic compounds. *Environ. Sci. Technol.*, **21**, 722–36.

Vonderheide A.P., Mueller-Spitz S.R., Meija J., *et al.*, 2006. Rapid Breakdown of brominated flame retardants by soil microorganisms, *J. Anal. Atom. Spect.*, **21**, 1232–9.

Wauchope R.D., Buttler T.M., Hornsby A.G., Augustijn-Beckers P.W.M. and Burt J.P., 1992. SCS/ARS/CES Pesticide properties database for environmental decisionmaking, *Rev. Environ. Contam. Toxicol.*, **123**, 1–157.

WHO, 1991. Chlorobenzenes other than hexachlorobenzene, *Environmental Health Criteria*, **128**, World Health Organisation, Geneva.

WHO, 1996. *Guidelines for Drinking-Water Quality*, 2nd ed. Vol.2. *Health Criteria and Other Supporting Information*. World Health Organization, Geneva.

Wuijts S., Schijven J.F., van der Aa N.G.F.M., Dik H.H.J., Versluijs C.W. and van Wijnen H.J., 2007. Components for a guidance document on groundwater protection. RIVM report 734301029, Bilthoven, the Netherlands.

Yu Z. and Smith G.B., 2000. Dechlorination of polychlorinated methanes by a sequential methanogenic-denitrifying bioreactor system, *Appl. Microbiol. Biotechnol.*, **53**, 484–9.

Zhang X. and Young L.Y., 1997. Carboxylation as an initial reaction in the anaerobic metabolism of naphthalene and phenanthrene by sulfidogenic consortia, *Appl. Environ. Microbiol.*, **63**, 4759–64.

3.3

Background Levels under the Water Framework Directive[1]

Ariane Blum[1], Hélène Pauwels[2], Frank Wendland[3] and Jasper Griffioen[4]

[1,2] *Bureau de Recherches Géologiques et Minières, Orléans cédex, France*
[3] *Institute of Chemistry and Dynamics of the Geosphere (ICG), Institute IV: Agrosphere, Research Centre Juelich, Juelich, Germany*
[4] *The Netherlands Organisation for Applied Scientific Research (TNO), Built Environment and Geosciences, Utrecht, The Netherlands*

---

[1] The views expressed in this chapter are purely those of the authors and may not in any circumstances be regarded as stating an official position of the European Commission.

Groundwater Monitoring Edited by Philippe Quevauviller, Anne-Marie Fouillac, Johannes Grath and Rob Ward
© 2009 John Wiley & Sons, Ltd

3.3.1   INTRODUCTION

3.3.1.1   Factors Influencing Groundwater Quality

The natural chemical composition of groundwater is derived from different factors such as:

- the petrographic properties of the aquifer;
- the chemical composition of rainfalls recharging the aquifer;
- the properties of the covering layers;
- the hydrodynamics properties of the aquifer including potential relationships with other aquifers (leakage) or with surface waters;
- the time of chemical interaction between the pore water and the solid matrix;
- redox conditions.

However, the petrographic properties of the solid matrix in the vadose and groundwater-saturated zones together with the regional hydrological and hydrodynamic conditions are the major geogenic influencing factors (Wendland *et al.*, 2007).

Today, groundwater composition of most European countries is largely influenced by human activities. Then, studying groundwater composition also means identifying anthropogenic sources of compounds such as diffuse source pollution (e.g. agriculture), point source pollutions (e.g. industries, waste water treatment plant), over-abstraction, mining activities, etc.

3.3.1.2   Definitions

Although an agreement exists about the above statements, there is no single definition of background levels. Here are three of them:

1. In 1999, the BaSeLiNe project adopted the following definition: 'groundwater quality baseline is the **concentration range** in water of a given present element, species or substance, derived from natural geological, biogenic, or atmospheric sources' (BaSeLine, 2003; Custodio and Manzano, 2007).

2. In 2006, the BRIDGE project aiming to support the derivation of threshold values in the context of the Groundwater Directive 2006/118/EC (GWD) adapted this definition to the Water Framework Directive (WFD) concepts and stated on the following one: 'the natural background level is the **concentration** of a given element, species or chemical substance present in solution which is derived by natural processes from geological, biological or atmospheric sources' (Müller *et al.*, 2006).

3. According to the GWD (Article 2.5), 'background level means the **concentration** of a substance or the value of an indicator in a body of groundwater corresponding to no, or only very minor, anthropogenic alterations to undisturbed conditions'.

Of course, all definitions are correct and the difference is explained by the use of 'background levels'. Natural quality of groundwater systems is variable (in time and space) and, from a scientific point of view, it makes no sense to define one single concentration for an aquifer as a whole. But to a policy point of view and referring to the WFD/GWD requirements, using an average concentration as a background level is helpful. As explained in Chapter 4.1 of the present book, under the WFD, background levels are not used to characterise groundwater bodies or aquifers but to derive threshold values. And threshold values do not reflect the status of groundwater bodies. Threshold values are just trigger values contributing to the assessment of groundwater bodies chemical status.

3.3.1.3   Why Assess Groundwater Background Levels?

Because many aquifers are contaminated, any groundwater management plan first requires the knowledge of the background levels of contaminants. Before the publication of the WFD, the main objective behind the assessment of background levels was the implementation and management of drinking water abstractions. In many European aquifers, high natural occurrences (i.e. exceeding drinking standards) of toxic elements (e.g. As, Se, Sb) are observed (BRIDGE, 2006) etc.). For this reason understanding background levels early became a major concern. In 1998 the publication of the Drinking Water Directive 98/60/EC confirmed the need to better characterise and predict the links between the geological context and groundwater chemical composition.

In 2000, the Water Framework Directive (WFD) offered an integrative approach to groundwater quality and, above all, made the assessment of groundwater background levels a legal obligation. On several occasions the WFD and the GWD have referred to groundwater background levels:

- To meet the WFD Article 5 requirements, 'Member shall carry out further characterisation of groundwater bodies at risk'. Where relevant, this characterisation shall include 'the chemical composition of the groundwater, including specification of the contributions from human activity. Member States may use typologies for groundwater characterisation when **establishing natural background levels** for these bodies of groundwater.'

- The GWD Annex II.A.1.d stipulates that 'when establishing threshold values, Member States will consider [...] hydro-geological characteristics including **information on background levels**'.

- Recital 10 of the GWD states that 'groundwater chemical status provisions do not apply to **high naturally-occurring levels** of substances or ions or their indicators, contained either in a body of groundwater or in associated bodies of surface water, due to specific hydro-geological conditions, which are not covered by the definition of pollution.'

Figure 3.3.1 summarises the relationships between groundwater quality and status under the WFD and clearly shows the need to distinguish background levels from human inputs for the assessment of groundwater bodies chemical status.

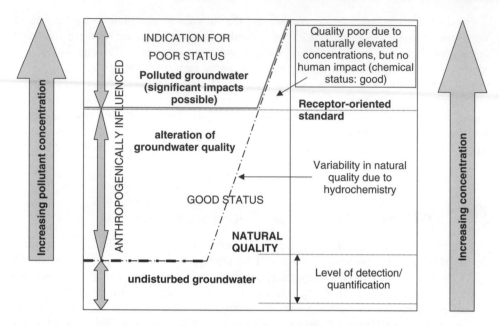

Figure 3.3.1   General relationship of groundwater quality and status under the WFD.

3.3.2   APPROACHES FOR DETERMINING GROUNDWATER BACKGROUND LEVELS

As explained before, the GWD clearly asks Member States to characterise background levels of groundwater bodies at risk. But how to make sure that it is feasible for the thousands European groundwater bodies at risk of failing poor chemical status? How to ensure that the assessment of background levels follows the same concepts and definition from one groundwater body to another and from one country to another? Does it mean Member States should adopt a common approach for the derivation of background levels? Would it be realistic to do so?

To answer these questions and to support Member States deriving threshold values, the BRIDGE project (2005–2006) has first studied the existing approaches adopted by Member States to characterise background levels (BRIDGE, 2006). A survey carried out in 12 countries[2] allowed giving an overview of existing approaches. The main conclusions of this survey are as follows:

- Before the WFD, many countries were already paying attention and were already carrying out studies to characterise their groundwater background levels. As a consequence, the assessment of background levels, as it is required by the WFD, must take this knowledge into account,

---

[2] Estonia, Finland, Belgium (Flanders), Hungary, United-Kingdom, Lithuania, Poland, Bulgaria, Denmark, Netherlands, Germany, Portugal.

- The level of knowledge is very different from one country to another and from one groundwater body to another (including in the same country).

For these reasons, and because each groundwater body has a unique chemical composition that is variable in time and space, when they are asked to assess background levels in response to the WFD, Member States are free to apply their own approach depending on existing studies and conceptual models of the groundwater bodies (European Commission, 2009).

3.3.3 PROPOSAL FOR A PROCEDURE TO DERIVE BACKGROUND LEVELS UNDER THE WFD/GWD

3.3.3.1 General Methodology

It is evident that Member States can use their own experience and their own approaches to derive background levels in cases where there is a high level of knowledge about geochemical transfers and processes and/or groundwater monitoring data is available for the whole country, specific groundwater typologies or for a specific groundwater body. Background levels, however, have also to be derivable in cases where only a medium or small level of knowledge about geochemical characteristics is available and/or only some groundwater monitoring data to precisely define the background levels are present. The background levels assessment in these cases is always a compromise between scientific validity and a general applicability throughout Europe. It became evident that the approach to derive background levels on a European level needs to be general applicable (and thus 'simplified') rather than 'sophisticated', but applicable only in well-monitored regions.

One issue of the BRIDGE project was the proposal for a procedure to derive background levels taking all these situations into account (Müller *et al.*, 2006). Figure 3.3.2 summarises this procedure. This one uses an aquifer typology established within the project (Wendland *et al.*, 2007; Chapter 2.2 of this book). Indeed, although many other aquifer typologies already exist, these are more often based on flow and structural concepts (karstic, fractured, granularity...). The criteria used in the BRIDGE project are the most relevant ones to explain the natural chemical signature of an aquifer. The main identified criteria are the lithology and the salinisation. Others such as the residence time have also been taken into account.

Depending on the level of knowledge and the available groundwater monitoring data the way to derive background levels can vary and be done:

- using own national or scientific approaches and experiences; or

- using a simplified (pre-selection) approach; or

- based on statistical reference results from other aquifers of the same type from the Member State or from another one.

The following section focuses on the pre-selection approach.

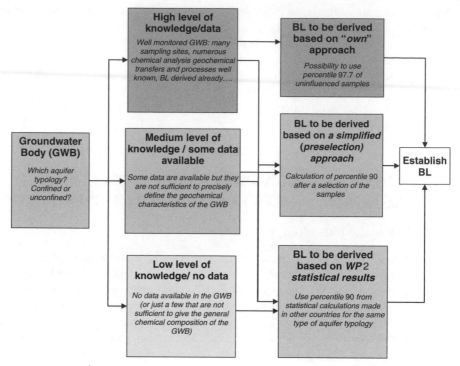

Figure 3.3.2   General procedure proposed by the BRIDGE project to derive background levels (BL).

3.3.3.2   Principles of the Pre-selection Approach

The basic idea of the pre-selection method is that there is a correlation between the concentration of certain indicator substances and the presence of anthropogenic influences.

Based on the experience in several French and German studies on background levels assessment (Wendland *et al*., 2005; Chery, 2006), the pre-selection approach recommends excluding:

1.   samples with purely anthropogenic substances (e.g. PAC, pesticides);

2.   samples, for which indicator substances for anthropogenic inputs (e.g. nitrate, ammonia, potassium) are exceeding a certain value. It is recommended to use NO3 > 10 mg/l as a criterion to exclude anthropogenic samples but depending on the groundwater body characteristics some other parameters can be relevant. This can be the case for potassium, chloride, sodium or ammonium. Potential impact of anthropogenic activity of groundwater composition can also be verified through dating with CFCs or tritium. Using other criteria can particularly be important for anaerobic aquifer. Because of denitrification processes, the $[NO_3^-] < 10$ mg/l criteria is indeed not sufficient to exclude all influenced samples. In case of denitrification, SO_4^{2-} or NO_2 concen-

trations (if available) could also be used to identify human influences but only in particular conditions. In all cases, denitrification,is revealed by excess of gaseous phases N_2, or N_2O. If additional sampling for analysis can be managed, it is recommended to proceed to the analysis of the three following gases in groundwater samples: N_2, N_2O and Ar.

By using criteria such as potassium, chloride, sodium, sulphates or ammonium, it is recommended to make sure that there are no high background levels of these parameters.

Background levels are finally calculated as the *90-percentile of the remaining samples*. In case samples from well defined aquifers are available and in case of being sure that no human influence exists, the 97.7-percentile may be used instead (BaSeLiNe, 2003).

An example of the application of this methodology is given is Chapter 5.2.

3.3.3.3 Limits of the Pre-selection Approach

In order to avoid bias in the results, the data-bases used for the background levels derivation should fulfil the following requirements:

- Samples with incorrect ion balance (exceeding 10%) should be removed.

- Samples of unknown depth and groundwater body type (see Chapter 3.2) should be removed.

- Data from hydrothermal aquifers should be removed.

- Data from salty aquifers (NaCl content of more than 1000 mg/l) (coastal or influenced by evaporates) should be removed. Salty aquifers must be considered as a separate type of aquifer and background levels should be, in this case, derived with a case by case approach.

- Data from aerobic aquifers should be separated from those of anaerobic aquifers. This can be done using the oxygen content: $>= 1$ mg/l O2 for aerobic and <1 mg/l O2 for anaerobic groundwater. If no reliable O2 data are available, Fe(II) and Mn (II) concentrations may be used for separating aerobic (Fe < 0.2, Mn < 0.05 mg/l) and anaerobic (Fe $>= 0.2$ and Mn $>= 0.5$ mg/l) groundwater) instead. Because of possible denitrification reactions in anaerobic groundwater, preselection to nitrate contents only may not separate samples with anthropogenic impacts.

- All the data available can be used (no restriction on the time series to consider) but time series should be eliminated by median averaging (in order to guarantee that all sampling sites contribute equally to the background levels derivation). After this data processing only one representative groundwater analysis remains for all the remaining monitoring stations.

- For traces elements, limits of quantification cannot be too high. In particular, limits of quantification equal to drinking water standards should be removed.

3.3.4 CONCLUSION AND PERSPECTIVES FOR GROUNDWATER MONITORING

With the adoption of the GWD in 2006, the *concept of background levels* has largely evolved. Although it is clear that each groundwater body (or each aquifer) has a unique chemical composition and that this one is variable in space and time, some concepts and proposals supporting a systematic assessment of background levels have risen. Now that Member States have to report on the chemical status of their groundwater bodies, assessing background levels became an obligation. But the WFD schedule, the large number of groundwater bodies across Europe (more than 7000) and the levels of knowledge (still low in many cases) do not allow carrying out in-depth studies of all groundwater bodies natural composition. In this context, the BRIDGE project proposes a practical procedure to assess background levels taking into account the level of knowledge.

However, the confidence in the results highly depends on the *quantity and on the quality of the data used*. According to the WFD Annex V.2.4.2, the surveillance control monitoring networks designed by Member States should 'provide information for use in the assessment of long term trends both as a result of *changes in natural conditions* and trough anthropogenic activities'. This objective of the WFD surveillance control is confirmed by the European Commission guidance document on groundwater monitoring (European Commission, 2006) in which it is written that 'the surveillance monitoring programme will also be useful for defining natural background levels [...] and characteristics within the groundwater body'. In principle, the WFD networks should hence be sufficient to assess background levels. But for the following reasons, the background levels assessment requires the use of all other relevant and available data within the groundwater body and in some cases further investigations:

- Most European groundwater bodies are influenced by human activities and applying the criteria of the BRIDGE pre-selection approach could exclude so many samples that the result would not be statistically representative. Therefore, all available data, if relevant and reliable (see Section 3.3.3.3), should be used.

- Although water–rock interactions are the main factor influencing groundwater natural chemical composition, other factors are involved and are locally significant (salinisation, confinement, residence time, leakage, dykes...). The identification and the understanding of these factors then become necessary and require further investigations and studies (including the acquisition of new data).

Therefore, the assessment of background levels cannot only be based on monitoring programmes. Where the existence of high background levels is recognised, data from these programmes and the application of a systematic approach like the one proposed by BRIDGE have to be completed and followed by an in-depth characterisation of the groundwater body chemical composition. This shall include the use of hydrodynamical modelling and geochemical tools supporting the understanding of water-rock interactions and all other factors influencing the chemical composition (leakage, salinisation, denitrification processes, redox process, time transfer, etc.).

Acknowledgments

The BRIDGE project was funded by the European Commission, DG Research, within the 6th Framework Programme under Priority 8 (contract No. 006538(SSPI)-Scientific Support to Policies). The authors would like to thank all the members of the BRIDGE consortium who contributed to this work: R. Kunkel (FZ-Jülich, GE), K. Walraevens, M. Van Camp, M. Coetsiers (LAGH-UGent, BE), R. Gorova (EEA, BG), R. Wolter (UBA, GE), K. Hinsby (GEUS, DK), A. Marandi (UT, EE), Z. Simonffy (BME, HU), K. Kadunas (LGT, LT), S. Witczak (DHWP, PL), J. Hookey (EA, UK), J. Gustafsson (SYKE, FI).

REFERENCES

BaSeLiNe, 2003. Natural baseline quality in European aquifers: a basis for aquifer management, www.bgs.ac.uk/hydrogeology/baseline/europe

BRIDGE (2006) Impact of hydrogeological conditions on pollutant behaviour in groundwater and related ecosystems, BRIDGE project, Deliverable D10, 3 vols, www.wfd-bridge.net.

Chery L. (ed.), 2006. Qualité naturelle des eaux souterraines. Méthode de caractérisation des états de référence des aquifères français, *Collection Scientifique et Technique*, BRGM éditions.

Custodio E. and Manzano M., 2007. Groundwater quality background levels. In Ph. Quevauviller (ed.), *Groundwater Science and Policy. An International Overview*, RCS Publishing, Chichester.

European Commission, 1998. Council Directive 98/83/EC on the quality of water intended for human consumption, *Official Journal of the European Communities* L 330, 5.12.98, p. 32.

European Commission, 2000. Directive 2000/60/EC of the European Parliament and of the Council of 23 October 2000 establishing a framework for Community action in the field of water policy, *Official Journal of the European Communities* L 327, 22.12.2000, p. 1.

European Commission, 2006. *Groundwater Monitoring, CIS Guidance Document* N°15, European Commission, Brussels.

European Commission, 2009. *Guidance on Groundwater Status and Trends Assessment*, CIS Guidance Document n°18, European Commission, Brussels.

Müller D., Blum A., Hart A., Hookey J., Kunkel R., Scheidleder A., Tomlin C. and Wendland F., 2006. *Final Proposal for a Methodology to Set up Groundwater Threshold Values in Europe*, Deliverable D18, BRIDGE project, 63p, available on www.wfd-bridge.net

Müller D. and Fouillac A.-M., 2007. Methodology for the establishment of groundwater environmental quality standards. In Ph. Quevauviller (ed.), *Groundwater Science and Policy. An International Overview*, RCS Publishing, Chichester.

Wendland F., Hannappel S., Kunkel R., Schenk R., Voigt H.-J. and Wolter R., 2005. A procedure to define natural groundwater conditions of groundwater bodies in Germany, *Water Sci. Technol.*, **51**(3-4), 249–57.

Wendland F., Blum A., Coetsiers M., *et al.*, 2007. European aquifer typology: a practical framework for an overview of major groundwater composition at European scale, *Environmental Geology*, DOI 10.1007/s00254-007-0966-5.

3.4

Quantitative Stresses and Monitoring Obligations

Emilio Custodio[1,4], Andrés Sahuquillo[2,4] and M. Rámon Llamas[3,4]

[1] *Technical University of Catalonia (UPC), Department of Geotechnics, Barcelona, Spain*
[2] *Technical University of Valencia (UPV), Department of Hydraulics, Valencia, Spain*
[3] *Complutense University of Madrid (UPM), Department of External Geodynamics, Madrid, Spain*
[4] *Royal Academy of Sciences, Spain*

3.4.1 INTRODUCTION

The role and behaviour of aquifers depend on climate, relief and geological conditions. In the temperate climates of Central and Northern Europe, in flat and low relief areas, the dominant shallow permeable formations are normally full of water which discharges into the more or less permanent surface water stream network. Losses by phreatic and phreatophyte evaporation are relatively small, and wetlands easily form in depressions and lowlands. Groundwater abstraction is often a small fraction of recharge, dominantly for human supply in urban and rural environments, so the groundwater flow pattern does not

Groundwater Monitoring Edited by Philippe Quevauviller, Anne-Marie Fouillac, Johannes Grath and Rob Ward
© 2009 John Wiley & Sons, Ltd

suffer significant changes. Groundwater quality problems, both natural and man–induced, represent a major concern, as well as the quality of groundwater–related surface water and habitats. This corresponds with the orientation and emphasis on water quality of the European Water Framework Directive (WFD, 2000/60/CE Directive to establish a community framework for action in water policy), and the recent Groundwater Daughter Directive (GWD, 2006/118/EC Directive on protection of groundwater against pollution and deterioration).

In arid and semiarid areas, such as those found in Southern Europe, around the Mediterranean Region, and especially in Central, Eastern and Southern Spain, groundwater plays important environmental roles and its dominant use is not for town and industrial supply, but to attend the important agricultural demand for irrigation of crops (EUWI, 2007). In some cases this amounts to more than 80% of total groundwater abstraction. This abstraction is a large fraction of the recharge, and in some areas may approach and even exceed annual mean recharge (MMA, 2000). In many areas this intensive development of aquifers (Llamas and Custodio, 2003; Custodio *et al*., 2005a, Sahuquillo *et al*., 2005; Llamas and Martínez–Santos, 2005) is more the rule than the exception, and this may involve serious ecological and water quality impacts, in addition to the problems of sustainability of groundwater abstractions and serious seawater intrusion in coastal groundwater bodies (Custodio, 2005b). Under intensive development, groundwater body quality problems and their ecological consequences continue to be important, but groundwater quantity issues are also important, and water scarcity becomes dominant in the mind of water users and society in many areas of the country. This is the case in most south-eastern and central areas of Spain. Only in areas of Catalonia and the humid NW human supply is the dominant concern. In any case quantity and quality issues are closely linked, although not always recognized by water administrators and society in general. When exploitation of water resources is high, agricultural return flows significantly increase water salinity and concentrations of non-desirable chemicals in aquifers. In water scarce areas, changes in groundwater bodies are already significant and restoration to background levels prior to intensive exploitation may be unrealistic in many cases. However, there are also other dependencies on groundwater including habitats and river systems (Box 3.4.1) that have to be taken into account, the more the more arid is the climate. They deserve protection, both in quantity and in quality, and at the same time taking into account their value and the social benefits from groundwater development (Box 3.4.2).

BOX 3.4.1 Role of groundwater

In Nature

- Contributing water and solutes to:
  - River base-flow
  - Springs and seeps
  - Groundwater dependent-wetlands
  - Phreatophyte areas

- ○ Coastal habitats depending on groundwater
- ○ Shore marine habitats

- Conditioning geotechnical properties of the ground
- Affecting the external and internal geodynamic behaviour of the Earth
- Influencing
  - ○ salinity and solute distribution in the ground
  - ○ mobilization of solutes (some of them may be noxious)

To supply human needs

- A reliable fresh water resource for:
  - ○ Drinking purposes
  - ○ Urban and rural supply
  - ○ Irrigated agriculture
  - ○ Animal raising
  - ○ Industry

- A strategic water resource in:
  - ○ Droughts
  - ○ Failures in supply systems
  - ○ Emergencies
  - ○ Criminal actions
  - ○ Conflict situations

A source of services

- from groundwater-dependent habitats
- stabilizing rural settlements and landscape preservation

BOX 3.4.2 Social benefits from groundwater development

Groundwater has proved its effectiveness to:

- Alleviate poverty in depressed areas
- Provide safe drinking water (in spite of some problems and failures)
- Dramatically improve population health
- Help developing an area
- Provide smooth social transition in developing poor areas
- Increase food production
- Improve employment and stabilizing population
- Increase security against droughts

- Guarantee water supply in extreme situations

- Reduce funds misuse and corruption

These social benefits are related to:

- Adequate quantity and quality

- Resilience to seasonality and inter-annual variation

- Easy location and distributed accessibility

- Readiness for use

- Low technical and administrative requirements

3.4.2   GROUNDWATER QUANTITY AND GOOD QUANTITATIVE STATUS

Groundwater quantity may be measured by the water storage in the ground when referring to reserves, and to recharge when referring to renewable flux. In a long-term stationary groundwater system under natural conditions average discharge equals average recharge. Discharge is produced in streams and springs as base–flow and part of it is evaporated in lakes, wetlands and phreatophyte areas, or transferred to other adjacent groundwater bodies.

The existence of groundwater abstractions may modify recharge to some extent, but especially reduces discharges. This means that stream and spring base-flow, and wetland and phreatophyte area decrease or may even be wiped out. These environmental impacts are due to the quantity of groundwater available, and also to quality changes due to aquifer chemical modifications, entry of pollutants and decreased dilution (Box 3.4.3).

BOX 3.4.3 Main impacts from developing groundwater

On water quantity:

- Groundwater level drawdown. This may mean:

  ○   increased water cost

  ○   early need to substitute wells, pumps and associated facilities

- Depletion of discharge to springs, rivers, lakes and wetlands

- Reduction of phreatophyte areas

- Impairment of groundwater dependent habitats

- Longer surface stream tracts where water infiltrates into the ground

On water quality:

- Water quality may change as a consequence of:
  - ○ Modifying the flow pattern
    - – displacement of low quality water bodies
    - – enhanced seawater encroachement in coastal aquifers
    - – more easy infiltration of degraded surface water
    - – penetration of shallow water into deep aquifers
    - – mobilization of noxious substances
  - ○ Changing the mixing of groundwater in wells, boreholes and springs
  - ○ Enhanced chemical reactions in the ground by
    - – altering the redox situation:
    - – dewatering, reatuation, allowing aerated water
    - – modifying flow velocity
    - – increasing flow through formations containing leachable minerals

On land

- Land subsidence due to decreased water pressure in the ground:
  - ○ more easy inundation
  - ○ changes in habitats
- Increased collapse rate in
  - ○ karstic areas
  - ○ areas around poorly constructed wells and drains
- Habitat modification due to water quantity and quality changes
- Impairment of landscape and scenic values

The good quantitative status of a groundwater body is more difficult to define than good chemical status, because the latter relies on easily measurable mineral solutes, suspended matter and microorganisms in water, or extensive water properties such as electrical conductivity, pH, redox potential and temperature. Good quantitative status should reflect the ability of groundwater bodies to maintain at least a part of their essential environmental roles as water sources, such as preserving river and spring base–flow, and a depth to the water-table compatible with pheatophyte and wetland subsistence. Also discharges into lakes, streams and littoral tracts need the chemical and biochemical environment that depends on groundwater outflow. Maintaining water head distribution prevents excessive land subsidence in order not to impair land drainage and flooding frequency, and, in coastal areas, marine inundation.

For the assessment of quantitative status, information on groundwater levels is needed, in conjunction with estimates of recharge. Spring-flow gauging and river base-flow evaluation are the more reliable methods of integrated groundwater recharge assessment, when this is possible. As indicated in EC (2007), an understanding of groundwater relationships with surface waters and terrestrial ecosystems is necessary for the development of the needed conceptual model of the hydrogeological system to groundwater resource and the assessment of groundwater quantitative and chemical status.

Often aquifers are complex systems consisting of several layers more or less linked vertically through aquitards. Good quantitative status not only refers to the upper layer, which interacts with surface water and the environment, but to the other layers through the influence they may have in the upper layer or in discharge areas.

Groundwater is transferred form high areas to lower areas, with lateral movement and vertical exchanges among the different layers. Groundwater flow is essentially three-dimensional, and two-dimensional representations are only simplifications for more easy treatment. In many cases these simplifications are acceptable for analyzing flow components and mass transfer, as stresses are likely to have significant effects, mostly in the upper aquifer layers. However, in other cases, a 2-D conceptual model may be inappropriate and even may lead to erroneous results. Often, cases of inappropriate conceptual models are due to the lack of wells or boreholes deep enough to determine the vertical component of the water head gradient, or poor measurements of the water-table elevation when exploitation is mostly by deep wells. However, in many other cases this is due to inadequate analysis of existing data.

In order to understand quantitatively an aquifer system the first step is to establish a conceptual model considering its characteristics and behaviour, and then check and quantify this model, or improve it. This may require hydraulic, hydrochemical and environmental isotope studies to obtain the parameters and boundary conditions to be fed into the appropriate calculation or numerical model(s). These may consist of simple calculations or sophisticated 2-D and 3-D flow and transport models.

In the WFD, Article 4.1,b,ii, requires EU Member States to protect, enhance and restore all groundwater bodies and ensure the equilibrium between abstraction and recharge of these waters in order to achieve good status for groundwater. A strict interpretation may be too limitative and in extreme situations may be used as an argument to exclude any groundwater development in many aquifers that are not permanently recharged by perennial surface water streams. Often this cannot be applied in dry climate areas. The limit should be acceptable ecological and quality impacts, although these are poorly defined concepts that largely depend on what is considered acceptable. Acceptability has complex ecological, economic and social components, that may change with time, and has to be agreed by the society.

A first appraisal of quantitative status is the decrease (as a fraction) of ecologically sensitive outflow, of wetland and phreatophyte surface areas, and of water salinity and quality. Water-table and piezometric level evolution is a relevant indicator that has to be monitored for the different groundwater bodies, taking into account the three-dimensional nature of the problem. Experience through well-studied cases is needed.

3.4.3   IMPACT OF GROUNDWATER DEVELOPMENTS

Any groundwater artificial development has some impact on the water balance components and often on water quality. The seriousness of the impacts increase with the intensity of such development (Custodio, 2001).

Groundwater development has many positive benefits from the hydrological, water quality, economic and social points of view (Custodio, 2005a; Llamas and Custodio, 2003). The social benefits are the more relevant to the WFD. Groundwater development is needed in many developing countries to attain the Millennium Declaration Goals (MDG) and to solve poor quality and scarcity of water (Ragone *et al.*, 2007), as recognized by the United Nations and supported by the International Association of Hydrogeologists. Possible drawbacks from groundwater development and unsustainable situations can be easily identified by the impact on water quantity, water quality and effects on land. Most of these drawbacks are the subject of the WFD related regulations.

In general terms, groundwater development had only a small impact on aquifers and related water bodies until 30–80 years ago, depending on the country or region. Since then, groundwater development has been almost exponential, especially in the semi–arid and arid areas of Southern Europe (Central, Eastern and southern Spain, south of France, areas of Italy, Greece) and around the Mediterranean sea, as well as in central and western USA, Australia, Middle East, India and China, and in many small islands (see Box 3.4.4). A large part of this largely uncontrolled development has been carried out by individuals and small communities producing what can be called a silent revolution (Fornés *et al.*, 2005; Llamas, 2005; Giordano and Willholth, 2007), mostly to develop irrigated agriculture. This has been a characteristic of groundwater development in the last 20–50 years (depending on the area) and will probably continue in the early part of present century.

A common although difficult concept is that of overexploitation (Custodio, 2002; Hernández-Mora *et al.*, 2001). Overexploitation puts the accent on the negative aspects, and not necessarily on environmental issues. The concept may be misused, for example, to try to halt aquifer exploitation, thus losing the large potential benefits from using groundwater. Groundwater mining is directly forbidden in the WFD except if it is clearly demonstrated that this may be a sound activity under special circumstances and for a limited time. An adequate study of intensive groundwater use, case by case, will provide answers to quantitative questions, provided they are carried out by trained personnel who do not yield to external interests, political pressure, media influence and self-interest (Llamas and Custodio, 2003).

3.4.4   DEFINITIONS OF GROUNDWATER QUANTITY

Groundwater quantity is mainly defined by:

1. Storage in the different groundwater bodies, that is to say, in aquifers and aquitards. These values are the saturated volume multiplied by total porosity. Only a part of this water can be abstracted since a part is retained in the formation by capillary forces. The volume is a fixed one in the case of confined aquifers or variable when the upper boundary is the water-table.

2. Recharge from outside the groundwater body. Under common circumstances most of the recharge comes from rainfall infiltration in the aquifer outcrops. In dry climates and mountaineous areas stream water infiltration may be significant and in some cases dominant. Also infiltration of snow-melt may be an important contribution to recharge. In shallow aquifers, especially in flat areas in dry climates, part of recharge goes back to the atmosphere through phreatic and phreatophyte area evaporation. What remains is net recharge.

3. Interaquifer leackage. Under many circumstances a groundwater body may receive groundwater from an adjacent (at the side, above, or below) water body. The outflow can be regarded as negative recharge. In this respect aquitards may play a very significant role.

BOX 3.4.4 Intensive exploitation of aquifers in southeastern Spain

South–eastern Spain is a semiarid area with a large water demand for irrigating high value crops that has induced a very intense exploitation of the region's aquifers. Often groundwater abstraction exceeds aquifer recharge, producing sustained water level drawdown and the drying of springs and of some wetlands. In other cases water salinity increases by seawater or saline water intrusion. Although in some cases groundwater exploitation has to be decreased or even abandoned, many aquifers continue providing valuable water resources at a relatively low cost.

Some of the more important aquifers have been modelled and different exploitation strategies have been simulated to foresee their future behaviour (Figure 3.4.1). The Guadalentín aquifer, with a surface area of $800\,km^2$ and a total recharge of 26×10^6 m^3/a, supports abstraction higher than 100×10^6 m^3/a, with a depletion of 3×10^9 m^3 in the last 30 years. From the aquifers of Jumilla–Villena ($340\,km^2$; recharge of 7×10^6 m^3/a) and Ascoy–Sopalmo ($420\,km^2$; recharge of 1.5×10^6 m^3/a) more than 50×10^6 m^3/a are pumped with a depletion of 1.5×10^9 m^3 in each of them. In some areas groundwater drawdown exceeds 160 m, but the average value is around 60 m to 80 m. In the Cingla–Cuchillos aquifer, with an average recharge of about 8.5×10^6 m^3/a, current pumpage is 28×10^6 m^3/a or a cumulative withdrawal since 1975 that exceeds recharge by 450×10^6 m^3. See the figure below.

Model simulations provide some interesting conclusions. If pumping is reduced to zero groundwater levels will still not recover to their original values within the next 30 or 40 years at current recharge rates. Former springs in Jumilla–Villena and Ascoy–Sopalmo areas will remain dry, and the former wetland in the Guadalentin valley will remain well above the aquifer water levels.

If pumping continues at current levels, sustained groundwater level drawdown will continue and in some areas wells will dry up during the next 10 to 15 years. This means that a further reduction in pumping will be needed. There will be no further environmental damage except some water salinity increase in the Guadalentin aquifer. This aquifer already contains high salinity water. It seems that these aquifers cannot be environmentally recovered from an economic and social point of view.

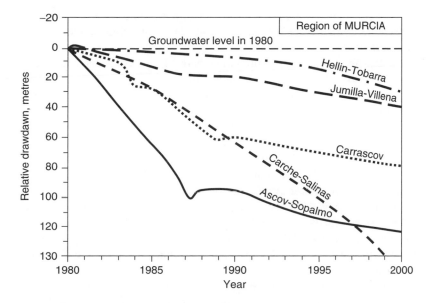

Figure 3.4.1   Aquifer modelling in the region of Murcia, Spain.

An understanding of the geometry of the groundwater body is needed to determine groundwater storage. However this is not always well known, and in many cases the lower limit (base) of the groundwater body may be quite speculative. Geological studies, drilling and geophysical surveys are the most commonly used means to gain knowledge. Total and drainable porosity (the part of porosity with water free to move under gravity) are also the result of studies and measurements.

Recharge is a key value required to evaluate quantitative status, and consequently to understand chemical status and its evolution, and make judgements on overexploitation and on the consequences of intensive use. However, recharge is one of the most difficult hydrogeological parameters to calculate. Numerous methods are available to make evaluations. Most of them refer to recharge by rainfall. They include soil water balance methods, which include evaluation of actual evapotranspiration by means of meteorological, runoff and plant effect calculations, measurement of flow in the unsaturated zone, chloride (or other conservative ion) balance in the soil, chemical and environmental isotope ground profiling and modelling (Lerner *et al.*, 1990; Simmers, 1988; Custodio *et al.*, 1997). Changes in water-table levels may be used, under favourable circumstances, to calculate and calibrate recharge. Recent studies include direct soil evaporation and phreatophyte evapotranspiration, in order to improve estimates in dry climates. Atmospheric chloride deposition on land and chloride balance in the soil is a fast, reliable method to estimate long-term average recharge under steady state conditions (Alcalá and Custodio, 2007; Custodio, 2009). Recharge varies with time and depends on location. Thus, the recent introduction of geographical information systems (GIS) allows more accurate evaluations provided that the significant parameters are known or can be calibrated. Advances are being made to consider the impact of land use activities and development on recharge. Additionally efforts are also being made to consider future climate and man-made land use change scenarios.

In situations where there is concentrated recharge through fissures, soil discontinuities and areas of runoff concentration evaluation of recharge is even more complex and uncertain. Recharge from surface waters, snow-melt and especially floods, which involve large volumes over a short timescale, irregular spatial and temporal patterns is still a subject of research, as well as recharge through fractured rocks,

Discharge is the result of recharge, after being smoothed and delayed by the effect of the large water storage in the aquifer system. It can be used as an averaged indicator of integrated recharge and to calibrate evaluation methods. This can be done by simple calculations or through aquifer system modelling.

However discharge is also prone to large uncertainties. Base-flow to streams has to be derived from hydrographs and/or chemical balances. However river hydrographs can be modified by aquifer abstraction and water intakes or inflows from other rivers. Evaporation from lakes and wetlands needs measurement stations and environmental correction factors.

Artificial abstraction can be determined using calibrated flow measuring devices. However these are often rare at sites where groundwater is used for irrigation. In these cases the irrigated surface area and the applied water depth have to be used as proxies. In many cases they are poorly known and even the number of operating wells and how they are pumped may be highly uncertain. Water intakes from rivers, directly or through canals, are also not always known and very often they are not metered. Similar quantitative uncertainties exist in water inflow to surface water from reservoirs, tributaries or other inputs. In dry climates fresh water is scarce and often attains high market values, so uncontrolled exploitation is often more the rule than the exception. Flows from artificial drainage of infrastructures and buildings in urban areas may be also highly uncertain.

Interaquifer flow has to be evaluated by hydraulic calculations and by means of hydrogeochemical and environmental isotope studies, and thus they may be also highly uncertain.

All the terms/parameters defining the aquifer system's quantitative behaviour must be used to calculate a water balance. Water balance calculation methods range from simple calculations to sophisticated methods employing numerical modelling of flow that can be reinforced by mass transport modelling when water chemistry is known with enough spatial and temporal detail. Long data series may greatly reduce uncertainty. This includes measuring chemical atmospheric deposition.

3.4.5 DELAYED EFFECTS IN AQUIFERS

The large storage associated with groundwater bodies with respect to annual flow introduces long delayed responses to external stresses, be they natural or artificial. This means that the impact produced by development at a given moment produces slow quantitative and chemical responses that may only diminish after years, decades and even millennia. Some large unconfined aquifers are still evolving after the large climatic changes at the end of last glacial period (16,000 to 10,000 years ago).

In a very simplified manner, aquifer response depends on the 'α parameter' where $\alpha = \beta L^2 S/T$. Parameter L is the aquifer size, S the storage coefficient, T the hydraulic transmissivity of a given aquifer and β a coefficient that depends on boundary conditions, but whose value is often between 1.5 a 2.5. While small, highly transmissive confined aquifers establish new steady state head conditions within days or weeks after a quantity perturbation, a large, unconfined low transmissivity aquifer may need millennia. This has important implications for defining the quantitative status of a groundwater body since the observed status may not be stationary but an evolving one. Thus the same observed groundwater levels may represent different situations, depending on the moment in time. Climatic variability can also affect simultaneously the aquifer, leading to delayed responses that need an adequate interpretation (see Box 3.4.5).

BOX 3.4.5 Time evolution of ground water. A simple example

Let us consider a hypothetical, oversimplified case of an unconfined groundwater body limited by impervious side and bottom boundaries and discharging into a river, thus contributing to base-flow and riparian habitats. Recharge is produced only on the aquifer outcrop. In a given moment diffuse exploitation is introduced, equivalent to a fraction of the recharge rate. The figure shows the shape of the time evolution of the depth to the water-table and the decrease of aquifer outflow to the river. The length of the time axis is decades to many hundreds of years in typical cases.

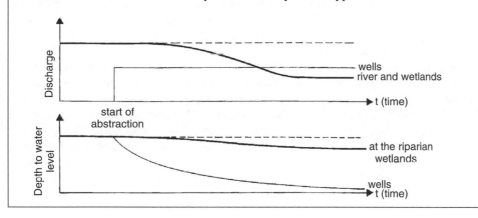

At the beginning the effect of abstraction seems to proceed at a very slow rate, but after some years it increases continuously and does not fade out until a long time (years to centuries). Consequently river base-flow and riparian habitats seem initially little affected, but afterwards they deteriorate fast, even when the groundwater body exploitation has not been changed for years. This may be improperly taken as overexploitation, when actually abstraction is less than recharge, but producing delayed groundwater level drawdown, decreased river discharge and riparian habitat deterioration.

These quantitative delayed effects may have a greater significance for the water quality status since groundwater moves slowly. Assuming piston flow in the aquifer (incoming groundwater fully displaces existing groundwater) the water velocity is given by the Darcy's law, $v = k \cdot i/m$, in which v is intergranular groundwater velocity, k is aquifer permeability (hydraulic conductivity) and m is dynamic porosity (close to total porosity in homogeneous sedimentary formations). Water and non-retarded solute movement in most cases varies from a fraction of a metre to a few hundreds of metres per year. This is much slower than changes in head and flow.

The result of a stress causing an impact may often be manifested in monitoring data after a considerable delay. For example, pesticide application to a wide area of land over a groundwater body may lead to increased concentrations in the groundwater many years after it was released. Practices to reduce nitrate inputs to groundwater in agricultural areas, will not be apparent years or decades later in deep parts of aquifers used for domestic supply. The movement (and temporal retention, and eventually the decay) of chemical changes in the unsaturated zone goes on unnoticed by normal monitoring, and the effects may be long delayed, and even appear after the stress has been suppressed.

3.4.6   MONITORING OF GROUNDWATER QUANTITATIVE STATUS

The key values to define groundwater body quantitative status – although not the only ones – are piezometric levels, both in space (in three dimensions) and along time. Information on groundwater levels and estimates of recharge should be used for the assessment of the quantitative status in conjunction with spring flows and the estimation of base-flow in rivers, when this is possible. Groundwater body parameters are not the subject of monitoring, but of studies. Such studies should be completed and extended every few years, especially if impacts are of concern.

The monitoring programmes should be designed to provide the information needed to evaluate the risk posed by existing stresses and to establish the magnitude and spatial and temporal distribution of any impacts (Condesso de Melo *et al.*, 2007), with a given strategy (Voigt *et al.*, 2007). Risk assessments for groundwater should be based on a conceptual model of the groundwater system and how stresses interact with that system. The conceptual model is necessary to design monitoring programmes and also to interpret the data provided by them, and hence assess the achievement of the monitoring objectives. The level of detail in any conceptual model needs to be proportional to the difficulty in judging the effects of the stresses, although a simple, generalized sketch of the groundwater system may be good to start. Monitoring should provide the information needed to test model performance and, where necessary, improve them so that an appropriate level of confidence can be achieved in the prediction and assessment of groundwater behaviour. When very costly restoration or enhancement measures are need in the case of aquifers failing to achieve good status, relatively complex models are likely to be required.

In the case of limited existing monitoring networks it may be convenient to iteratively build or enlarge them to the extent needed to test or develop the needed conceptual models.

An understanding of groundwater relationships with surface waters and terrestrial ecosystems is necessary for the development of the conceptual model of the hydrogeological system, the determination of the available groundwater resources and the assessment of groundwater quantitative and chemical status (EC, 2007).

A given groundwater body may be not represented – and often it is not – by one monitoring point. This depends greatly on aquifer characteristics and shape, degree of development, density of links with surface water and the importance of groundwater-dependent ecological and water quality situations. The more spatially variable the groundwater flow system or the stresses on it, the greater the density of monitoring points needed. The amount of monitoring required also depends on the extent of existing information on water levels and on the groundwater flow system. Where this information is adequate and reliable, it may not be necessary to extend monitoring programmes. A guide could be to monitor sufficient points to validate the conceptual model. In case of having a consistent conceptual and validated model and an adequate knowledge of aquifer recharge and stresses, the monitoring network could even be reduced.

The measurement frequency should allow short-term and long-term level variations within the groundwater body to be detected, and to distinguish short and long-term variations in recharge from impacts of abstraction and discharges. For formations in which the natural temporal variability of groundwater level is high or in which the response to stresses is rapid, more frequent monitoring will be required than will be the case for groundwater bodies that respond slowly to short-term variations in precipitation or stresses. The number of points in the monitoring network can be reduced where measurements are made frequently, from weekly or fortnightly to every two months, but a more dense network is needed where data are taken more sparsely, once or twice per year. Installing some continuously recording devices is recommended in some cases, especially when groundwater body behaviour is being studied. In order to better use the technical and scientific capacity of the monitoring organization, the strategy of using more sparse monitoring networks between periods of simple monitoring of a particular groundwater body, and a denser network for periods of deeper analysis, allow a more efficient follow up of different groundwater bodies, which alternate in the strategy. Designing and operating integrated groundwater and surface water monitoring networks will also produce cost-effective monitoring information for assessing the achievement of the objectives of the WFD (EC, 2007) for groundwater bodies.

Groundwater level measurement points – piezometers – are not simply open holes in to the groundwater body, as is unfortunately the case in many situations. Although existing wells can be used – provided their characteristics, especially depth and length of the screen or open section are known – dynamic or residual pumping levels may impact on what is being monitored. Purpose-drilled piezometers are preferred. These should have relatively short screens at the appropriate depths, and be properly constructed. This means taking care during construction, for example by ensuring grouting, tube joints waterproofing and properly installed filter packs. Often several piezometers, clustered or nested, are needed at a given point to separately monitor the groundwater body in depth, including deep layers and aquitards. These may be expensive to construct and maintain, so a cost-benefit analysis should be carried out. The representativity of what is being monitored may also inform the need for investment. The design of a piezometric level monitoring network is a complex task that has to be carried out by trained hydrogeologists.

Not any variable of interest to define groundwater quantity status is for general purpose monitoring, although they may be needed to understand the aquifer behaviour or to help in model design and calibration. Such is the case of discharge quantity. This means gauging significant springs and river tracts to obtain base-flow. It should be considered that the needed points may not coincide – and often do not – with surface water quality monitoring needs.

Monitoring of abstraction is an easy task when wells, water galleries and drains have flow measuring and recording devices. But often this is not the case and then indirect means of measurement are needed, based on pumping energy consumption, well yield and hours of functioning, irrigated surface, served population, etc. This should be part of the monitoring operation, at the appropriate level.

Aquifer outflow through lakes, wetlands and phreatophyte areas should also be monitored by repeated surveys. Modern satellite and air-borne remote sensing devices are interesting monitoring tools for some variables, such as evapotranspiration and vegative cover, if properly calibrated. Advances are continuously being made to help in making these methods easily available to monitoring agencies.

3.4.7 AQUIFER MANAGEMENT AS A TOOL FOR GROUNDWATER QUANTITY STATUS COMPLIANCE

Aquifer management is one of the most challenging aspects of water resources management. Generally there are a large number of unrelated agencies and stakeholders often with a poor or non-existent understanding that the groundwater body is a common asset. This is more acute in dry areas, where irrigated agriculture is an important economic and social factor, even when in developed countries this accounts only for a small percentage of a region's total gross income. Generally, governments and regulating institutions have not been able to cope with this relatively new situation, and often they do not have the appropriate staff, economic resources, and understanding.

In dry areas irrigation and water supply consumes the whole groundwater recharge over large surfaces. This may lead to too intensive or excessive groundwater development. This development is difficult to control using classical means when the cost of water relative to crop value is low. Normally farmers will do all they can to get groundwater, even if they deplete it, as in a classical 'Tragedy of the Commons' situation. Only, with an understanding of the common asset will there be capability to establish the limits to development (López-Gunn and Martínez-Cortina, 2006). Efficient tools to engage and involve farmers in the management and collective stakeholder participation are needed (López-Gunn, 2007; Schlager and López-Gunn, 2006). To achieve this, groundwater quantity monitoring is a necessary tool, and where stakeholders are not able or prepared to carry it out the responsibility rests with public institutions, to do this subsidiarily at the beginning and to prepare and help users to carry out this jobs by themselves and to produce data and reports for the Water Authorities to be able to carry out general and management planning, but not the local tasks.

3.4.8 GROUNDWATER QUANTITY STATUS AND THE WFD – CONCLUDING REMARKS

The long history, high population density and concentration of activities in Europe have degraded and even wiped out many environmental assets and degraded water quality. This explains the programme for preservation and restoration of water-related environmental values in the European Union. These rely on quality parameters, but are closely linked to the quantitative situation.

For groundwater the concerns are:

- water-table depletion, which induce decay of wetlands, and reduction of phreatophyte areas and riparian tracts;

- stream base-flow and spring flow reduction;

- seawater contamination in coastal groundwater bodies;

- modification of lake functioning and even their disappearance;

- land subsidence in some areas, with increased risk of flooding, and modified inundation periods; and

- abstraction of poor quality groundwater that is later on disposed into the environment. This is mostly saline water, but also water with nitrates and pesticides from artificial contamination and some heavy metals, etc.

Often the actual situation is complex due to the specific characteristics of each aquifer, and their heterogeneity. A further complication is the long delayed behaviour of aquifer systems that produces non-steady situations of long duration, both in terms of quantity and quality. The chemical changes are often more important and difficult of identify. This means that the evaluation of an actual situation must be carried out in a dynamic context, and monitoring results should be interpreted in the same way. This may not be obvious, especially for non-specialists, and so well designed and continuously updated information is needed. This is not explicitly mentioned in the WFD, but must be considered.

Stakeholders have an important role as collaborators and partners with public institutions. This often has to be through their own representative bodies given the often very large number of stakeholders involved, especially when irrigated agriculture is important. They should contribute to monitoring in a given water body and any other related water bodies and use the results to draw their own conclusions after careful independent studies. It is not unusual that well informed stakeholders conclusions may be at odds with those of the public administration. An effective forum to resolve conflicts is a goal to be achieved, using common monitoring, and comparable evaluation tools.

REFERENCES

Alcalá F.J. and Custodio E., 2007. Recharge by rainfall to Spanish aquifers through chloride mass balance in the soil. In: *Groundwater and Ecosystems*. Proc. XXXV IAH Congress, Lisbon (L. Ribeiro, A. Chambel and M.T. Condesso de Melo, eds.). CD printing. ISBN: 978–989.95297–3–1.

Condesso de Melo T., Custodio E., Edmunds W.M. and Loosly H., 2007. Monitoring and characterization of natural groundwater quality. In: *The Natural Baseline Quality of Groundwater* (eds W.M. Edmunds and P. Shand). Chap 7. Blackwell, Oxford: 155–77.

Custodio E., 2001. Effects of groundwater development on the environment. *Bol. Geolog. Minero*, Madrid, 111(6): 107–20.

Custodio E., 2002. Aquifer overexploitation, what does it mean? *Hydrogeology Journal*, 10(2): 254–77.

Custodio E., 2005a. Groundwater as a key water resource. In *Water Mining and Environment. Libro Homenaje al Profesor D. Rafael Fernández Rubio*. Instituto Geológico y Minero de España, Madrid: 68–78.

Custodio E., 2005b. Coastal aquifers as important natural hydrogeological structures. In: *Groundwater and Human Development* (ed. E. Bocanegra, M. Hernández and E. Usunoff). Intern. Assoc. Hydrogeologists, Selected Papers no. 6. Balkema, Lisse: 15–38.

Custodio E., 2009. Estimation of aquifer recharge by means of atmospheric chloride eposition balance in the soil. *Contributions to Science* (in press).

Custodio E., Llamas M.R. and Samper J. (eds), 1997. *La evaluación de la recarga a los acuíferos en la planificación hidrológica*. Assoc. Intern. Hidrología Subterránea–Grupo Español/Instituto Geol ogico y Minero de España, Madrid: 1–455.

Custodio E., Kretsinger V. and Llamas, M.R., 2005. Intensive development of groundwater: concept, facts and suggestions. *Water Policy*, 7: 151–62.

EC, 2007. *Guidance on groundwater monitoring*. Common Implementation Strategy for the Water Framework Directive (2000/60/EC), Doc. 15. Technical Report 002–2007. European Commission (Environment), Publications Office. Brussels: 1–52.

EUWI, 2007. *Mediterranean Groundwater Report*. Mediterranean Groundwater Working Group. http//www.semide.net/iniciatives/medeuwi/SP/GroundWater

Fornés J.M., de la Hera A. and Llamas M.R., 2005. The silent revolution in groundwater intensive use and its influence in Spain. *Water Policy*, 7(3): 253–68.

Garrido, A., Martínez–Santos, P., Llamas, M.R., 2006. Groundwater irrigation and its implications for water policy in semiarid countries: The Spanish experience. *Hydrogeology Journal*, 14(3): 340–49.

Giordano M. and Willholth, K.G. (eds), 2007. *The Agricultural Groundwater Revolution: Opportunities and Threats to Development*. CAB International, Wallingford, UK: 1–336.

Hernández-Mora N., Llamas M.R. and Martínez Cortina L., 2001. Misconceptions in aquifer over-exploitation: implications for water policy in Southern Spain. In *Agricultural Use of Groundwater: Towards Integration between Agricultural Policy and Water Resources Management* (ed. C. Bori). Kluwer Acad. Publ., Doordrecht: 107–25.

Lerner D.N., Issar A.S. and Simmers I., 1990. *Groundwater Recharge*. Intern. Assoc. Hydrogeologists, International Contributions to Hydrogeology, 8. Heisse, Hannover: 1–345.

Llamas M.R. 2005. La revolución silenciosa de uso intensivo del agua subterránea y los conflictos hídricos en España. In: Water Mining and Environment: Libro Homenaje al Profesor D. Rafael Fernández Rubio. Instituto Geológico y Minero de España, Madrid: 79–86.

Llamas M.R. and Custodio E., 2003. *Intensive use of groundwater: challenges and opportunities*. Balkema, Lisse: 1–478.

Llamas M.R. and Martínez-Santos, 2005. Intensive groundwater use: silent revolution and potential source of social conflicts. *J. Water Resources Planning and Management*: 337–41.

López-Gunn E. 2007. Groundwater management in Spain: Self-regulation as an alternative for the future? In: *The Global Importance of Groundwater in the 21 Century* (ed. S. Ragone). The National Ground Water Association Press, Westerville, Ohio: 351–7.

López-Gunn E. and Martínez-Cortina L. 2006. Is self–regulation a myth? Case study on Spanish groundwater user association and the role of higher-level authorities. *Hydrogeology Journal*, 14(3): 361–75.

MMA, 2000. *Libro blanco del agua en España*. Secretaría de Estado de Aguas y Costas, Ministerio de Medio Ambiente. Madrid: 1–637.

Ragone S., Hernández-Mora N., de la Hera A., Bergkamp G. and McKay J. (eds), 2007. The global importance of groundwater in the 21th century. *Proc. Intern. Symp. Groundwater Sustainability*. Nat. Groundwater Assoc. Press, Westerville, Ohio: 1–382.

Sahuquillo A., Capilla J., Martínez–Cortina L. and Sánchez-Vila X., 2005. *Groundwater intensive use*. Intern. Assoc. Hydrogeologists, Selected Paper Series 7. Balkema, Leiden: 1–450.

Schlager E. and López-Gunn, E. 2006. Collective systems for water management: is the tragedy of the commons a myth? In: *Water Crisis: Mith or Reality* (eds M.R. Llamas & P. Rogers). Balkema, Amsterdam: 43–58.

Simmers I. (ed.), 1988. *Estimation of natural groundwater recharge*. NATO Advanced Workshop ASI Series C, V. 222. Reidel Publ. Co., Dordrecht: 1–510.

Voigt H-J., Nitsche C., Tamás J., Biró T., Broers H.P. and Kozel R., 2007. Strategies and effectiveness of groundwater monitoring systems for different aims. WAPO Congress. Intern. Assoc. Hydrogeologists. *Selected Papers on Hydrogeology*. Balkema, Lisse, NL. (under evaluation).

Part 4
Groundwater Quality Standards and Trend Assessment

4.1

Threshold Values and the Role of Monitoring in Assessing Chemical Status Compliance

Ariane Blum[1], Hélène Legrand[2], Johannes Grath[3], Andreas Scheidleder[4], Hans-Peter Broers[5], Cath Tomlin[6] and Rob Ward[7]

[1] *Bureau de Recherches Géologiques et Minières, Orléans cédex, France*
[2] *Ministère de l'Ecologie, du Développement et de l'Aménagement durables, Direction de l'Eau – PREA, Paris, France*
[3,4] *Umweltbundesamt GmbH, Wien, Austria*
[5] *The Netherlands Organisation for Applied Scientific Research (TNO), Built Environment and Geosciences, Utrecht, The Netherlands*
[6] *Environment Agency – England and Wales, Apollo Court, Hertfordshire, United Kingdom*
[7] *Environment Agency – England and Wales, Olton Court, West Midlands, United Kingdom*

Groundwater Monitoring Edited by Philippe Quevauviller, Anne-Marie Fouillac, Johannes Grath and Rob Ward
© 2009 John Wiley & Sons, Ltd

4.1.1   INTRODUCTION

What is water of 'good quality'? There is no single answer to this question as water quality can be defined according to its *function* or to its *use*. The quality of the groundwater required is different depending on whether the water is used for drinking-water supply or for industrial uses. These requirements or objectives are even stricter if the water quality assessment is by comparison to its natural and original status (hereafter called 'background level'; see Chapter 2.3 for further details). The key therefore to understanding what is groundwater of 'good quality' is knowing the purpose of assessing groundwater quality.

For some groundwater quality assessments, specific threshold values or standards exist for the different objectives. As an example, when referring to nitrates in water intended for human consumption, 'good quality' water is that in which the nitrate concentration is below 50 mg/L (European Commission, 1998), but when referring to its natural status 'good quality' water should have a nitrate concentration below 10 mg/L.

Consequently, for the same set of samples, the assessment of groundwater quality can lead to totally different results depending on the assessment and therefore the standards used. This is illustrated by Figure 4.1.1 that shows an assessment of French groundwater contamination by nitrates in 2002. While only 53% of the monitoring stations recorded 'good quality' when compared to background levels, 89% of them were 'good quality' when the assessment was based on drinking-water standards.

The Water Framework Directive 2000/60/EC (WFD) and its Daughter Directive 2006/118/EC (GWD) on groundwater protection (European Commission, 2000 and 2006), for the first time combined different requirements and objectives and set an

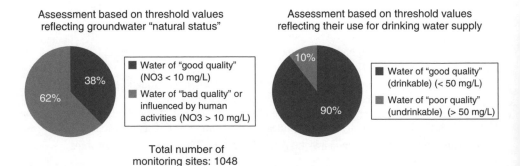

Figure 4.1.1   Assessment of the quality of French metropolitan groundwaters in 2002 using two types of criteria: groundwater 'natural status' and drinking-water supply. Adapted from Blum, 2004.

integrated approach for the assessment of Chemical Status. Criteria for the identification of groundwater chemical status assessment, including groundwater-quality standards and threshold values, are given in the *'Groundwater daughter directive' 2006/118/EC* on the protection of groundwater against pollution and deterioration (GWD). Member States are required to define threshold values 'having particular regard to its impact on [...] associated surface waters and directly dependent terrestrial ecosystems and wetlands'. These threshold values shall also 'take into account human toxicology and ecotoxicology knowledge' (GWD, Article 3.1). Although the environmental objectives set by the WFD for groundwater status cover both *quantitative and chemical* aspects, this chapter only discusses chemical status.

In order to ensure a coherent and harmonious implementation of the WFD, the European Member States, Norway and the European Commission agreed on a *Common Implementation Strategy* (CIS). In order to clarify methodological questions enabling a common understanding to be reached on the technical and scientific implications of the WFD, several guidance documents were published, e.g. the CIS guidance document no. 15 on 'Groundwater monitoring' (European Commission, 2006). However, in view of the preparation of the First River Basin Management Plan (RBMP), the main objective of the CIS Working Group C on Groundwater during its 2007–2009 mandate was the *development of a common methodology for establishing groundwater threshold values and of a guidance document on status assessment and trend assessment*. Supported by the outcomes of the FP6 project BRIDGE (Müller *et al.*, 2006), a guidance document on 'Groundwater Status and Trend Assessment' has been published (European Commission, 2009). The following sections are largely based on this guidance document.

4.1.2 THRESHOLD VALUES AND CHEMICAL STATUS ASSESSMENT UNDER THE WFD/GWD: LEGAL BACKGROUND

This section provides the legal background of the chemical status assessment procedure, including the threshold values derivation.

4.1.2.1 Chemical Status Assessment

The definition of chemical status is set out in WFD Annex V 2.3.2. It states that good groundwater chemical status is achieved when:

'The chemical composition of the groundwater body is such that the concentrations of pollutants:

- as specified below, do not exhibit the effects of saline or other intrusions,

- do not excced the quality standards applicable under other relevant Community legislation in accordance with Article 17 WFD;[1]

---

[1] This corresponds to the WFD requirement leading to the GWD adoption.

- are not such as would result in failure to achieve the environmental objectives specified under Article 4 for associated surface waters nor any significant diminution of the ecological or chemical quality of such bodies nor in any significant damage to terrestrial ecosystems which depend directly on the groundwater body.

    Changes in conductivity are not indicative of saline or other intrusion into the groundwater body.'

According to Article 4(2) of the GWD a groundwater body is considered to be of good status when:

- Annex V 2.3.2 (WFD) conditions have been met;

- relevant threshold values (Article 3 and Annex II GWD) or groundwater quality standards – GW-QS (Annex I GWD) have not been exceeded at any monitoring point; or

- a threshold value or GW-QS has been exceeded at one or more monitoring points but appropriate investigations (Annex 3 GWD) confirm that:

    1. pollutant concentrations do not present a significant environmental risk taking account, where appropriate, the extent of the groundwater body which is affected;

    2. other conditions for good status of Annex V 2.3.2 (WFD) are being met in accordance with paragraph 4 of Annex III GWD;

    3. no deterioration in quality of waters for human consumption (DWPA) in accordance with paragraph 4 of Annex III GWD; and

    4. no significant impairment of human uses.

Therefore, the use of threshold values or GW-QS is just the first step of the overall process when assessing the chemical status of a groundwater body. When data from the WFD monitoring networks show threshold values or GW-QS have been exceeded, this does not automatically imply that the groundwater body is in poor status. Exceedance of a threshold value just acts as a trigger value leading to an in-depth investigation of the groundwater body characteristics. Figure 4.1.2 summarises the general procedure when assessing the chemical status of a groundwater body

4.1.2.2 Background and Requirements

Article 3 of the GWD lays down criteria for assessing groundwater chemical status:

'1. For the purposes of the assessment of the chemical status of a groundwater body [. . . .] Member States shall use the following criteria:

(a) groundwater quality standards as referred to in Annex I,
(b) threshold values to be established by Member States in accordance with the procedure set out in Part A of Annex II [. . .]'.

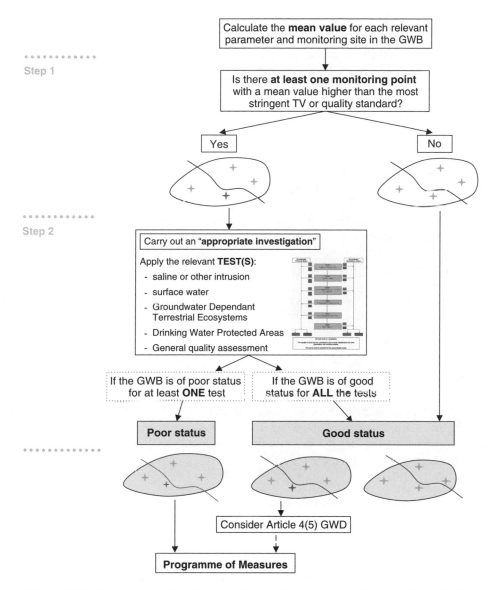

Figure 4.1.2   General procedure for assessing the chemical status of a groundwater body.

The GWD defines groundwater quality standards for nitrate and pesticides, see Table 4.1.1.

However, if these groundwater quality standards are not adequate for achieving the environmental objectives set out in Article 4 of the WFD, e.g. if concentrations in groundwater that are lower than quality standards could result in failure to achieve the environmental objectives for associated surface waters, or in any significant diminution of the ecological or chemical quality of such bodies, or in any significant damage to

Table 4.1.1   Groundwater quality standards (GWD Annex I).

Pollutant	Quality standards
Nitrates	50 mg/l
Active substances in pesticides, including their relevant metabolites, degradation and reaction products	0.1 µg/l 0.5 µg/l (total)

associated terrestrial ecosystems, then more stringent values have to be applied. These new values become 'threshold values' and the procedure to define them follows Article 3 and Annex II of the GWD and the specifications described in the CIS guidance document.

Article 3.1(b) of the GWD requires Member States to derive 'threshold values' for other relevant parameters that are causing a groundwater body to be at risk of not meeting the WFD Article 4 objectives. Member States need to take into account at least the list of substances in Annex II.B that are:

- 'Substances or ions or indicators which may occur both naturally and/or as a result of human activities': As, Cd, Pb, Hg, NH_4^+, Cl^-, SO_4^{2-};

- 'Man-made synthetic substances': Trichloroethylene, Tetrachloroethylene;

- 'Parameters indicative of saline or other intrusion': Conductivity or Cl^- and SO_4^{2-} depending on Member States decision.

'Taking into account. . .' does not mean that deriving threshold values for all parameters in Annex II.B is obligatory. Deriving threshold values for other substances/parameters that are not on the list, but which cause the groundwater body to be at risk, is an obligation.

Criteria for Establishing Threshold Values

As mentioned in the GWD Annex II.A, 'the determination of threshold values should be based on:

- 'The extent of interactions between groundwater and associated aquatic and dependent terrestrial ecosystems;

- The interference with actual or potential legitimate uses or functions of groundwater; [. . .]

- Hydrogeological characteristics including information on background levels'.

Moreover, it is also written in Annex II.A of the GWD that 'threshold values will be established in such a way that [. . .] this will indicate a risk that one or more of the conditions for good groundwater chemical status referred to in Article 4.2.c.(ii), (iii) and (iv) are not being met'. The latter Article refers to:

- the definition of good groundwater chemical status (WFD Annex V 2.3.2);

- protected areas used for the abstraction of drinking water (WFD Article 7);

- the ability of a groundwater body to support human uses.

Based on these elements, two criteria can be considered when deriving threshold values:

- **Environmental criteria:**
  - ○ Threshold values that aim to protect associated aquatic ecosystems and groundwater-dependent terrestrial ecosystems.

- **Usage criteria**
  - ○ Threshold values that aim to protect drinking water in Drinking Water Protected Areas (DWPA); and

  - ○ Other legitimate uses of groundwater: crop irrigation, industry, etc...[Only uses involving a significant surface (or volume) of the groundwater body compared to the whole surface (or volume) of the groundwater body should be considered.]

Scale for Setting Threshold Values

Depending on the type of pollutant, the risk to groundwater, and the observed concentrations, Member States can derive threshold values at different scales: groundwater body (or group of groundwater bodies), river basin district, national part of an international river basin district, or national level (Article 3.2 GWD). The groundwater body is the most detailed scale allowed for threshold values derivation.

For example, where a purely anthropogenic pollutant (e.g. trichloroethylene) is commonly observed at very low levels, Member States may set a threshold value at the national level as long as the achievement of environmental objectives in any individual groundwater body is not compromised. Alternatively, for parameters with natural concentrations that vary from one groundwater body to another (e.g. As, Cl^-, SO_4^{2-}, NH_4^+ and metals), it is highly recommended to establish threshold values at the groundwater body scale.

Timetable and Revision

Threshold values had to be established by Member States for the first time by 22 December 2008 (Article 3.5 of the GWD) and published in the first River Basin Management Plan, i.e. by 22 December 2009 (Article 13 WFD).

However, the threshold value derivation process is an on-going process and Member States can add, remove or re-insert threshold values for any substance whenever necessary (Article 3.6 of the GWD). Any changes will depend upon 'new information' about the parameters derived from new scientific knowledge and understanding.

4.1.3 METHODOLOGY TO DERIVE THRESHOLD VALUES UNDER THE WFD/GWD

The general methodology for establishing threshold values in a groundwater body is summarised in Figure 4.1.3. As explained before, two types of criteria should be considered when establishing the threshold values: *environmental criteria* and *usage criteria*.

Threshold values (TV) will be set by Member States by comparing the background level (BL) to the criteria value (CV). The criteria value is the concentration of a pollutant, not taking into account any natural background concentrations, that, if exceeded, may lead to a failure of the relevant 'good status' criterion. CVs should take into account risk assessment and groundwater functions.

When BLs and CVs are compared, two situations may arise:

- Case 1: BL is below CV_i. In this case, Member States will define the TV according to national strategies and a risk assessment, which may result in establishing a TV above the BL, provided it can be clearly justified.

- Case 2: BL is higher than CV_i. In this case, the TV should be equal to the BL.

However, in order to integrate the concept of sustainable development and allow for the growth of economic activities (especially existing activities). Member States may consider a small addition to the BL which represents an acceptable amount of human influence as long as this is considered not to be harmful in protecting the relevant receptors.

4.1.3.1 Determining Threshold Values for the 'Associated Aquatic Ecosystems and Dependent Terrestrial Ecosystems' Criteria

When groundwater and surface waters are linked and especially when surface waters or dependent terrestrial ecosystems are fed by groundwater, the criteria value(s) relevant for the protection of associated surface water or GWDTE, will be derived using environmental quality standards (EQS) for surface water (or any other relevant ecotoxicological value). For priority substances and other pollutants listed in the 2008/105/EC Directive (European Commission 2008a), EQS values set in this text may be used. Any other EQS derived locally or nationally by a Member State using for example ecotox-test-results of aquatic organisms may also be applied.

Because the concentration of a parameter varies between the aquifer and the river, a dilution factor (DF) or an attenuation factor (AF) may be applied for deriving an appropriate criteria value.

The calculation of attenuation and dilution factors will depend on the level of knowledge of groundwater/surface-water interaction, the conceptual model and the position of monitoring points in the groundwater system relative to the receptor. Member States are free to set the value(s) of a dilution factor (DF) and an attenuation factor (AF) for each groundwater body according to its own approach and knowledge. The BRIDGE project proposals for calculating DF and AF may also be used (*Müller et al.*, 2006).

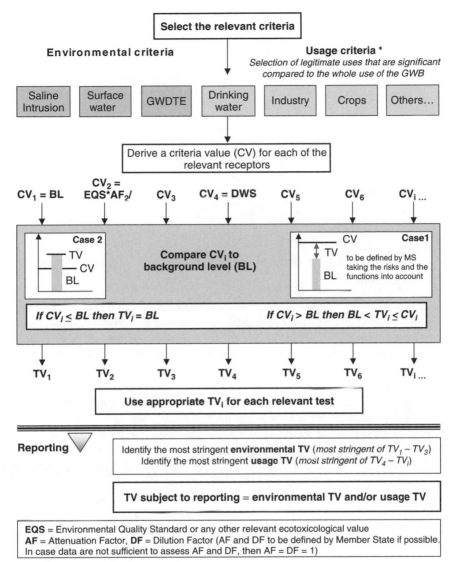

Figure 4.1.3 General methodology of deriving groundwater threshold values (TV). European Environment Agency http://www.eea.europa.eu/.

The relevant criteria value is hence equal to:

$$CV = EQS^*AF/DF.$$

Dilution and attenuation should not be included when monitoring takes place in the receptor. In this case:

$$DF = AF = 1. \text{ Therefore } CV = EQS_{\text{surface water}}$$

The use of AF and DF require a good understanding of the groundwater system and its relationship to surface water. Where this understanding is lacking, the use of AF and/or DF may not be possible. In this case a precautionary approach may be taken in the first instance; i.e. $CV = EQS$

4.1.3.2   Determining Threshold Values for the 'Legitimate Uses' Criteria

Where a groundwater body has uses, some other relevant values may need to be derived. For instance, if groundwater is used for drinking-water supply, crop irrigation or the food industry, then a criteria value can be derived for these uses. However, such a value will only be derived and considered if the total surface or volume of the polluted area related to such 'legitimate use' is 'significant' when compared to the whole surface or volume of the groundwater body. In the case of drinking-water supply, drinking-water standards (DWS) should be considered when deriving criteria values. For other uses such as crop irrigation and industry, a case-by-case approach is recommended.

If the abstraction point is not the monitoring point where compliance with the threshold value is to be assessed, it may be appropriate to also take into account dilution and attenuation when deriving the threshold values and criteria values for the usage criteria which is to be decided by Member States.

It should be noted, however, that the compliance regime for Drinking Water Protected Areas (DWPA) does not only account for a check on exceeding of the threshold values, but also relies on testing whether there will be no need for a (further) increase of water treatment measures as required by Article 7.3 of the WFD.

4.1.3.3   Determining Threshold Values for the 'Saline or Other Intrusions' Criteria

The relevant threshold value for saline or other intrusions will be the BL for key parameters (i.e. those indicative of intrusion), as this is the most appropriate environmental value when examining if there has been any intrusion caused by anthropogenic activities.

4.1.4 THRESHOLD VALUES, CHEMICAL STATUS COMPLIANCE AND MONITORING: RELATIONSHIPS AND LIMITS

With the support of the CIS guidance documents no. 7 and no. 15 'Groundwater Monitoring', Member States have put emphasis on the design of harmonised monitoring networks. But despite all efforts, groundwater monitoring programmes for assessing the chemical status show differences from one country to another. Data reported by Member States in response to the WFD Article 8[2] on monitoring, clearly illustrate this gap (Figures 4.1.3 and 4.1.4). The *diversity of approaches* was already pointed out by the BRIDGE consortium (Scheidleder, 2005), which underlined that 'the monitoring network, the assessment methodology and the thresholds are tightly connected to each other'; the author concluded 'that the methodology of establishing threshold values has to bring several variables under one hat'.

This difference also concerns the *delineation of groundwater bodies*. An analysis of the WFD Article 5 reports clearly highlights that European groundwater bodies (i.e. units for groundwater chemical status assessment) have very different sizes. The average size of groundwater bodies ranges from $10 \, \text{km}^2$ (Slovenia) to $4411 \, \text{km}^2$ (Estonia). Nevertheless, the groundwater body remains the most detailed scale for threshold values derivation; for each body there is a single 'environmental threshold value' and a single 'usage threshold value', which have to be relevant for all WFD monitoring stations in the groundwater body.

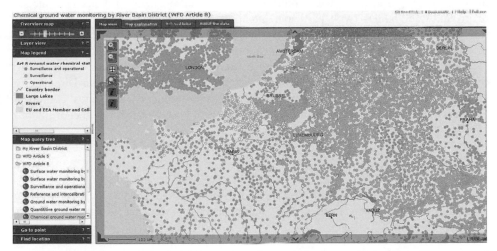

Figure 4.1.4 WFD Article 8 monitoring stations for groundwater monitoring surveillance control and operational controls.

---

[2] http://www.eea.europa.eu/themes/water/mapviewers/art8-gw.

Another bias is introduced by the *performance of methods used for sampling and analysis*. Inter-laboratory testing for the analysis of chemical contaminants in water has shown major uncertainties in the results. For instance, for most pesticides in the priority substances list (European Commission, 2008a), the coefficient of variation ranges from 30% to 40% (Coquery *et al*., 2005). Such uncertainties have a direct consequence on the assessment of water chemical status and low-grade monitoring can lead to misjudgements on the chemical status of a groundwater bodies of Member States.

4.1.5   CONCLUSIONS AND PERSPECTIVES

Setting and applying threshold values for the assessment of groundwater status is a real challenge. Assessing the chemical status of groundwater bodies is not only limited to the comparison of chemical data to groundwater threshold values or groundwater quality standards. The GWD clearly triggers further *investigations* to be carried out by Member States in case such threshold values or quality standards are exceeded. This is a key step in the assessment, where experts are invited to combine their knowledge and carry out further detailed studies on the links between groundwater, surface water and wetlands, on the potential impact on human uses of exceeding such values, especially drinking-water supply, and on the potential for intrusion of seawater or from other groundwater bodies

The implementation of the WFD is an on-going process and improvements are continuous thanks to *interactions between science and policy*. Many studies carried out at local, national or European levels have contributed to the improvement of conceptual models of groundwater bodies (with particular attention to groundwater/surface-water interactions, pollutant behaviour in the vadose zone, etc...). This will certainly contribute to the revision of groundwater bodies delineation and of the monitoring networks in 2013.

During the past years, emphasis has also been put on the quality and comparability of analytical results, in particular with the publication of the Commission Directive on 'Technical specifications for chemical analysis and monitoring of water status' (European Commission, 2008b) and with the FP6 research project EAQC-WISE (Held *et al*., 2008).

Acknowledgments

The authors would like to thank Philippe Quevauviller and the members of Working Group C for their involvement in the establishment of the concepts presented in this chapter.

REFERENCES

Blum A., 2004. *The state of groundwater resources in France. Quantitative and qualitative aspects*, Editions de l'Institut Français de l'Environnement (IFEN), Coll. 'Etudes et travaux' no. 43, 16p., available on www.ifen.fr

Coquery M., Morin A., Bécue A., Lepot B., 2005. Priority substances of the European Water Framework Directive: analytical challenges in monitoring water quality, *Trends in Analytical Chemistry*, 24(2).

Dahlström K., Müller D., 2006. *Report on national methodologies for groundwater threshold values*, Deliverable D14, BRIDGE project, 20p, available on www.wfd-bridge.net

European Commission, 1998. Council Directive 98/83/EC on the quality of water intended for human consumption, *Official Journal of the European Communities* L330, 5.12.98, p. 32.

European Commission, 2000. Directive 2000/60/EC of the European Parliament and of the Council of 23 October 2000 establishing a framework for Community action in the field of water policy, *Official Journal of the European Communities* L 327, 22.12.2000, p. 1.

European Commission, 2003. *Monitoring under the Water Framework Directive*, CIS Guidance Document No. 7, European Commission, Brussels.

European Commission, 2006. *Groundwater Monitoring*, CIS Guidance Document No. 15, European Commission, Brussels.

European Commission, 2007. *Guidance on preventing or limiting direct and indirect inputs in the context of the Groundwater Directive 2006/118/EC*, CIS Guidance Document No. 17, European Commission, Brussels.

European Commission, 2008a. Directive 2008/105/EC of the European Parliament and of the Council of 16 December 2008 on environmental quality standards in the field of water policy, amending and subsequently repealing Council Directives 82/176/EEC, 83/513/EEC, 84/156/EEC, 84/491/EEC, 86/280/EEC and amending Directive 2000/60/EC of the European Parliament and of the Council, *Official Journal of the European Communities* L 348, 24.12.2008, p. 84.

European Commission, 2008b. *Commission Directive laying down, pursuant to Directive 2000/60/EC of the European Parliament and of the Council, technical specifications for chemical analysis and monitoring of water status*, draft version, May 2008.

European Commission, 2009. *Guidance on groundwater status and trends assessment*, CIS Guidance Document no. 18, European Commission, Brussels.

Held A., Emons H., Taylor P., 2008. *EAQC-WISE (European Analytical Quality Control in support of the Water Framework Directive via the Water Information System for Europe), Final report of the project: The blue print*, deliverable D25, December 2008, 58p.

Müller D., Blum A., Hart A., Hookey J., Kunkel R., Scheidleder A., Tomlin C., Wendland F., 2006. *Final proposal for a methodology to set up groundwater threshold values in Europe*, Deliverable D18, BRIDGE project, 63p, available on www.wfd-bridge.net

Müller D., Fouillac A.-M., 2007. Methodology for the establishment of groundwater environmental quality standards. In Ph. Quevauviller (ed.), *Groundwater Science and Policy. An International Overview*. RCS Publishing, Chichester.

Scheidleder A., 2005. *Summary report on groundwater quality monitoring network designs for groundwater bodies*, Deliverable D13, BRIDGE project, 48p, available on www.wfd-bridge.net

4.2

Assessing and Aggregating Trends in Groundwater Quality

Hans Peter Broers[1], Ate Visser[2], John P. Chilton[3] and Marianne E. Stuart[4]

[1] *The Netherlands Organisation for Applied Scientific Research (TNO), Geological Survey of The Netherlands, Utrecht, The Netherlands*
[2] *Faculty of Geosciences, Utrecht University, Utrecht, The Netherlands*
[3] *International Association of Hydrogeologists, Reading, United Kingdom*
[4] *British Geological Survey, Wallingford, United Kingdom*

4.2.1   LEGISLATIVE REASONS FOR PERFORMING TREND ANALYSIS

'Groundwater is a valuable natural resource and as such should be protected from deterioration and chemical pollution. This is particularly important for groundwater dependent ecosystems and for the use of groundwater in water supply for human consumption'

Groundwater Monitoring Edited by Philippe Quevauviller, Anne-Marie Fouillac, Johannes Grath and Rob Ward
© 2009 John Wiley & Sons, Ltd

(after EU, 2006). The EU Water Framework Directive (EU, 2000) aims for integrated management of surface water and groundwater. It is the most advanced regulatory framework for the protection of all natural waters in the European Union, seeking to achieve 'good status' for all water bodies by the end of 2015. In relation to groundwater quality, the WFD requires Member States to:

- delineate groundwater bodies and characterise them according to the anthropogenic pressures in order to identify groundwater bodies at risk of failing to meet their environmental objectives and that may fail to meet the criteria for 'good status';

- establish a groundwater monitoring network to provide a comprehensive overview of the chemical and quantitative status of the groundwater body. This was required to be operational by the end of 2006.

The recently adopted EU Groundwater Directive (GWD) (EU 2006) on the protection of groundwater against pollution in the EU Member States better defines the environmental objectives of the WFD for groundwater. Thus, the GWD is based on three pillars:

- specific criteria for defining 'good chemical status' (Article 3);

- criteria for the detection of significant and sustained long term anthropogenic induced upward trends in the concentrations of pollutants (Article 5) as well as the definition of starting points for trend reversal and requirements on the implementation of measures necessary to reverse any significant and sustained upward trends;

- Preventing and limiting the inputs of pollutants to groundwater (Article 6).

So, the identification of sustained upward pollution trends and their reversal is the second 'pillar' of the new directive, which stipulates that trends must be identified for any pollutant putting the groundwater 'at risk'. This links to the analysis of pressures and impacts carried out under the WFD (Article 5). The issue of 'significance' is clarified in Annex IV of the GWD. Trends must be both statistically significant (mathematical) and environmentally significant. Environmental significance relates to potential future impact of the identified upward trends.

The trend reversal obligation requires that any significant and sustained upward trend will need to be reversed when reaching 75% of the values of EU-wide groundwater quality standards and/or threshold values (Figure 4.2.1). Trend reversal has to be achieved through establishing the programmes of measures defined by the WFD (Annex VI).

The Water Framework Directive also requires monitoring and trend assessments at individual groundwater drinking water abstraction sites. Under Article 7.3 of the WFD, Member States shall ensure the necessary protection for groundwater bodies identified as Drinking Water Protected Areas 'with the aim of avoiding deterioration in their quality in order to reduce the level of purification treatment required in the production of drinking water'.

By implementing groundwater protection measures that are technically feasible and proportionate, Member States need to use their best endeavours to ensure that groundwater quality does not deteriorate at the point of abstraction for drinking water supply, and, so that there is no need to increase the level of purification treatment (EU, 2007b).

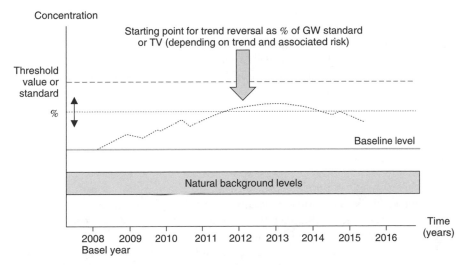

Figure 4.2.1 Principle of the identification and reversal of statistically and environmentally significant upward trends (after Quevauviller 2008).

Member States should ensure that raw water quality monitoring is representative and sufficient to ensure that significant and sustained changes in groundwater quality due to anthropogenic influences can be detected and acted upon. Compliance points must be set at appropriate locations to detect such changes. This objective may be achieved by groundwater protection measures (which may be focused using safeguard zones) and the monitoring of raw groundwater quality to demonstrate significant and sustained improvements (trends).

4.2.2 SCOPE AND TREND DEFINITION

In this chapter we explore different methodological aspects of trend analysis in relation to the new WFD and GWD legislation. Part of the approach presented originates from the 6th EU Integrated Research and Technology Development Framework Programme (FP6) Project Aquaterra. Within this project, work package TREND 2 (Visser *et al.*, 2008) was dedicated to the development of operational methods to assess, quantify and extrapolate trends in groundwater systems. Trend analysis techniques were tested on data from a wide range of European environmental settings including unconsolidated lowland deposits in the Netherlands and Germany, chalk aquifers in Belgium and a fractured aquifer with a thick unsaturated zone in France. In addition, this chapter presents recent developments in trend analysis on data from abstraction sites in the UK. This is particularly relevant for the Drinking Water Protected Area (Article 7.3) requirements of the WFD.

We define a trend as 'a change in groundwater quality over a specific period in time and over a given region, which is related to land use or water quality management'. Trend analysis for the Groundwater Directive is dedicated to distinguishing these anthropogenic changes from natural variation 'with an adequate level of confidence and precision' (GWD, Annex IV, Article 2(a)(i)).

Temporal variations due to climatological and meteorological factors have the potential to complicate trend detection. Also, spatial variability is an additional complicating factor, especially when aggregating trends at the groundwater body scale. The requirement to aggregate trends is defined in European Union WFD Common Implementation Strategy guidance (EU, 2008). Relevant factors influencing spatial variation include:

- flow paths and travel times;

- pressures and contaminant inputs; and

- the chemical reactivity of groundwater bodies.

These variations result in variable and different trend behaviour across the scale of the groundwater body, because some monitoring points might be along flow paths which originate from areas with high contaminant inputs to groundwater and others along flow paths that originate from areas of low input.

Trend analysis techniques aim to reduce the variability which is not related to anthropogenic changes themselves. Therefore, trend detection becomes more efficient when the aforementioned spatial and temporal variability are reduced by taking into account the physical and chemical temporal characteristics of the body of groundwater, including flow conditions, recharge rates and percolation times (GWD, Annex IV, (2(a)(iii)). Several statistical techniques, modelling techniques and combinations of both are available for trend analysis and some of the promising techniques have been tested in the TREND2 work package and at UK abstraction sites, including age dating and transfer-function approaches (Visser *et al*., 2008; Stuart *et al*., 2007).

4.2.3   TRENDS IN RELATION TO PRESSURES, MONITORING STRATEGIES AND PROPERTIES OF GROUNDWATER SYSTEMS

The Aquaterra comparative approach showed that there is no unique approach which works under all hydrogeological conditions and for all monitoring systems across Europe. However, reducing variability by including information on pressures, hydrology and hydrochemistry did help to improve the detection of relevant trends in each of the hydrogeological settings studied. Specific conclusions included:

- Grouping of wells is recommended to improve trend detection efficiency.

- Grouping is preferably done according to pressures (often land use related), hydrologic vulnerability (travel time frequency distributions, unsaturated zone depth) and chemical characteristics such as rock type and organic matter contents (Figure 4.2.2).

- Grouping should also consider the depth dimension because groundwater generally becomes older with depth (Figure 4.2.3) and trends at depth might be completely different from trends in the shallower parts of the aquifer.

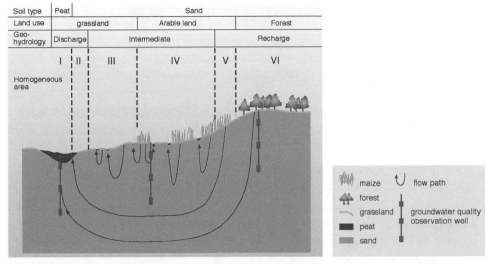

Figure 4.2.2 Grouping of wells according to pressures (land use), hydrologic vulnerability (hydrogeological situation) and chemical vulnerability (soil type). The resulting combinations were called homogeneous areas and used for determining trends and assessing chemical status (Broers and van der Grift, 2004).

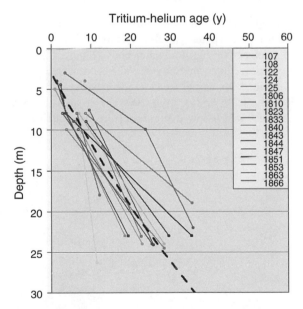

Figure 4.2.3 Increase of groundwater age with depth as determined by an analytical equation (dashed line) and tritium-helium age dating in 14 multi-level observation wells (separate colours for each well).

- It is essential to distinguish abstraction wells and springs from observation wells which are not pumped or naturally flowing;

  o *Pumping wells and springs* normally have water mixed from different layers and the resulting water quality reflects a broad range of travel times. As a further complicating factor, the contributions of young and old water in the mixture may change with time.

  o Water quality measured in *observation wells* is normally related to a distinct groundwater age, and the time series can be related to a specific infiltration period once the age has been determined.

  If different monitoring types occur in a groundwater body, trend detection is best done by grouping similar types of monitoring point together.

- Unsaturated zone thickness is one of the controlling variables when considering the choice of trend analysis technique. Thick unsaturated zones lead to long response times which can lead to difficulties in early detection of trends related to anthropogenic inputs from the land surface.

4.2.4   AGGREGATION OF TRENDS AT THE GROUNDWATER BODY SCALE

Although grouping of wells according to pressures and monitoring depths helps to identify trends (previous section), large spatial variability is also often observed in trend direction (up/down) and trend slope across a groundwater body (Figure 4.2.4). The implementation of the GWD requires a procedure where the trend assessment results at individual monitoring points are combined (or aggregated) to identify significant and sustained trends at the groundwater body scale body' (EU, 2008). Two possible ways of aggregating individual trends are illustrated below using data from the Dutch monitoring network in Noord-Brabant.

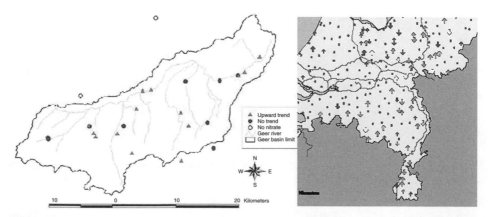

Figure 4.2.4   Spatial variability in trends in the Geer basin, Belgium (left) and southeast Netherlands (right).

The monitoring network comprises standardized monitoring wells with fixed screens at specific depths. The wells consist of purpose built nested piezometers with a diameter of 50 mm and a screen length of 2 m at a depth of about 8 and 25 m below surface (Broers, 2002). The subsurface of Noord-Brabant consists of fluvial unconsolidated sand and gravel deposits from the Meuse River, overlain by a 2–5 m thick cover of Middle- and Upper-Pleistocene fluvio-periglacial and aeolian deposits consisting of fine sands and loam. Noord-Brabant is a relatively flat area with altitudes ranging from 0 m above Mean Sea Level (MSL) in the north and west to 30 m above MSL in the south-east. Groundwater tables are generally shallow, usually within 1–5 m below the surface.

As a first step in aggregating trends it was recommended to group monitoring wells on the basis of pressures/vulnerability and hydrological properties such as the probable travel time distribution in the groundwater body (previous section). Two methods for aggregating the trends are possible:

1. a statistical method, for example by defining the median trend slope and the corresponding confidence interval;

2. a deterministic method, for example using age dating to aggregate time series along a standardized X-axis showing recharge time.

Both approaches are illustrated below using Aquaterra results.

Example 1: Aggregation Using Median Trend Slopes

First, all trend slopes of individual monitoring points were determined, through linear regression or a Kendall-Theil robust line (Helsel and Hirsch, 1992). Aggregated trends were then determined by taking the median of all trend slopes to test whether this median differs significantly from zero (Broers and van der Grift, 2004). A significant upward aggregated trend for the group of wells is established when the 95% confidence level of the median is completely above the zero slope line (Figure 4.2.5). A downward trend is identified if the complete confidence interval is below the zero slope line. Confidence intervals around the median slope were determined non-parametrically following the method of Helsel and Hirsch (1992) and using a table of the binomial distribution. In the example shown in Figure 4.2.5, significant upward trends (filled symbols) were detected for OXC and Sumcat in the lower graphs which represent the deeper screens, and downward trends for Sumcat and OXC in the upper graphs which represent the shallow screens. The results indicate reversal of trend direction with depth, with improving conditions in the shallow subsurface due to action programmes which effectively reduced the pollutant inputs, while the old pollution front still leads to deteriorating conditions in deeper groundwater.

It should be noted that trends can often have reversed directions at different depths in the aquifer, due to differences in groundwater age and the corresponding contaminant inputs during the period of infiltration (see for example Figure 4.2.5). One of the conclusions of aggregating trends in a statistical manner is that often a relatively large number of observation wells (20–40) is necessary to statistically demonstrate trends because of the observed large temporal and spatial variability which is inherent in groundwater quality datasets.

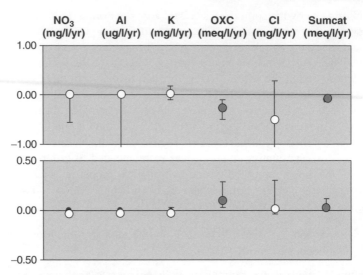

Figure 4.2.5 Aggregated median trend slopes for agricultural recharge areas in the province of Noord-Brabant for 6 chemical indicators for shallow screens (upper graph) and deeper screens (lower graph). *Source*: Visser *et al.*, 2005. (OXC = oxidation capacity, Sumcat = sum of cations).

Example 2: Aggregation Based on Recharge Time Using Age Dating

A new and promising aggregation technique is to use age dating to determine the recharge period of the groundwater and relate the measured concentration data to the derived recharge time. This technique proved to work well for monitoring systems based on multi-level observation wells in areas with porous aquifers. In this example, tritium-helium ages were used to determine the travel time to the monitoring screens. These travel times were used to relate the time-series of measured concentrations to the time of recharge, instead of the time of sampling (Figure 4.2.6). In this example, the aggregated time series shows a sustained upward trend with higher concentrations in recently infiltrated groundwater.

Subsequently, the results of all 28 time series in the 'intensive agricultural land use in recharge areas' type were aggregated in one graph and analysed using LOWESS smoothing (Cleveland, 1979) and ordinary linear regression approaches (Figure 4.2.7). The method successfully identified statistically significant ($P < 0.005$) trend reversal of nitrate concentrations and oxidation capacity for this area type.

The observed trend compares well with the known input history of agricultural pollutants based on historical data series of the production and use of fertilizer and manure under various crop types. Trend reversal is generally most easily demonstrated for conservative solutes and indicators, such as 'oxidation capacity' (Visser *et al.*, 2007). Downward trends in the most recent groundwater could also be demonstrated for reactive solutes such as nitrate, which is transformed to nitrogen when it encounters denitrification by reactive organic matter or sulfides at some depth in the subsurface.

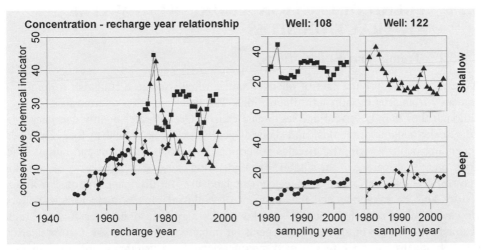

Figure 4.2.6 Translating time series measured in individual observation multi-level wells at shallow depth (10 m −sl) and deep (25 m −sl) into an aggregated time series plot using recharge year as X −axis after age dating using tritium-helium (Visser *et al*. 2007). The aggregated time series shows a sustained upward trend with higher concentrations with recharge time.

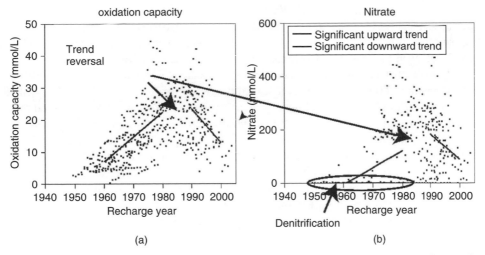

Figure 4.2.7 Aggregation by using age dating to determine recharge year corresponding to the measured concentrations.

4.2.5   TREND DETECTION AT DRINKING WATER ABSTRACTION SITES

4.2.5.1   Trends and Fluctuations

A trend is an underlying rate of change, and is often used to distinguish a long-term tendency from erratic, short-term fluctuations. The latter are often referred to as 'noise', although they may have real and legitimate causes. Groundwater quality can vary over time scales ranging from tidal and daily cycles, seasonal or annual to longer periods, depending on the varying time scales governing the sources and input and output functions, and the properties of the aquifer. The underlying cause of these patterns of variation may reflect 'regional' changes in catchment land use, fertiliser applications, pollution history and the evolution and development of pollutant plumes and climatic factors. However, solute concentrations in samples of discharging groundwater from a single abstraction borehole or spring also depend on numerous 'local' or site factors such as borehole depths and open or screened interval, depths and lengths of groundwater flow paths, possible groundwater quality stratification in the aquifer and changes (at various timescales) in groundwater levels, directions of flow and pumping regime. These sources of variation are often superimposed on one another in a time series of an individual parameter, such as nitrate or chloride, at a drinking water abstraction site and their resolution can be a challenging task (Stuart *et al*., 2007).

Short-term peaks in solute concentration at abstraction sites are a particular problem for water supply utilities, and may compromise their ability to meet groundwater quality obligations under the Drinking Water Directive and the Water Framework Directive. Such variations may be qualitatively understood to be related to, for example, seasonal responses to groundwater recharge. However, the precise nature of the variations and the hydrogeological processes and pollutant transport mechanisms controlling them may be difficult to identify and quantify and the timing and scale of the peaks correspondingly difficult to predict. If observed groundwater quality time series are reasonably well described by a statistical model which accounts for both trends and seasonal variability, then there is a good prospect for determining trends and predicting groundwater quality within the timescales envisaged by the Water Framework Directive and Groundwater Directive.

In many cases the monitoring itself can introduce its own characteristics which may make it difficult to assess the presence and significance of trends. These characteristics include the sampling frequency, the amount of missing data and its distribution within the time series, the length of the monitoring period and the presence of uncontrolled variables such as intermittent abstraction, varying abstraction rates and unrecorded pumping regime.

4.2.5.2   Using Abstraction Site Monitoring for Trend Detection

Many, perhaps most, national groundwater quality monitoring programmes, especially those that have developed gradually over time, depend to a large extend on sampling of groundwater at water supply sites (EU, 2007a). Of these, public supply boreholes have one major advantage of being operated and discharging more or less continuously. Purging is not normally required, the discharging groundwater represents water from within the aquifer, although sometimes from uncertain and varying locations within the aquifer, and sampling the discharge may be easy and relatively inexpensive. Private domestic, industrial and irrigation boreholes are also widely used, but may be operated less regularly.

It is, therefore, not surprising that such boreholes and, where suitable, springs form the backbone of many networks. There are, however, some pitfalls and limitations of abstraction sites which can affect the assessment of trends in water quality, and the regional and local factors referred to above must be understood in the interpretation of the monitoring results. At the simplest operational level, it is critical to obtain the sample from the supply pump or directly at the spring, and at the same point each time, before any treatment, storage or blending processes. Groundwater quality rarely changes extremely rapidly and if in examining closely-spaced time series data from abstraction boreholes there are sharp excursions of individual points (either upward or downward), these should be examined carefully. They are unlikely to be 'real' groundwater responses. Single individual outliers may be an analytical error; repeated individual outliers or very noisy plots are likely to represent local operational factors.

Particular problems can occur at multiple borehole sites, wellfields and multiple spring sources. It might be expected that variations in solute concentration with time for a cluster of boreholes within metres or tens of metres of each other at one site would be similar and related. However, individual abstraction points may have different concentrations and trends. This may be due to stratification of groundwater quality in the aquifer combined with differences in borehole depths, water levels, abstraction rates and inflow levels, or to differences in direction of groundwater flow, capture zone, soils and protective geological cover and land use (Stuart *et al.*, 2007). In such situations, complex operating regimes, with rotating duty and standby boreholes mean that pumping from one may affect the quality of the others, and individual capture zones may be disturbed by the regular rotation of pumping. In some cases, abstraction from one borehole at such a site has been discontinued because of high nitrate concentrations, only for the adjacent low-nitrate borehole to experience a sharp step rise in nitrate as it captures more of the available high-nitrate water (Figure 4.2.8).

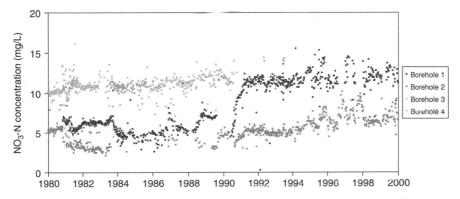

Figure 4.2.8 At this Triassic sandstone site, the shutdown of borehole 3 in 1990 due to high concentrations of nitrate leads to the transfer of high nitrate water to the next nearest borehole (1) within a few months (after UKWIR, 2004). (See Plate 2 for a colour representation).

4.2.5.3 Approaches to Trend Detection

There is a long history of the application of statistical methods to water quality data, particularly to surface waters (Helsel and Hirsch, 1992; Peters, 1996), but also to ground-water (Frapporti, 1994; Beeson and Cook, 2004; Broers and van der Grift, 2004; Grath *et al.*, 2001; Stuart *et al.*, 2007). Many of the classical statistical procedures for analysing time series, such as autoregressive integrated moving average (ARIMA) methods, require regular sampling intervals. Although missing data can sometimes be accommodated by these methods, groundwater quality data from most water supply abstraction sites are so irregular as to preclude these types of analysis.

The trends of interest include monotonic, linear, cyclic (seasonal) and step changes. Grath *et al.* (2001) proposed statistical tests for each of these, although their robustness against outliers, missing data and censoring varies. The Spearman rho and Mann-Kendall tau methods have been used to test for the presence of monotonic trends (Yue and Wang, 2002; Broers and van der Grift, 2004) and the Spearman rank correlation coefficient was also used by Yue and Wang (2002). Seasonal responses can be detected by methods such as periodograms, Students *t*-test (Helsel and Hirsch, 1992); Mann-Witney rank-sum test, analysis of variance, Kruskall-Wallis test, periodic functions, seasonal Mann-Kendall (Helsel and Hirsch, 1992) and spectral analysis (Fleming *et al.*, 2002). However, seasonal patterns in groundwater quality are often complicated by the variation between years of the length, timing and scale of responses to climatic factors and the associated modifications to operational abstraction regimes related to increases or decreases in water demand and water availability.

Stuart *et al.* (2007) summarise a simple semi-automated approach to trend estimation which incorporates a series of descriptive and statistical tests to determine the regularity and frequency of sampling, whether the data show a significant linear trend with time, whether there is any seasonality in the data, whether the data show any unusually large deviations from the assumptions made in the statistical tests used, and whether there is any evidence of a change in trend or a trend reversal. The 'R' statistical programming language (R Development Core Team, 2005) is used because of its powerful built-in graphical features, its ability to deal with large numbers of data sets in 'batch' mode and its facility for summarising the results of these tests. The approach has been extensively applied to groundwater nitrate time series data from public water supply sites in the major Chalk, limestone and sandstone aquifers of the UK (UKWIR, 2003; Stuart and Kinniburgh, 2005; Stuart *et al.*, 2007).

The steps employed in this approach comprise descriptive, statistical and trend tests. The descriptive tests include graphical methods and summary statistics, and do not involve any estimation or testing of hypotheses. Five descriptive plots are automatically produced by this method:

- Plot 1: a raw data scatterplot of concentration versus date;

- Plot 2: a step plot showing the gap between successive samples in days, annotated with the mean and standard deviation of the gap, to illustrate the regularity of sampling;

- Plot 3: a histogram of the gap to show sampling interval;

- Plot 4: a box and whisker plot of concentrations binned into calendar months to show the range of monthly values. Cyclical behaviour on an annual timescale indicates seasonality (see for example Figure 4.2.9a)

- Plot 5: a smoothed trend based on a LOESS smoother plotted onto the raw data.

These plots provide a quick summary of the amount, range and quality of the time series data at each abstraction point, with information about the regularity of sampling, the presence and importance of outliers, the degree of seasonality and the 'smoothness' of the data. Where the quality of the time series data is poor, this may be all that is possible or appropriate.

Following this, two standard plots are produced to show the results of statistical tests. Plot 6 shows the raw data overlain with linear trend lines determined by three regression-based methods (for example Figure 4.2.9b, Stuart *et al.*, 2007). The plot is annotated with potential outliers, trend values, probability of significant seasonality, and the root mean square error (r.m.s.e). If a change or reversal of trend is detected, a 'broken stick' plot is included. Where there is variation that cannot be accounted for by a linear model, a warning is included that 'additional' structure exists in the time series data. Plot 7 illustrates the results of standardised residuals tests in the form of a scatterplot of standardised residuals against date based on the seasonal or non-seasonal model and influential points and possible outliers are highlighted.

Figure 4.2.9 (Stuart *et al.*, 2007) illustrates results obtained for nitrate in groundwater from an abstraction source in the UK Chalk aquifer which is subjected to strongly seasonal influence. As a consequence, while the overall upward trend is rather modest, the seasonal peaks have provided non-compliant groundwater nitrate concentrations since 1995, except in the dry years of 1996 and 1997. The very good correlation between nitrate concentrations and groundwater levels in a nearby observation borehole (Figure 4.2.9) is clear, and was maintained in these dry years, suggesting a fundamental, process-based connection between the two.

4.2.5.4 Aggregation of Data from Abstraction Sites

Time series and trends from single drinking water abstraction sites are essential for determining whether there is deterioration in groundwater quality in the safeguard zones and Drinking Water Protected Areas (DWPAs) established under the Water Framework Directive and Groundwater Directive (EU, 2007b). As is the case for observation boreholes described in Section 4.2.4 above, aggregation of data from abstraction sites is needed for assessing trends in groundwater bodies. Two approaches have been tested for a selection of groundwater bodies in the UK (Kinniburgh *et al.*, 2004):

- an 'average of averages' approach in which an average of all data within the groundwater body for each time interval is taken, and the trend with time of these averages taken – spatial average and then time trend;

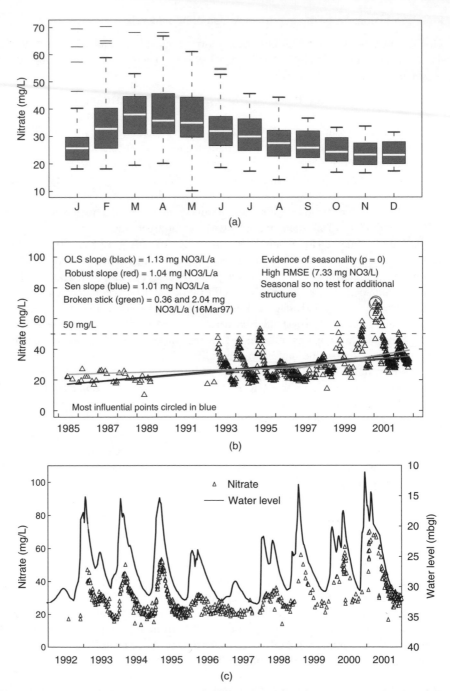

Figure 4.2.9   Seasonal data from a site in the Chalk aquifer: (a) range of monthly values (Plot 4), (b) trend fitting (Plot 6), (c) correspondence with water level for part of the data series (1993–2001). (Stuart *et al.*, 2007). Reproduced from the Quarterly Journal of Engineering Geology and Hydrogeology, by permission of The Geological Society Publishing House, vol 40 pp 361–376.

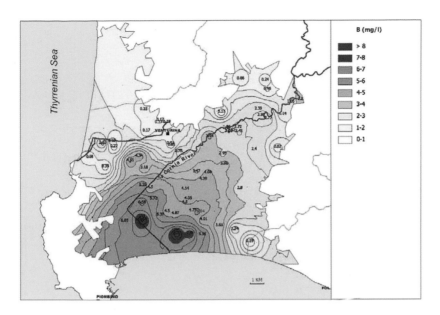

Plate 1

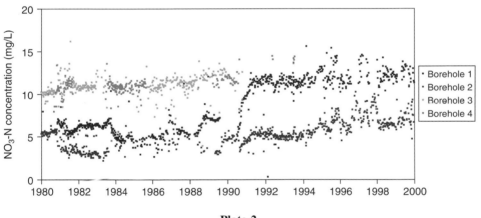

Plate 2

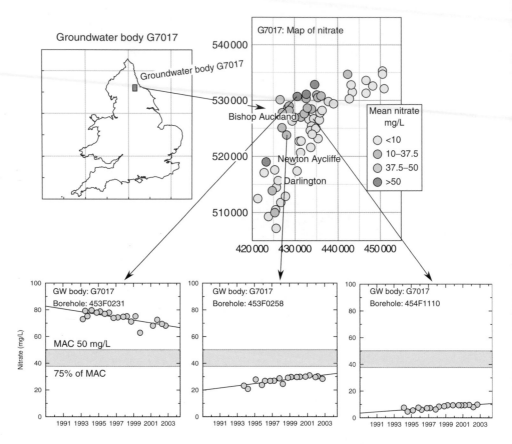

Plate 3

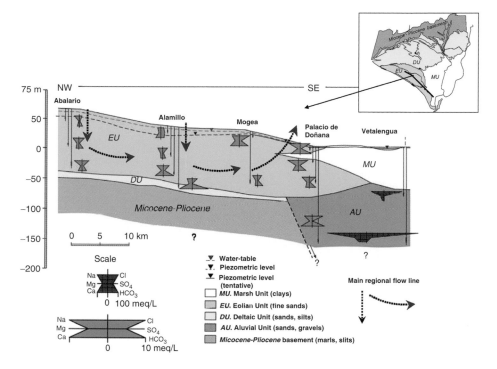

Plate 4

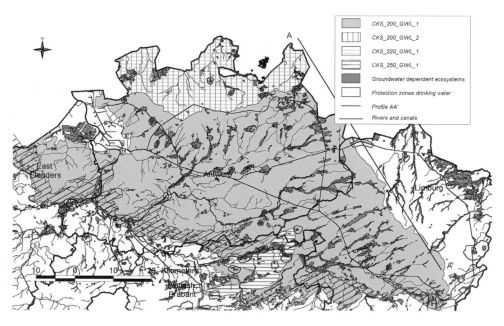

Plate 5

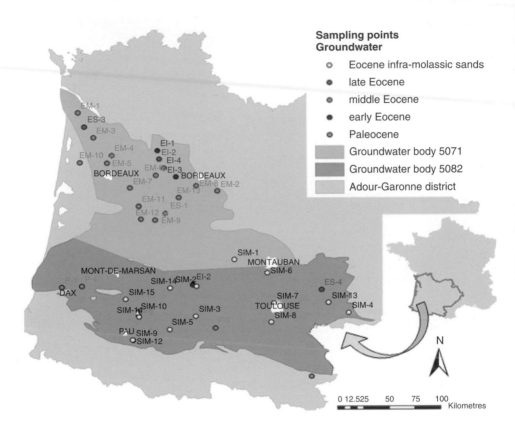

Sampling points
Groundwater

- Eocene infra-molassic sands
- late Eocene
- middle Eocene
- early Eocene
- Paleocene

Groundwater body 5071
Groundwater body 5082
Adour-Garonne district

Plate 6

- a 'median trend' approach in which the data for each site are used to determine an individual trend, and then the median of these individual trends is used to provide a trend for the groundwater body – time trend and then spatial average. This is similar to the approach described in Section 4.2.4.

The former method is that proposed by Grath *et al.* (2001). This works well if the sampling is indeed very regular and there are few missing data points. However, where there is both a large amount of systematic variation within the groundwater body and many missing data, the average of averages approach becomes less robust, as sites are counted in and counted out for different time intervals. The median trend approach is less susceptible to outliers and missing data. Moreover, as the trend at the individual site is in any case needed for DWPAs and, for nitrate, also for the Nitrates Directive and probably for other purposes, it may make more sense to take this approach. In addition, trends for individual abstraction sites may respond to both the regional and local factors outlined above, about which there may be considerable knowledge and information. There is also likely to be detailed construction and operational detail about the site, and the trend information provided may greatly assist both the regulatory agency and the water supply operator in managing groundwater quality, before being incorporated into a broader assessment at groundwater body level.

Where groundwater bodies are of substantial size and there are considerable numbers of monitored abstraction boreholes, differing concentrations and trends may be observed. These may vary systematically across the body (Figure 4.2.10, Stuart *et al.*, 2007), in a broadly similar way to the relationship with depth illustrated in Figure 4.2.5. This ground-water body in the north-east of England comprises part of a productive Permian limestone aquifer, dipping from west to east beneath younger confining strata (Figure 4.2.11). The outcrop receives recent recharge from relatively nitrate-rich infiltration from agricultural land. As the groundwater move eastwards along flowpaths down the dip of the aquifer (Figure 4.2.11), nitrate concentrations decrease, either because the water is older recharge with less nitrate, or some of the nitrate is removed by denitrification in changing redox conditions beneath the confining strata. Evidence of chemical denitrification in the hydro-geological setting shown in Figure 4.2.11 has been widely detected in Chalk, sandstone and limestone aquifers in the UK.

Trends in nitrate concentration also vary systematically across this groundwater body (Figure 4.2.10). Thus the highest nitrate concentrations at the outcrop tend to be decreasing, which is likely to reflect the beneficial impact of agricultural control measures, as was inferred in the example shown in Figure 4.2.5. Further down the groundwater flowpath, nitrate concentrations are still increasing towards 50 mg/l – a statistically and environmentally significant trend. The lowest concentrations are also increasing (Figure 4.2.10) but, although clearly statistically significant these trends are not yet really significant environmentally. It should be noted that the reversing directions of trends along a flow path are conceptually similar to the trend direction reversal shown in Section 4.2.4, Example 1, which deals with observation wells screened at multiple depths at the groundwater body scale.

Applying both of the aggregation approaches described above, the average of averages (Grath *et al.*, 2001) suggests an almost imperceptible downward trend of 0.08 mg/l/a (Stuart *et al.*, 2007) and the median trend approach an upward trend of 0.08 mg/l/a. Although these are different directions of change, the magnitude is small

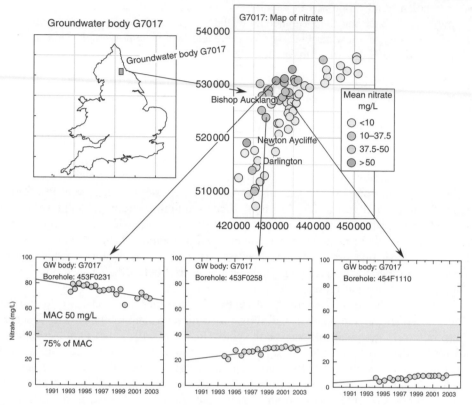

Figure 4.2.10   Variation in mean nitrate concentration in a groundwater body, and differing trends across the groundwater body. Reproduced from the Quarterly Journal of Engineering Geology and Hydrogeology, by permission of The Geological Society Publishing House, vol 40 pp 361–376. (See Plate 3 for a colour representation).

and both suggest there is very little overall trend within the groundwater body. Of the two, the median trend approach provides a better indication of how the situation varies across the groundwater body, and where groundwater quality management and pollution control measures should most effectively be targeted. These findings of course raise the question as to whether it would be more appropriate under the Water Framework Directive to take account of such major differences in chemical quality status by sub-dividing the groundwater body. The soundness of hydrogeological definition of the groundwater body and the integrity of the groundwater flowpath, however, suggest that as a management unit it should remain as it is.

4.2.6   CONCLUSIONS

The trend analysis results presented in this chapter show that it is feasible to detect trends and demonstrate trend reversal both at the individual abstraction site and groundwater body scale, and to assess the corresponding level of confidence. The results show

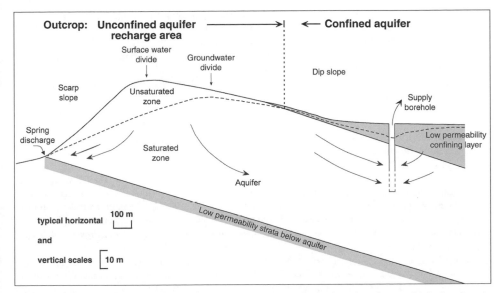

Figure 4.2.11 Simplified conceptual sketch of hydrogeological setting from which the data in Figure 4.2.10 are taken.

that trend detection is preferably tuned to pressures to the groundwater system, to the monitoring set-up and to the hydrological and chemical properties of the system. It also illustrates how groundwater age dating can improve trend detection.

REFERENCES

Batlle Aguilar J., Orban P., Dassargues A. and Brouyère S., 2007. Identification of groundwater quality trends in a chalk aquifer threatened by intensive agriculture in Belgium. *Hydrogeology Journal*, **15**(8): 1615.

Beeson S. and Cook M., 2004. Nitrate in groundwater: a water company perspective. *Quarterly Journal of Engineering Geology and Hydrogeology*, **37**, 261–70.

Broers H.P., 2002. Strategies for regional groundwater quality monitoring. Netherlands Geographical Studies no. 306, Ph.D. Thesis University of Utrecht, the Netherlands.

Broers, H.P. and van der Grift, B., 2004. Regional monitoring of temporal changes in groundwater quality. Journal of Hydrology, **296**(1–4): 192–220.

Cleveland W.S., 1979. Robust locally weighted regression and smoothing scatterplots. *J. Amer. Statist. Assoc.* **74**, 829–36.

EU, 2000. Directive 2000/60/EC of the European Parliament and of the Council of 23 October 2000 establishing a framework for Community action in the field of water policy, *Official Journal of the European Communities*, L327, 22.12.2000, p. 1.

EU, 2006. Directive 2006/118/EC on the Protection of Groundwater against Pollution and Deterioration.

EU, 2007a. Common Implementation Strategy for the Water Framework Directive(2000/60/EC). Guidance Document No. 15. Guidance on Groundwater Monitoring.

EU, 2007b. Common Implementation Strategy for the Water Framework Directive(2000/60/EC). Guidance Document No. 16. Guidance on Groundwater in Drinking Water Protected Areas. Technical Report – 2007 – 010.

EU, 2008. Common Implementation Strategy for the Water Framework Directive(2000/60/EC). Guidance on Groundwater Status and Trend Assessment – Final Draft 2.0, 15 October 2008, Working Group C – Groundwater, Activity WGC-2, 'Status Compliance & Trends'.

Fleming S.W., Lavenue A.M., Aly A.H. and Adams A., 2002. Practical applications of spectral analysis to hydrologic time series. *Hydrological Processes*, **16**, 565–74.

Frapporti G., 1994. Geochemical and statistical interpretation of the Dutch national groundwater quality monitoring network. University of Utrecht.

Gourcy, L. Dubus, I.G. Baran, N. Mouvet C. Gutierrez A. 2005. First investigations into the use of environmental tracers for age dating at the Brévilles experimental catchment (Ch. 4). In: *Report on Concentration-depth, Concentration-time and Time-depth Profiles in the Meuse Basin and the Brévilles Catchment* (Deliverable T2.3), ed. Broers H.P. Visser A. Utrecht, The Netherlands. (http://www.attempto-projects.de/aquaterra/21.0.html)

Grath J., Scheidleder A., Uhlig S., Weber K., Kralik M., Keimal T. and Gruber D., 2001. The EU Water Framework Directive: Statistical aspects of the identification of groundwater pollution trends, and aggregation of monitoring results. Final Report. No.41.046/01-IV1/00 and GZ 16 2500/2-I/6/00, Austrian Federal Ministry of Agriculture and Forestry, Environment and Water Management and European Commission, Vienna

Helsel D.R. and Hirsch R.M., 1992. Statistical methods in water resources, *Studies in Environmental Science* 49, Elsevier, Amsterdam.

Kinniburgh D.G., Chilton P.J. and Cooper D.M., 2004. Identification and reversal of trends in groundwater pollution: Part 2 – possible approaches and their implications with some trial assessments using existing monitoring data. British Geological Survey Commissioned Report, CR/04/207C.

Loftis J.C., 1996. Trends in groundwater quality. *Hydrological Processes*, **10**: 335–55.

Peters N.E. (ed), 1996. Trends in water quality. Special Issue. *Hydrological Processes*, 10.

Pinault J.L. and Dubus I.G., 2008. Stationary and non-stationary autoregressive processes with external inputs for predicting trends in water quality. *Journal of Contaminant Hydrology*, in press.

Quevauviller Ph. (2008) European regulatory framework of integrated groundwater management. Theory versus realities. In: *Proceedings EU Groundwater Policy Developments Conference, 13-15 November 2008, Unesco, Paris, France.*

Stuart M.E. and Kinniburgh D.G. 2005. Nitrate trends in groundwater. *British Geological Survey Internal Report*. IR/05/137R.

Stuart M.E., Chilton P.J., Kinniburgh D.G. and Cooper D.M., 2007. Screening for long-term trends in groundwater nitrate monitoring data. *Quarterly Journal of Engineering Geology and Hydrogeology*, **40**, 361–76.

UKWIR, 2003. Implications of changing groundwater quality for water resources and the UK water industry. Phase 2: Trend detection methodology and improved monitoring and assessment programmes: Main report. UK Water Industry Research Report 03/WR/09/06.

UKWIR, 2004. Implications of changing groundwater quality for water resources and the UK water industry. Phase 3: Best practice guidelines for investigating processes controlling groundwater quality. UK Water Industry Research Report 04/WR/09/07.

Visser A., Broers H.P., Van der Grift B. and Bierkens M.F.P., 2007. Demonstrating Trend Reversal of Groundwater Quality in Relation to Time of Recharge determined by 3H/3He. *Environmental Pollution*, **148**(3): 797–807.

Visser A., Broers H.P., Dubus I.G., *et al.*, 2008. Comparison of methods for the detection and extrapolation of trends in groundwater quality, submitted to *Journal of Environmental Monitoring*.

Yue S. and Wang C.Y., 2002. The influence of serial correlation on the Mann-Whitney test for detecting a shift in median. *Advances in Water Resources*, **25**, 325–33.

Part 5
Case Studies for Groundwater Assessment and Monitoring in The Light of EU Legislation

5.1

Groundwater Monitoring in Denmark and the Odense Pilot River Basin in Relation to EU Legislation[1]

Klaus Hinsby and Lisbeth Flindt Jørgensen

Geological Survey of Denmark and Greenland, Copenhagen K, Denmark

5.1.1   INTRODUCTION

The EU Water Framework and Groundwater directives stipulates that groundwater status assessments have to be based on environmental standards and objectives for actual or potential groundwater legitimate uses (e.g. drinking water) as well as associated aquatic and dependent terrestrial ecosystems. The groundwater status assessments requires a considerable amount of monitoring data, which often are not available, especially when based

---

[1] The views expressed in this chapter are purely those of the authors.

Groundwater Monitoring Edited by Philippe Quevauviller, Anne-Marie Fouillac, Johannes Grath and Rob Ward
© 2009 John Wiley & Sons, Ltd

on environmental objectives for and interaction with ecosystems. This chapter describes the general design and principles of the Danish groundwater monitoring programmes, and briefly discusses these in relation to the requirements of the EU directives and guidelines, while showing examples of monitoring results for selected priority substances in groundwater (nitrate, total P, As, selected pesticides etc.). Examples of nitrate data for the unsaturated zone, drainage tiles and streams are also shown. Compiled national data from all sand and gravel groundwater bodies are used to illustrate the Danish groundwater status based on groundwater legitimate uses, in this case by comparing average values from monitoring points with EU drinking water standards. Data and studies from the Odense Pilot River Basin are then used to illustrate and discuss the more detailed and extensive data requirements for the groundwater status assessment based on environmental objectives for an associated aquatic ecosystem, the Odense Fjord Estuary (Hinsby *et al*., 2008).

Examples of results from groundwater monitoring wells located in sand and gravel aquifers in the Odense Pilot River Basin is compared to the national average from all groundwater monitoring wells in Danish sand and gravel aquifers and to average results from monitoring in agricultural catchments for selected priority substances of the EU Groundwater Directive. These priority substances have to be considered for groundwater chemical status assessment and derivation of groundwater threshold values in EU member states.

5.1.2   HISTORY OF THE DANISH GROUNDWATER QUALITY MONITORING PROGRAMME

The water supply in Denmark has been based mainly on groundwater resources for many decades (at present ∼99 %). Hence the location of the subsurface water resources and the evolution of the quantity and quality of these have been of high priority at national and regional policy levels for many years. The first Danish Water Supply Law was adopted in 1926 and stipulated that all new water abstraction wells had to be reported to the Geological Survey of Denmark and include a description of the collected data. The current national groundwater monitoring database at the Geological Survey of Denmark and Greenland ('Jupiter') contains specific well information from more than 240,000 wells on technical development, geology, pump tests and groundwater chemistry etc. that dates back to the late nineteenth century. The database has an increasing annual number of groundwater chemistry entries from a few in the 1890s to presently more than 10,000 analysis packages from monitoring wells, and nearly twice this number of analysis packages on drinking water, i.e. mixed groundwater, which has received only simple treatment (filtration and aeration) before leaving the water works. Today each analysis package includes ∼100 compulsory parameters analysed on groundwater samples from specific wells or from mixed water at the water works.

Instrumentation of dedicated groundwater and agricultural catchment monitoring sites ('GRUMO' and 'LOOP', respectively) were initiated by the Geological Survey of Denmark and the National Environmental Research Institute in 1988.

National Danish groundwater monitoring comprises three different monitoring activities ('GRUMO', 'LOOP' and 'PLAP'), which are briefly described in the following. The location of the monitoring sites is shown in Figure 5.1.1.

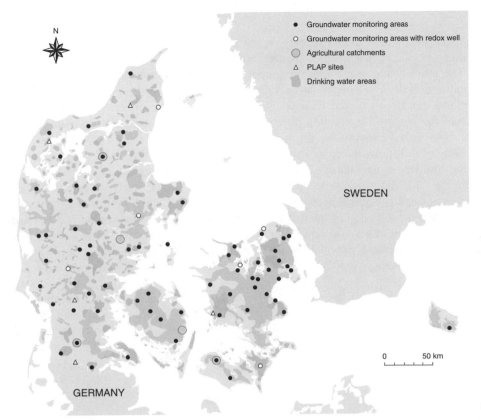

Figure 5.1.1 Location of national groundwater monitoring sites in Denmark (modified from Jørgensen and Stockmarr, 2009).

The GRUMO and LOOP monitoring programmes are two of more monitoring programmes on the aquatic environment initiated as a result of a political agreement from 1987 on the 'First Action Plan for the Aquatic Environment', a plan that has been revised twice since in the second (1998) and third (2004) action plan. The aim of the first plan was to reduce N- and P-losses to the aquatic environment by 50% and 80%, respectively. The action plan entailed a number of measures such as restriction on the use of nutrients in agricultural practice and more efficient waste water treatment before discharge into streams and coastal areas to improve the status of the aquatic environment. The third action plan considers the requirements of the EU Water Framework Directive and the EU Habitat Directive and stipulates that N- and P-leaching from agriculture must be reduced by a further 13% and 50%, respectively, by 2015. The monitoring programmes were established to monitor whether the measures have the expected effect on the status of the aquatic environment. Kronvang *et al.* (2008) evaluates the effects of the action plans and policy measures on nitrate pollution and concentration trends for the period 1989–2004 in Danish rivers in 86 catchments, and compares the nitrate concentration trends with trends in rivers in 16 other European countries. In Sections 5.1.4 and 5.1.5 we show some selected results to illustrate the effects of the policy measures implemented in Denmark.

GRUMO

The Danish Groundwater monitoring programme (GRUMO) dates back to the late 1980s (Czako, 1994; Jørgensen and Stockmarr, 2009)

The groundwater monitoring programme including new dedicated specially installed monitoring wells was operational in 1989 and the first national reporting of results was published in 1990. At that time 66 monitoring areas were established spread evenly around the country representing different types of geology and land use. Almost 1000 wells (screens) were available for sampling.

Figure 5.1.2 shows the annual number of analysis packages from water works reported to the 'drinking water' database at GEUS and from specific dedicated monitoring wells from the GRUMO programme for the period 1970–2007. It clearly illustrates the large increase of hydrochemical analyses reported to the groundwater monitoring database at GEUS after the initiation of the first action plan for the aquatic environment in Denmark and the installation of a new network of dedicated groundwater monitoring wells in 1988. The observed increase around 1980 in the number of drinking water analyses is a result of a new water supply legislation with more stringent rules on compulsory analyses and reporting.

As the main focus of the action plan in the first years was to reduce the emission of nutrients, analyses were focussed on nitrate and phosphorous and some main chemical elements such as chloride, fluoride, sodium, sulphate, iron and pH to characterise the spatial and temporal variation of some important groundwater constituents in the aquifers

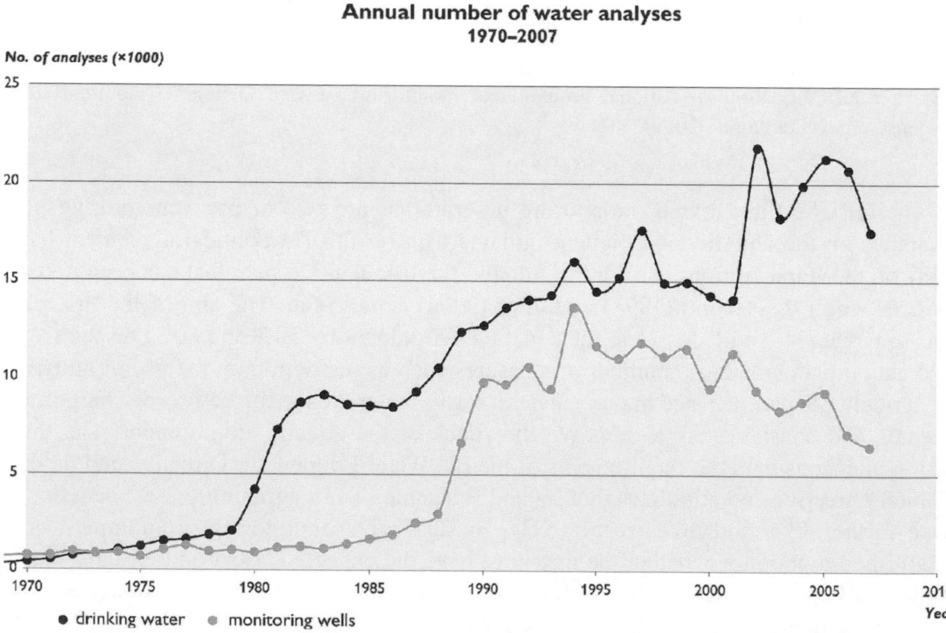

Figure 5.1.2   Annual number of analysis packages reported to the Danish groundwater monitoring database for the period 1970–2007. Each analysis package may contain analyses of up to more than 100 different chemical substances.

in each area. However, soon other parameters of importance for drinking water quality were included such as inorganic trace elements, organic micro-pollutants and pesticides.

Gradually more areas and more screens were included in the programme; old wells were taken out and replaced by new more adequate ones. Further the programme has been extended with five new multi-level monitoring wells in different type areas, 'redox wells', with down to few cm distance between screens. These wells were installed to follow the progression of the redox front and the evolution of the hydrochemistry around the redox front, which was estimated and investigated in an earlier study (Postma *et al.*, 1991).

In the period from 1998–2003 the analytical programme were quite comprehensive both in terms of parameters and annual analyses, see Table 5.1.1.

Today there are 73 groundwater monitoring areas in Denmark with a total of almost 1500 screens. The number of parameters as well as the frequencies has decreased in recent years due to reductions in the programme budget (Table 5.1.1, Figure 5.1.2). With the latest revisions the programme has gradually been adjusted to comply with the objectives of the European Water Framework Directive. This has resulted in about 330 new wells to monitor the shallow upper groundwater, and new sampling frequencies. Previously the strategy was to analyse young groundwater (CFC age younger than 1950) more often than older groundwater, but currently the frequency is determined based on the groundwater quality in each individual well; that is the higher ('closer to a environmental standard' or threshold value) the concentration of a given parameter the more often analyses will be carried out.

LOOP

Besides from the 73 groundwater monitoring areas shallow groundwater quality is monitored in five agricultural catchments (the LOOP monitoring programme) with a total

Table 5.1.1 Number of analysed parameters and sampling frequencies in the Danish groundwater monitoring programme during the period 1998–2009. Y = years.

Group	1998–2003		2004–2006		2007–2009	
	No. of parameters	Frequency	No. of parameters	Frequency	No. of parameters	Frequency
Main chemical elements	27	1 per 1–2 y	26	1 per 1–6 y	20	1 per 1–6 y
Inorganic trace elements	23	1 per 1–6 y	14 (8)[a]	1 per 2 y	8	1 per 1–2 y
Organic micro pollutants	24	1 per 1–3 y	22[b]	1 per 3–6 y	18	1 per 1–6 y
Pesticides and metabolites	45	1 per 1–6 y	34[b]	1 per y	24	1 per 1–3 y

[a] 14 parameters in young groundwater, 8 in old groundwater, - young groundwater is defined as groundwater with a CFC age younger than 1950 (e.g. Hinsby *et al.*, 2001; Hinsby *et al.*, 2008).
[b] Only analysed in young groundwater.

of about 100 screens in the uppermost groundwater, $1\frac{1}{2}$ to 5 m below surface. In this monitoring programme the impact from conventional agricultural practice is followed in the unsaturated zone, drainage tiles, monitoring wells in shallow groundwater, and streams (Rasmussen, 1996; Grant, 2007).

PLAP

A supplementary monitoring programme focusing on pesticide leaching research, The Danish Pesticide Leaching Assessment Programme (PLAP), regularly gives input to the list of pesticides and metabolites in both the groundwater monitoring programme and the compulsory groundwater control by the water supply companies. The PLAP encompasses five test sites, two sandy and three clayey, established during 1998–2000 (Kjær, 2008). The sites are representative of the dominant soil types and climatic conditions in Denmark. Each site is instrumented with both horizontal and vertical wells supplemented by drainage systems and suction cups for soil water sampling. Cultivation of the PLAP sites is in line with conventional agricultural practice in the vicinity and the pesticides in the programme are applied in the maximum permitted dosage and in the manner specified in the regulations. The pesticides included in the PLAP are selected by the Danish Environmental Protection Agency based on expert judgements. At present, 36 pesticides and several of their degradation products are included in the PLAP (Kjær, 2008).

MONITORING OF WATER SUPPLY WELLS

Finally, results from groundwater control carried out by the Danish water supply companies also add to the picture of the groundwater quality in Denmark. By licensing the water companies are obliged to carry out compulsory groundwater control analyses in each individual well every 3rd to 5th year depending on the amount of groundwater abstracted. The list of parameters is similar to that of the GRUMO programme.

5.1.3   PRINCIPLES OF DANISH GROUNDWATER MONITORING SITES

The principles of the GRUMO monitoring areas are illustrated in Figure 5.1.3. The monitoring areas are selected to represent all regions and main aquifer types in Denmark (primarily unconsolidated sand and gravel, chalk and limestone). The groundwater monitoring areas typically consists of 15–25 monitoring points located in the caption zone of a large volume monitoring water supply well. Figure 5.1.4 illustrates the monitoring of shallow groundwater and the unsaturated zone in the agricultural catchments monitoring areas (LOOP). Besides samples and analyses from suction cups in the unsaturated zone and shallow groundwater monitoring wells (1.5–5 m below surface), analyses on shallow groundwater from tile drainage systems (~1 m below surface) and streams are also available.

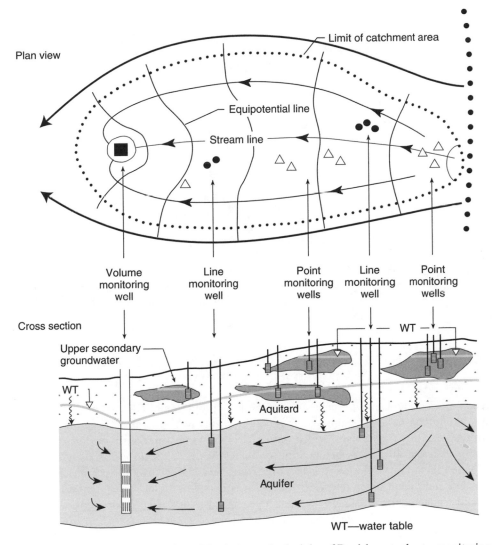

Figure 5.1.3 Schematic illustration of the design and principles of Danish groundwater monitoring sites in the 'GRUMO' programme (modified from Jørgensen and Stockmarr, 2009).

The EU Water Framework and Groundwater directives stipulate that member states have to assess groundwater chemical status for the protection of human health and the environment based on environmental standards for groundwater legitimate uses and associated aquatic and dependent terrestrial ecosystems. In Sections 5.1.4 and 5.1.5 we briefly describe and discuss the Danish monitoring programme in relation to groundwater legitimate uses and associated aquatic ecosystems, respectively, according to the EU Water Framework and Groundwater directives.

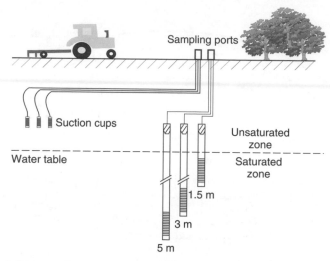

Figure 5.1.4 Schematic illustration of subsurface monitoring points in the Danish 'LOOP' programme. Besides monitoring points (piezometers and suction cups) drainage tiles located at around 1 m below surface in clayey subsoils are also monitored (Rasmussen, 1996, Grant, 2007), Figure 5.1.6.

5.1.4   GROUNDWATER MONITORING FOR PROTECTION OF GROUNDWATER LEGITIMATE USES

The design of the Danish groundwater monitoring programme (GRUMO) combined with the monitoring in water supply wells generally seems well suited for assessing threats to groundwater legitimate uses, which primarily is drinking water abstracted at an average depth of about 40 meters below surface. Tables 5.1.2, 5.1.3 and 5.1.4 show results of selected important pollution indicators measured in the GRUMO programme in screens located in different depth intervals. Nitrate, arsenic and pesticides are the most frequently found contaminants/substances in Danish groundwater that exceeds drinking water standards (Thorling, 2007), see Figure 5.1.5 and Tables 5.1.2 to 5.1.4.

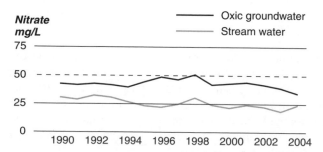

Figure 5.1.5   Annual mean concentrations of nitrate in oxic groundwater (Thorling, 2007) and streams (flow-weighed average of streams in 86 catchments) in Denmark. (Kronvang *et al.*, 2008).

Table 5.1.2 Mean values and 80th percentiles[a] of selected main chemical elements and pollution indicators from Danish groundwater monitoring wells in sand and gravel aquifers, 1998–2007. All concentrations are in mg/L. Underlined bold figures indicate values above EU standards – bold figures indicate values above Danish standards.

Depth	N[b]	NO_3	NO_3(p80)	SO_4	SO_4(p80)	$P_{tot} \times 10^{-2}$	$P_{tot} \times 10^{-2}$ (p80)	Cl	Cl (p80)
0–10	278	23	45	51	79	6.2	10	46	57
10–20	413	29	45	65	113	6.5	12	54	67
20–30	254	29	**<u>57</u>**	60	90	6.6	9.7	40	42
30–40	131	15	<u>35</u>	58	86	9.8	15	38	48
40–50	83	6.3	0.69	41	64	15	**20[d]**	49	40
50–60	33	2.5	0.33	29	38	13	**16[d]**	31	43
60–70	27	1.7	1.6	27	36	14	**20[d]**	30	34
70–80	4	0.28	0.38	46	63	8.2	9	28	33
80–90	7	0.11	0.15	16	19	15	**19[d]**	20	22
EU/DK DWS[c]		50/50		250/250		–/15		250/250	

[a] p80 are used since the EU guideline stipulate that groundwater bodies with >20% (by gwb volume or number of monitoring wells) above standards or threshold values does not comply with the good status objectives of the WFD (European Commission, 2009).

[b] n is number of monitoring points with top of screen in defined depth interval.

[c] EU and DK drinking water standards.

[d] The high P values have a natural source and hence does not result in poor status – further, a major part of the dissolved and particulate P is retained in the filter sands at the water works.

Table 5.1.3 Mean values and 80th percentiles of some of the most frequently found pesticides and metabolites in Danish sand and gravel aquifers. All concentrations are µg/L. Bold figures indicate values above EU and DK standards.

Depth	n[a]	BAM[b]	p80 BAM	$\sum$ Triazines[c]	p80 $\sum$ Triazines
0–10	278	**0.11**	**0.12**	0.054	0.10
10–20	413	**0.32**	**0.61**	0.10	0.075
20–30	254	**0.14**	**0.23**	0.035	0.039
30–40	131	0.043	0.05	0.060	0.083
40–50	83	0.072	0.09	0.015	0.017
50–60	33	0.020	0.02	0.048	0.048
60–70	27	0.032	0.042	0.057	0.057
70–80	4	0.012	0.012	-	-
80–90	7		-	-	-
EU/DK DWS[e]		0.1/0.1		_[d]	-

[a]n is number of monitoring points with top of screen in defined depth interval.
[b]= 2,6 Dichlorbenzamid (metabolite of Dichlobenil).
[c](atrazine + dieethyl-atrazine+deisopropyl-atrazine).
[d]The standard for sum of all pesticides is 0.5.
[e]EU/Danish drinking water standard.

From Table 5.1.2 it is apparent that nitrate values in Danish groundwater is quite high and e.g. that more than 20% of the monitoring screens in the interval between 20 and 30 meters exceeds the EU (and DK) drinking water standard of 50 mg/l. Table 5.1.3 show that the number of wells with pesticide metabolites exceeding drinking water standards are even higher. Hence nitrate and pesticides clearly is a large threat to the Danish drinking water resources, and they are the main reason why a very large part of the shallow Danish groundwater (0–30 m below surface) does not comply with the good status objectives of the Water Framework and Groundwater directives. Trace elements such as As and Ni (Table 5.1.4) may locally also exhibit concentrations above the drinking water standards, but these elements almost always originate from natural sources in the subsurface, and they have a more local impact on the groundwater quality than diffuse pollution. A sound knowledge of the natural background levels in various types of aquifers is however important, when the sources of trace elements in groundwater needs to be identified (Hinsby and Rasmussen, 2008; Wendland *et al.*, 2008).

By comparing the GRUMO monitoring results with results from the LOOP monitoring programme (Figure 5.1.6) it becomes obvious that the GRUMO monitoring in the upper groundwater (0–10 m below surface) does not reflect the important variation in nitrate concentrations within this zone. This has important consequences for the assessment of groundwater status based on environmental standards and/or objectives for associated aquatic and dependent terrestrial ecosystems as illustrated in the next section.

Table 5.1.4 Mean values and 80th percentiles of selected trace elements and pollution indicators from Danish groundwater monitoring wells in sand and gravel aquifers. All concentrations are µg/L. Underlined bold figures indicate values above EU standards- bold figures indicate values above Danish standards.

Depth	n[a]	As	As (p80)	Ni	Ni (p80)	Pb	Pb (p80)	Hg x10^{-3}	Hg x10^{-3} (p80)
0–10	278	1.5	1.9	9.1	9.8	0.37	0.22	2.7	2.5
10–20	413	1.6	2.6	5.3	3.6	0.28	0.25	1.6	1.8
20–30	254	2.6	2.5	2.8	2.8	0.23	0.17	1.4	2.1
30–40	131	1.9	2.0	2.1	2.2	0.15	0.18	2.8	1.7
40–50	83	3.5	5.3	1.1	1.4	0.12	0.23	1.8	2.7
50–60	33	5.0	**<u>12</u>**	0.67	0.93	0.15	0.12	0.9	1.2
60–70	27	3.7	**<u>5.6</u>**	1.6	0.80	0.09	0.1	0.8	0.9
70–80	4	3.9	4.4	0.11	0.11	0.23	0.31	0.6	0.86
80–90	7	1.8	2.6	2.8	4.9	0.13	0.21	0.1	0.1
EU/DK DWS[b]		10/5		20/20		10/5		1000/1000	

[a] n is number of monitoring points with top of screen in defined depth interval.
[b] EU and DK drinking water standards

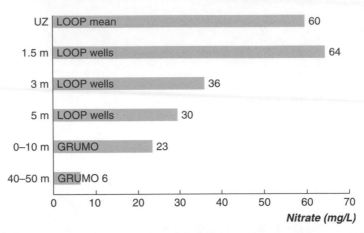

Figure 5.1.6 Mean nitrate concentrations at different depths and depth intervals. Data are compiled from LOOP (Grant, 2007) and GRUMO (Thorling, 2007) monitoring databases. In total there are about 100 LOOP monitoring wells (piezometers) at the depths of approximately 1.5, 3 and 5 meters below the surface. The number of GRUMO wells in sand and gravel aquifers in the intervals 0–10 and 40–50 m below surface are 278 and 83, respectively (Table 5.1.2).

5.1.5 GROUNDWATER MONITORING FOR PROTECTION OF ASSOCIATED AQUATIC ECOSYSTEMS – THE CASE OF THE ODENSE PILOT RIVER BASIN (OPRB)

Unfortunately, no LOOP monitoring sites or similar detailed monitoring of shallow groundwater exist in the OPRB, and only very few monitoring wells exist in the upper oxic groundwater. This is a problem when e.g. the sources and source areas for the nitrogen loads to the Odense river and ultimately the Odense Fjord Estuary have to be identified. Therefore, it is necessary to use average data from similar settings outside the OPRB in order to understand the relative importance of nitrogen loads via different flow paths. Here we assume that average nitrate concentrations in the uppermost oxic groundwater in the OPRB are similar to or slightly less than values from the LOOP programme (Figure 5.1.6). This seems reasonable as the average stream flow weighted nitrate concentrations in Odense river is close to the national average in streams in Denmark (~25 mg/L nitrate, Figure 5.1.5), and as the attenuation factor between the root zone and Odense river has been estimated to approximately 0.5 in several studies (e.g. Hinsby *et al.*, 2008).

Figure 5.1.6 shows that annual mean values of nitrate concentrations in drainage tiles in LOOP monitoring sites with clayey subsoils generally compares quite well to mean values of monitoring points (suction cups) in the unsaturated zone and of the shallow groundwater monitoring wells at 1.5 meter below the surface. They all have about 60 mg/L nitrate. This is to be expected as the drainage tiles generally are located at depths around 1 m in between the suction cups and the shallow monitoring wells (Figure 5.1.6). These nitrate levels result in a N transport in drains directly to streams of around half of the total N leaching from the root zone in catchments with clayey soils (Grant, 2007). The runoff in drains accounts for about 61% and 52% of the runoff in streams in the two clayey LOOP monitoring sites. Hence the major part of the remaining discharge to the streams is from anoxic groundwater where nitrate has been reduced, e.g. by iron sulphides in the groundwater bodies (Postma *et al.*, 1991; Grant, 2007; Hinsby *et al.*, 2008). These numbers compares fairly well to what was found in studies for the Odense Pilot River Basin (Hinsby *et al.*, 2008), indicating that the situation in the OPRB is comparable to the situation in the two clayey LOOP monitoring sites.

The data in Table 5.1.5 clearly demonstrate that the monitoring of the shallow groundwater quality ($\sim$<10 m below surface) in the OPRB is insufficient to assess the evolution of shallow groundwater quality and estimate the groundwater pollutant loads to associated aquatic and terrestrial ecosystems in this area. When comparing data in Table 5.1.5 with data in Table 5.1.2 and Figure 5.1.5 and 5.1.6, it is clearly seen that the average nitrate value measured in the upper 0–10 m significantly underestimate the probable nitrate concentration in the uppermost oxic groundwater. The reason for this is that the redox boundary is located quite shallow in the OPRB (on average $\sim$3 m below surface), and hence that nitrate has been reduced (degraded) by iron sulphides in shallow glacial sediments (Hinsby *et al.*, 2008). There are practically no monitoring wells in the upper oxic and nitrate containing zone in the OPRB, however, data from the Odense river compared to other Danish streams (Hinsby *et al.*, 2008; Larsen *et al.*, 2008; Kronvang *et al.*, 2008) indicate that nitrate concentrations in shallow oxic groundwater in the OPRB probably are quite similar to average concentrations in upper shallow oxic groundwater in Denmark), and that drainage tiles are responsible for a major part of the nitrate load to the Odense river. Data indicate that about half of the N leakage from agricultural soils in the OPRB is transported directly to the Odense river by drains, as was the case too for the clay sites in the LOOP programme. The other half is diluted or degraded (attenuated) primarily below the drainage tiles in the OPRB, hence a combined dilution and attenuation factor of 0.5 has been suggested for the OPRB (Hinsby *et al.*, 2008).

Table 5.1.5 Mean values of selected main chemical elements and pollution indicators from OPRB groundwater monitoring wells (GRUMO), 1998–2007. Each parameter has been analysed between 1 and 36 times in the different screens during the period. All concentrations are in mg/L. Bold figures indicate values above derived groundwater threshold values.

Depth	n^a	O_2	NO_3	SO_4	Ptot	Cl
0–10	7	0,94	17	99	**0,085**	67
10–20	10	0,79	6,0	91	0,061	24
20–30	8	0,75	0,72	84	0,070	56
30–40	4	0,88	0,17	69	0,042	55
40–50	10	0,70	0,06	60	0,068	21
50–60	3	0,67	0,06	38	0,060	19
EU/DK DWS[b]			50/50	250/250	-/0.15	250/250
TV			18[c]		0,08[c]	

[a] n is number of monitoring points with top of screen in defined depth interval,
[b] EU and Danish drinking water standards (EU/DK DWS).
[c] Groundwater threshold value derived in Hinsby *et al.* (2008).

5.1.6   DISCUSSION

The groundwater monitoring in the OPRB and in Denmark in general, is comprehensive, and includes regularly monitoring of more than 10,000 wells nationally, of which currently about 1500 are dedicated monitoring wells in the GRUMO programme (Jørgensen and Stockmarr, 2009). For the OPRB the total numbers of wells and dedicated monitoring wells are approximately 200 and 50, respectively. The existing groundwater monitoring network provides quite detailed quality trend assessments of relatively deep groundwater used for drinking water purposes, which on average is abstracted at depths of about 40 m below the surface. However, the shallow oxic groundwater, which is the most polluted and in general affects the groundwater dependent ecosystems the most, is closely monitored in just about 100 dedicated shallow monitoring wells, nationally. These wells are all installed 1–5 m below the water table in the five different LOOP monitoring sites below agricultural areas, and most of them are located in the important oxic groundwater zone. For the OPRB no such wells exist and only 7 GRUMO wells are located within the upper 10 meters below the surface (Table 5.1.5). The monitoring programmes in Denmark generally provide extensive data for a general national overview of the chemical status. However, the data on chemical composition of shallow groundwater and the knowledge about physical and chemical interaction between groundwater and surface water is often insufficient when chemical status assessment is performed for protection of specific ecosystems (e.g. Natura 2000 areas) or ecosystems in regional river basins such as the OPRB. The Odense PRB case clearly demonstrate that the present groundwater monitoring programme is insufficient for evaluation of the pressures on and the protection of e.g. specific EU Natura 2000 terrestrial and marine habitats and the coastal and transitional waters outlined in the Water Framework Directive (Hinsby *et al.*, 2008). Many of these local and regional systems do not have sufficient chemical data on the shallow oxic groundwater types, which generally affect the associated aquatic and dependent terrestrial ecosystems the most. Further, the groundwater monitoring is inadequate for the identification and planning of relevant and cost-efficient measures for restoration of good chemical and ecological status in specific ecosystems. For this task it is necessary to integrate the design of the groundwater and surface water monitoring networks based on available spatially distributed data on groundwater and surface water chemical status as well as land use and geology e.g. guided by a typology for groundwater/surface water interaction (Dahl and Hinsby, 2008; Dahl *et al.*, 2007). Further, integrated hydrological modelling provides valuable information for optimised design of monitoring programmes (Hojberg *et al.*, 2007; Jorgensen *et al.*, 2007).

5.1.7   CONCLUSIONS, PERSPECTIVES

The data from the Danish groundwater monitoring programme clearly show that the groundwater in Denmark is highly affected by human activities with agriculture being the single most important and severe pollution source. Generally, shallow oxic groundwater bodies below agricultural areas do no comply with good status objectives of the EU directives.

The Danish groundwater monitoring programme is extensive and provides a wide range of important data for groundwater quality and quantity status assessments. However, the groundwater monitoring programme needs to be continuously revised and developed to face new requirements and challenges with respect to e.g. new emerging pollutants, optimised location of monitoring points, climate change impacts, and better integration of groundwater and surface water quality and quantity monitoring programmes. The current programme is quite well suited for protection of legitimate uses (primarily drinking water) as requested by the Water Framework and Groundwater directives. Still, it needs to be developed further in order to be able to perform ground-water status assessments based on good status objectives for the associated aquatic and dependent terrestrial ecosystems as also requested by the EU directives, i.e. more monitoring points need to be established in order to follow the impact on and plan the protection of ecosystems. Important issues which have to be considered more efficiently for the latter objectives include: (1) the biogeochemical environments in groundwater bodies and riparian areas; (2) the shallow groundwater quality (depth depends on size of the groundwater and surface water bodies); (3) quantitative and chemical groundwater/surface water interaction; (4) dilution, attenuation and flow paths from recharge to discharge areas; (5) spatial and seasonal variation of pollutant concentrations in shallow groundwater and surface water bodies (the more detailed information the better the possibility for detailed land use planning); (6) spatial and seasonal variation in water tables, discharge and pollution loads from different flow components and stream runoff; (7) identification of surface water quality standards and sustainable loads; and (8) finally, climate change impacts must also be considered and simulated in order to establish a scientific sound basis for future adaptive groundwater management and land use planning.

The good status objective for all water bodies in Denmark as well as in the rest of the EU in 2015 is an enormous challenge. This challenge will become even larger in the future considering that climate change scenarios indicate significant change in temperature and precipitation during the 21st century, with considerable geographical variation. These changes will significantly affect e.g. stream runoff and nutrient loads even within relatively small and weakly affected geographical areas such as Denmark. Therefore, research on optimised integrated monitoring and modelling of water quantity and quality, integrated characterisation and description of mainly the shallow groundwater bodies and their quantitative and chemical interaction with associated ecosystems is strongly needed. The research needs to be performed by groundwater and surface water research institutions in close cooperation with ecosystem scientists, economic scientists, agricultural associations and research institutes, and relevant authorities in order to plan sustainable land use and efficiently protect water resources, human and environmental health, including coastal and marine ecosystems and resources, and establish a sound basis for policy making.

REFERENCES

Czako T., 1994. Groundwater Monitoring Network in Denmark – example of results in the Nyborg Area, *Hydrological Sciences Journal-Journal des Sciences Hydrologiques*, **39**, 1–17.

Dahl M. and Hinsby K., 2008. GSI typology – Typology of groundwater/surface water interaction 2008. In: *Proceedings of the Conference 'EU Groundwater Policy Developments - Good Status Objectives and Integrated Management Planning', UNESCO Paris, 13–14 November 2008*.

Dahl M., Nilsson B., Langhoff J. H. and Refsgaard J. C., 2007. Review of classification systems and new multi-scale typology of groundwater–surface water interaction, *Journal of Hydrology*, **344**, 1–16.

European Commission, 2009: Guidance on Groundwater Status and Trend Assessment. Guidance Document No 18. Technical Report 2009-026. ISBN 978-92-79-11374-1. European Communities, Luxembourg.

Grant R. (ed.), 2007. Landovervågningsoplande 2006. Report no. 640, National Environmental Research Institute, Univ. Aarhus.

Grant R., Nielsen K. and Waagepetersen J., 2006. Reducing nitrogen loading of inland and marine waters - Evaluation of Danish policy measures to reduce nitrogen loss from farmland, *Ambio*, **35**, 117–23.

Hansen B. and Thorling L., 2008. Use of geochemistry in groundwater vulnerability mapping in Denmark, *Geological Survey of Denmark and Greenland Bulletin*, **15**, 45–8.

Hinsby K. and Rasmussen E. S., 2008. The Miocene Sand Aquifers, Jutland, Denmark. In W. M. Edmunds and P. Shand (eds), *Natural Groundwater Quality*, Wiley-Blackwell, London.

Hinsby K., Condesso de Melo M. T. and Dahl M., 2008. European case studies supporting the derivation of natural background levels and groundwater threshold values for the protection of dependent ecosystems and human health. Science of the Total Environment **401**, 1–20.

Hinsby K., Purtschert R. and Edmunds W. M., 2008. Groundwater age and quality. In P. Quevauviller (ed.), *Groundwater Science and Policy - An International Overview*. The Royal Society of Chemistry, RSC Publishing, London/Cambridge.

Hinsby K., Edmunds W. M., Loosli H. H, Manzano M., Melo M. T. C. and Barbecot F., 2001. The modern water interface: recognition, protection and development - Advance of modern waters in European coastal aquifer systems. In: Edmunds and Milne (eds), *Palaeowaters in Coastal Europe: Evolution of Groundwater since the Late Pleistocene*. Geol. Soc., London,Spec. Publ., **189**, 271.

Jørgensen L. F. and Stockmarr J., 2009. Groundwater monitoring in Denmark – review, status and perspectives, *Hydrogeology Journal*, **17**, 827–842.

Jørgensen L. F., Refsgaard J. C. and Højberg A. L., 2007. Joint use of monitoring and modelling, *Water Science and Technology*, **56**, 21–29.

Kjær J. (ed.), 2008. The Danish Pesticide Leaching Assessment Programme. Monitoring results May 1999–June 2007. Report, Geological Survey of Denmark and Greenland, http://www.pesticidvarsling.dk.

Kronvang B., Andersen H. E., Borgesen C., *et al.*, 2008. Effects of policy measures implemented in Denmark on nitrogen pollution of the aquatic environment, *Environmental Science & Policy*, **11**, 144–52.

Larsen M. A. D., Søgaard H. and Hinsby K., 2008. Temporal trends in N & P concentrations and loads in relation to anthropogenic effects and discharge in Odense River 1964-2002, *Hydrology Research*, **39**, 41–54.

Postma D., Boesen C., Kristiansen H. and Larsen F., 1991. Nitrate reduction in an unconfined sandy aquifer – water chemistry, reduction processes, and geochemical modeling, *Water Resources Research*, **27**, 2027–45.

Rasmussen P., 1996. Monitoring shallow groundwater quality in agricultural watersheds in Denmark, *Environmental Geology*, **27**, 309–19.

Refsgaard J. C., Jørgensen L. F., Højberg A. L., Demetriou C., Onorati G. and Brandt G., 2008. Joint modelling and monitoring of aquatic ecosystems. In: Ph. Quevauviller, U. Borchers, K. Clive Thompson and T. Simonart (eds), *The Water Framework Directive – Ecological and Chemical Status Monitoring*. Wiley, Chichester.

Thorling L. ed., 2006. Groundwater – status and trends 1989–2006. Report, Geological Survey of Denmark and Greenland, www.grundvandsovervaagning.dk, in Danish (English summary).

Wendland F., Blum A., Coetsiers M., *et al.*, 2008. European aquifer typology: A practical framework for an overview of major groundwater composition at European scale, *Environmental Geology*, **55**, 77–85.

5.2
Upper Rhine

Ralf Kunkel[1], G. Berthold[2], Ariane Blum[3],
H.-G. Fritsche[2] and Frank Wendland[1]

[1] Research Centre Juelich, Institute of Chemistry and Dynamics of the Geosphere (ICG),
Institute IV: Agrosphere, Juelich, Germany
[2] Hessian Agency for Environment and Geology (HLUG), Wiesbaden, Germany
[3] Bureau de Recherches Géologiques et Minières, Orléans cédex, France

5.2.1   CASE STUDY AREA

The 'Upper Rhine basin', as described in the characterisation for the Water Framework Directive, belongs to the river basin district Rhine and includes the natural landscapes Upper Rhine Valley (which is in the centre of examination) as well as the low mountain ranges Vosges and Pfälzerwald (in the west) and the Black Forest and the Odenwald (in the east) up to their watersheds, which are oriented more or less south to north and an altitude up to 1500 m. The total area of the catchment area Upper Rhine is 21,700 km², 8200 km² of which is French, the rest German with a small part in Switzerland. For this case study, only the groundwater bodies of the Upper Rhine Valley, comprising a total area of 9290 km², are examined (see Figure 5.2.1).

Geologically, the Upper Rhine Valley is the result of the building of the Upper Rhine Rift structure since the Middle Eocene and its filling with sediments up to 4000 m thickness. The Pliocene and lower quaternary sediments of the Upper Rhine Valley consist of

Groundwater Monitoring Edited by Philippe Quevauviller, Anne-Marie Fouillac, Johannes Grath and Rob Ward
© 2009 John Wiley & Sons, Ltd

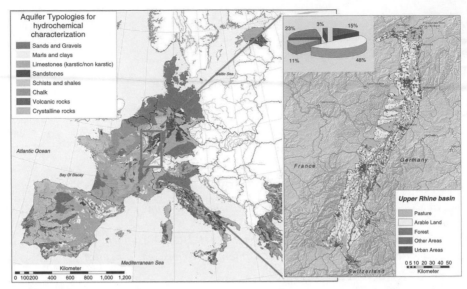

Figure 5.2.1 Location and land use distribution of the case study area 'Upper Rhine Valley' within the Aquifer typology scheme for hydrochemical characterisation (Wendland *et al*., 2008). Reproduced from Environmental Geology, F. Wendland, 55, 1, 2008, with kind permission from Springer.

alternating layers of silt, clay and sand. In the overlying strata, the quaternary continues with variations of gravely-sandy and clayey-sandy sediments, all building pore aquifers with up to 15% porosity. The floodplain is erosive and is embedded in the intermediate horizon. Due to the tectonic lateral tilting of the pre-quaternary basement the thickness of porous rocks increases towards the east. In general, the hydraulic conductivity decreases from south to north along the flow direction which is associated with a decrease in the grain size of the alluvial sediments. With regard to the aquifer typologies developed in the BRIDGE project for hydrochemical characterisation (Wendland *et al*., 2008) the Upper Rhine Valley belongs to the 'Fluviatile deposits of major streams'.

The Upper Rhine Valley is one of the most important drinking water reservoirs in Europe. At the same time it is a good example for anthropogenic impacts on groundwater quality and quantity due to intensive agriculture, industry, salt mining and water withdrawal. The hilly outliers, for example, are particularly suitable for viniculture and fruit growing. Woodlands occur predominantly in the adjacent grounds of the low mountain ranges. As a result of the manifold and intensive pressures on groundwater occurrence, the water quality and water quantity in the Upper Rhine Valley is thoroughly monitored by means of a lot of monitoring sites and special monitoring networks. For the purpose of reaching an integrative, transboundary management of groundwater, a common transboundary data base with regard to groundwater composition has been developed. This data base is used here for the derivation of natural background levels and threshold values.

5.2.2 DATA BASE

In the last years a lot of joint German-French-Suisse projects dealing with the implementation of the EU – Water Framework Directive have been carried out. In this framework BRGM (France) developed a joint groundwater quality data base, fed by different German, French and Suisse State authorities (BRGM Alsace, LUA Baden-Württemberg, LUA Rheinland-Pfalz, HLUG Hessen, Bâle Ville, Bâle Campagne). In total the data base consists of about 1700 monitoring points, of which 67 samples originating from Switzerland, 734 from France and 910 from Germany (535 from Baden-Württemberg, 209 from Rhineland-Palatinate, 166 from Hesse) from the years 2002 and 2003 (see Figure 5.2.2, left part). For each of the monitoring stations one sample containing the solution contents for integral and chemical environment parameters (el. conductivity, O_2, pH, temperature, DOC, TOC, hardness), characteristic major and minor parameters (B, Ba, Ca, Fe, Mg, Mn, Na, K, Cl, SO_4, HCO_3), and pollutants mentioned in the Water Framework Directive (As, NH_4, NO_3, NO_2, PO_4) was available. In total, natural background levels could be derived for 23 parameters.

Before evaluating the data, the whole database was subjected to several plausibility checks:

- samples with an error larger than 10% on the ionic balance were removed;

- data from hydrothermal aquifers were removed;

- data from salty aquifers were removed;

- time series were eliminated by median averaging.

For statistical purpose two-thirds of the detection limit was used for samples with analytical results below detection limit.

5.2.3 DERIVATION OF NATURAL BACKGROUND LEVELS

Apart from the 'natural' factors groundwater quality is influenced by anthropogenic inputs mainly from diffuse sources (e.g. agriculture, atmosphere). Whereas some of these inputs (e.g. pesticides) are a direct indicator of human impact, most inorganic contents occurring in the groundwater originate both from natural and anthropogenic sources. This makes it difficult to decide whether the observed groundwater solution contents in a certain area is influenced by diffuse pollution intakes or still represents an (almost) natural state. To simulate groundwater solution content patterns and to separate the 'natural' groundwater concentration pattern from the influence of pollutant intakes, different procedures may be used.

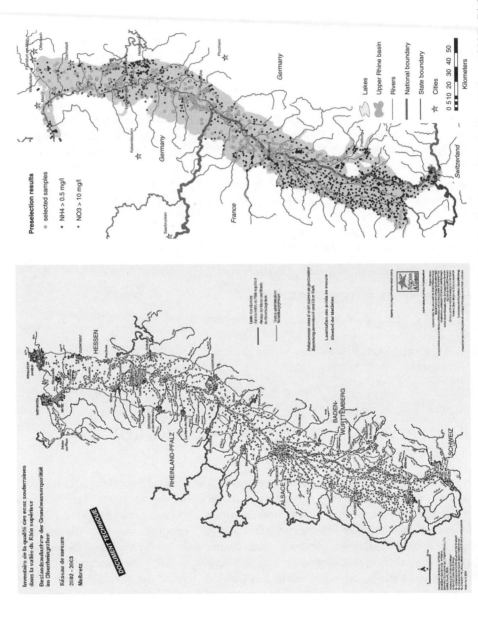

Figure 5.2.2 Left: Sampling sites of groundwater in the Upper Rhine Valley (Direction Regionale de l'Environment Alsace, 2005); Right: Sampling sites preselected for the derivation of natural background levels.

An evaluation of existing approaches for natural background level assessment (Kunkel *et al.*, 2004; Blum *et al.*, 2006) has shown that statistical methods, which separate the 'natural' groundwater concentration pattern from the influence of (diffuse) pollution intakes to groundwater, are appropriate and widespread to derive natural background levels, especially for large area considerations. On one hand a number of methods exist which separate the observed concentration distributions of a certain groundwater parameter into two distribution functions, describing the natural and the influenced component (Kunkel *et al.*, 2007). The natural situation as well as the influence of (diffuse) pollution is characterized by confidence intervals of the individual distribution function. Although the applicability and good performance of this procedure has been proven on a national scale in Germany already (Wendland *et al.*, 2005), its applicability on a European scale was regarded as doubtful due to the fact that 'deeper knowledge' about statistical analysis is a prerequisite for an application.

On the other hand, there are statistical preselection methods available, which derive natural background values based on the exclusion of samples in case the concentrations of indicator substances exceed a certain concentration. An evaluation of existing preselection approaches (HLUG, 1998; Christensen *et al.*, 2000; Kunkel *et al.*, 2004; Wendland *et al.*, 2005; Chery, 2006) has shown that they are appropriate to derive natural background levels on the level of aquifer typologies as defined by (Wendland *et al.*, 2008). Furthermore, they fulfil the requirements to be applicable on a European scale since they require 'no deeper knowledge' about statistical analysis, can be applied by non-experts and to groundwater bodies, for which only few samples are available.

The basic idea of preselection methods is that a correlation between the concentration of certain indicator substances and the presence of anthropogenic influences is assumed to be present. Consequently, the natural background levels are derived only from groundwater samples, where the concentrations of these indicator substances are below a defined level, indicating that the samples can be regarded as not being influenced by anthropogenic activities. Although preselection methods have the disadvantage that the relation between high concentration of indicator substances and anthropogenic induced high concentrations of other substances is not true in all cases (Kunkel *et al.*, 2004), this kind of approach is particularly representative and interesting for elements for which a considerable number of samples are available.

Starting point of the analysis is the frequency distribution of the observed concentrations of a groundwater parameter. A preselection approach, already described in Chapter 3 has been used to identify anthropogenic influenced groundwater samples:

- samples displaying purely anthropogenic substances;

- samples displaying nitrate concentrations above $10\,mg\ NO_3/l$ in oxidized aquifers ($O_2 > 2\,mg/l$ and Fe (II) $< 0.2\,mg/l$);

- samples displaying ammonia concentrations above $0.5\,mg\ NH_4/l$ in reduced aquifers ($O_2 \leq 2\,mg/l$ and Fe (II) $\geq 0.2\,mg/l$).

The natural background values are defined subsequently for all groundwater parameters as the concentration range between the 10th and 90th percentiles of the concentration distributions from the remaining samples.

5.2.4 RESULTS AND DISCUSSION

The method for natural background levels derivation has been applied to the database for the Upper Rhine Valley Preselection according to Nitrate ($NO_3 < 10\,mg/l$) has lead to the exclusion of 1094 samples (64%). Evaluation of the redox status of groundwater has shown that 188 samples indicate reduced aquifer conditions. 35 of those samples were excluded from the natural background level derivation because of their high NH_4-contents ($NH_4 > 0.5\,mg/l$). In the end 594 groundwater samples remained (see Figure 5.2.2, right part). For those samples it was assumed that they are to a large extent free from human impact. Hence, they are regarded as being appropriate for the natural background level derivation in the Upper Rhine Valley.

The results of the statistical evaluation of the monitoring data are presented in Table 5.2.1. For all available groundwater parameters the number of remaining samples, the minimum and maximum values, the 50th percentile (median), the 10th and 90th percentiles, used for background assessment, and the 2.3rd and 97.7th percentiles as a reference for the spread of data values are printed in the table. As can be seen from the table, the natural background levels obtained choosing the 10th to 90th percentile range and the 2.3rd to 97.7th percentile range differentiate considerably for the same parameter. For all parameters the range of values derived from the 2.3rd–97.7th

Table 5.2.1 Results of the statistical evaluation and natural background levels for the investigated groundwater parameters in the Upper Rhine Valley. DL denotes the detection limit. The natural background levels are defined here as the concentration range between the 10th and 90th percentiles (P_{10} and P_{90}).

Parameter		Values	Min	Median	Max	P_{10}	P_{90}	$P_{2.3}$	$P_{97.7}$
B	mg/L	518	<DL	0.03	0.69	**0.01**	**0.1**	0.01	0.17
Ba	mg/L	397	<DL	0.10	2.9	**0.04**	**0.26**	0.02	0.43
Ca	mg/L	517	6.7	92	1910	**52**	**166**	19.0	239
Cl	mg/L	580	0.8	32	10470	**8.6**	**84**	6.0	139
Fe (II)	mg/L	460	<DL	0.15	18	**<DL**	**3.6**	0.2	6.4
H	μg/L	578	0.0003	0.05	3.6	**0.02**	**0.18**	0.02	0.71
HCO₃	mg/L	469	6	275	647	**142**	**421**	36	299
K	mg/L	517	0.4	2.3	417	**1.18**	**7.2**	0.67	26.4
Mg	mg/L	577	1.1	10.6	308	**5.5**	**25**	2.8	46
Mn (II)	mg/L	523	0.001	0.15	10	**<DL**	**0.82**	0.002	1.53
Na	mg/L	577	0.90	15.5	4130	**6.2**	**41**	4	75
SO₄	mg/L	514	1.67	42	620	**21**	**173**	10.1	339
DOC	mg/L	512	<DL	1.4	17.9	**0.45**	**3.8**	0.2	6.4
TAC	mg/L	542	0.1	4.6	10.6	**2.4**	**7.0**	0.57	8.3
O₂	mg/L	576	<DL	1.7	10.3	**0.20**	**6.9**	0.10	9.0
LF	μS/cm	580	5	544	23746	**305**	**951**	149	1276
As	μg/L	336	0.6	1	45	**<DL**	**4**	<DL	11.9
NH₄	mg/L	544	<DL	0.05	4.1	**<DL**	**0.39**	<DL	0.79
NO₂	mg/L	544	<DL	0.01	1.5	**<DL**	**0.04**	<DL	0.11
NO₃	mg/L	583	<DL	2.2	10.0	**<DL**	**8.2**	<DL	9.5
PO₄	mg/L	538	<DL	0.04	6.8	**<DL**	**0.17**	<DL	0.58

Table 5.2.2 Comparison of derived natural background levels from different studies for the Upper Rhine Valley.

Parameter		Derived natural background levels			
		This study	(Berthold & Toussaint, 1999)	(Kunkel *et al.*, 2004,a)	(Kunkel *et al.*, 2004,b)
Samples		594	320	953	351
B	mg/L	0.1		0.035	0.051
Cl	mg/L	84	75.7	99	59
Fe (II)	mg/L	3.6	6.17	3.3	3.7
K	mg/L	7.2	4.2	4.7	3.5
Mg	mg/L	25	23.8	33	22
Mn (II)	mg/L	0.82	0.61	0.6	0.5
Na	mg/L	41	36	19	31
SO$_4$	mg/L	173	157.3	249	119
LF	μS/cm	951		1296	840
As	μg/L	4	5.6	3.6	4.1
NH$_4$	mg/L	0.39		0.04	0.4
NO$_2$	mg/L	0.04			
NO$_3$	mg/L	8.2	5.8	1.2	3.5

percentiles are often 2–4 times higher than those of the 10th–90th percentiles. This is due to the strong skewness of the concentration distributions for some groundwater parameters. Especially for individual aquifers, where the origin of the groundwater data is known very well, the 2.3rd–97.7th percentiles are used as a reference to derive natural background levels. However, because there is no objective way to proof the general preference of one of the two percentile ranges, both of them can be used. Due to the size of the case study area, the associated variability in local conditions and the heterogeneity of the data sources we used the 10th–90th-percentile value range as a reference for natural background level assessment in the Upper Rhine Valley.

As has been mentioned already, natural background levels for the Upper Rhine Valley have been derived previously in several other investigations, each using different approaches. In Berthold and Toussaint (1999) and Kunkel *et al.* (2004), denoted as Kunkel *et al.*, 2004b, background levels are derived based on a preselection approach. In contrast to this study, different data bases and slightly different methodologies with respect to the preselection conditions and the used percentile ranges have been used. In Kunkel *et al.* (2004), denoted as Kunkel *et al.* (2004a), a different approach based on the separation of the natural and influenced components (Kunkel *et al.*, 2007). In Table 5.2.2 the results of the different studies are compared to each other. Due to the differing derivation methodologies, sampling sites, sampling periods and monitoring networks, however, the derived natural background values can only be compared restrictively in terms of comparability.

In spite of the heterogeneities with regard to the evaluated samples and the methodology the derived natural background level derived for the investigated parameters are relatively similar and lie in the same range for most of the parameters. This can be regarded as a proof for the validation of the general methodology used here to derive natural background level in the Upper Rhine Valley. At the same time it is a proof for the homogeneity of the different water bodies in the Upper Rhine Valley with regard to their hydrogeochemistry. Finally, it can be stated that the natural background value derivation methodology developed for the application on an EU-scale leads to results which compare sufficiently to more detailed studies on a regional scale, at least for the Upper Rhine Valley.

REFERENCES

Berthold G. and Toussaint B., 1999. *2. Grundwasserbeschaffenheitsbericht 1998, Handbuch Teil III – Grundwasser*. Hydrologie in Hessen.

Blum A., Kunkel R. and Wendland F., 2006). Natural background levels: State of the art and review of existing methodologies. BRIDGE – Background Criteria for the Identification of Groundwater thresholds: Impact of hydrogeological conditions on pollutant behaviour in groundwater and related ecosystems, Project report, D 10, 85–158.

Chery L., 2006. Qualité naturelle des eaux souterraines. Méthode de caractérisation des états référence des aquifères francais., Rapport BRGMeditions – Orléans, 238 p.

Christensen T. H., Bjerg P. L., Banwart S. A., Jakobsen R., Heron G. and Albertsen H.-J., 2000. Characterization of redox conditions in groundwater contaminant plumes. *Journal of Contaminant Hydrology*, 45, 165–241.

Direction Regionale de l'Environment Alsace, 2005. Internationales Bearbeitungsgebiet Oberrhein, Bericht zur Bestandsaufnahme, Strasbourg, France.

HLUG, 1998). *Grundwasserbeschaffenheit in Hessen, Auswertung von Grund- und Rohwasseranalysen bis 1997*. Hessische Landesanstalt für Umwelt, Umweltplanung, Arbeits- und Umweltschutz, 102 pp.

Kunkel R., Voigt H. J., Wendland F. and Hannappel S., 2004. *Die natürliche, ubiquitär überprägte Grundwasserbeschaffenheit in Deutschland*. Schriften des Forschungszentrums Jülich, Reihe Umwelt/Environment. Forschungszentrum Jülich GmbH, Jülich, Germany.

Kunkel R., Wendland F., Hannappel S., Voigt H. J. and Wolter R., 2007. The influence of diffuse pollution on groundwater content patterns for the groundwater bodies of Germany. *Water Science and Technology*, 55(3), 97–105.

Wendland F., Hannappel S., Kunkel R., Schenk R., Voigt H. J. and Wolter R., 2005. A procedure to define natural groundwater conditions of groundwater bodies in Germany. *Water Science and Technology*, 51(3–4), 249–57.

Wendland F., Blum A., Coetsiers M., *et al*., 2008. European aquifer typology: a practical framework for an overview of major groundwater composition at European scale. *Environmental Geology*, 55(1), 77–85.

5.3

The Colli Albani Volcanic Aquifers in Central Italy

Alfredo Di Domenicantonio, Manuela Ruisi and Paolo Traversa

Tevere River Basin Authority, Roma, Italy

5.3.1 Introduction
5.3.2 Characterisation of Groundwater Bodies
5.3.3 Pressures and Impacts
5.3.4 First Assessment of Groundwater Chemical Status
5.3.5 Conclusion
References

5.3.1 INTRODUCTION

The management of aquifers subject to intense overexploitation for household, agricultural and industrial water supply uses requires specific and complex planning and management measures. This case study of the Colli Albani volcanic area provides a good example. Other aspects of interest in this area are the presence of protected areas and groundwater dependent terrestrial and aquatic ecosystems. Moreover the volcanic nature of the area results in a complex distribution of natural hydrochemistry from which different qualities of groundwater are derived. The identification of threshold values, especially in relation to natural background concentrations, will be required to demonstrate the achievement the good status objectives of the Water Framework Directive (2000/60/EC) and its daughter Groundwater Directive (2006/118/EC).

Groundwater Monitoring Edited by Philippe Quevauviller, Anne-Marie Fouillac, Johannes Grath and Rob Ward
© 2009 John Wiley & Sons, Ltd

5.3.2   CHARACTERISATION OF GROUNDWATER BODIES

The Colli Albani volcanic structure is situated south of the city of Rome. It has an important value from a landscape, historical and cultural point of view because it has been widely exploited since the Roman epoch. It comprises important drinking water protected areas and natural protected areas of local, national and European interest. It is characterised by an isolated relief with a characteristic truncated cone shape that surmounts the Roman countryside with an altitude between 0 and 970 m above sea level.

The Colli Albani district is one of the many volcanic structures which developed along the Lazio and Toscana regions' continental platform in the north-eastern part of the Tevere river basin. Including the hydraulically linked coastal strip, it covers an area of about 1950 km^2.

Volcanic activity in Colli Albani area started between 600,000 to 20,000 years ago. It was characterised by long alternating explosive and effusive phases, accompanied by diffuse eccentric activity, culminating in a series of violent phreatomagmatic explosions, which concluded the cycle of the entire complex.

The outcrops are constituted by pyroclastic products, lava flows, volcanic scoriae and ash, sand and lapilli strata. The overall thickness of these formations varies from a few meters to over a kilometre. The chemical and petrographic characteristics of the pyroclasts and lava are typical of a magmatic series rich in potassium (Capelli *et al.*, 2005).

The water circulation is sustained by 500 to 600 m thick marine clay deposits. The substratum is constituted of the carbonate series of the Apennine ridge, where deep water circulation takes place (Capelli & Mazza, 2005).

The Colli Albani structure's water circulation radiates from the centre of the structure to the periphery following complex patterns, and it is characterised by a strong interaction between groundwater and surface water circulation.

The geological setting established an upper aquifer in the central area, sustained by low permeability volcanic rocks, and a regional aquifer, sustained by marine pre-volcanic clay deposits contained in the more ancient volcanic rocks. Water recharge to the upper aquifer is characterised by the presence of vast semiendorheic areas. The upper complex feeds the lakes and drains into the basal aquifer. The perennial surface water circulation is fed, at the base of the riverbed, by the regional aquifer through linear springs at an altitude between 50 and 70 m above sea level.

The water circulation has been subdivided into four water bodies (Autorità di Bacino del Fiume Tevere, 2003) delimited by piezometric levels. For each it was possible to carry out water balance calculations (Figure 5.3.1). The main parameters of the hydrogeological balance are summarized in Table 5.3.1 (Capelli *et al.*, 2005).

The depth of the aquifer from the surface varies from a few meters in the periphery, where the aquifer is mainly unconfined, up to 500 m in the central area of the volcanic structure, where the aquifer is mostly confined.

Studies commissioned by the Tevere River Basin Authority, the Lazio Regional River Basins Authority and the Hydrogeology Department of Roma Tre University were carried out in order to develop a conceptual model the water circulation (Autorità di Bacino del Fiume Tevere, 2003). These studies included the interpretation of hydrogeological maps and a measurement campaign on 500 wells deeper than 100 m.

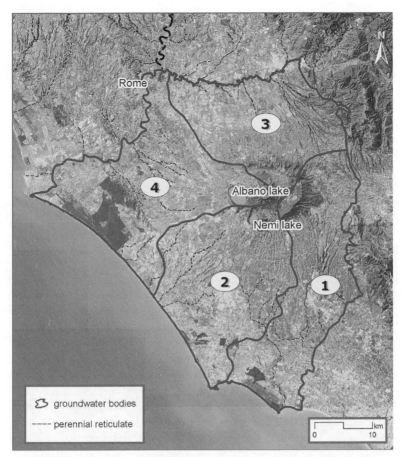

Figure 5.3.1 Map of the four groundwater bodies in the Colli Albani volcanic hydrogeological structure.

Table 5.3.1 Annual average water balance components in the Colli Albani hydrostructure (averages for 1997–2001).

Hydrogeological system of the Colli Albani structure	mm/year	l/s	Mm³/year	% of P
Precipitation[*]	731	45 925	1 448	100.0
Evapotranspiration	346	21 659	683	47.2
Runoff	138	8 610	272	18.7
Effective Infiltration	245	15 364	485	33.5
Base flow in the riverbed (from in situ measurements)	62	3 893	123	8.5

[*]Comprises 5.7 Mm³/year of precipitation in the lakes.

The basal aquifer's piezometric level was derived from these measurements. The interpretation of piezometric data in this area is particularly complex due to the presence of numerous perched aquifers that are less relevant but nevertheless hydraulically linked to the rest of the structure. In the upper part of the aquifer the water circulation is distributed homogeneously in all directions, whereas under the caldera area there is a strong influence from pre-volcanic morphology. In this area the paleovalleys filled with intensely fractured lava flows have become the main groundwater flow paths (Figure 5.3.2) (Capelli *et al.*, 2005).

Before the 1980s it was possible to identify springs with different geo-chemical characteristics in the Albano aquifer (Figure 5.3.3) (Di Domenicantonio *et al.*, 2007). The chemical composition of the water reflecting the complex interaction between the circulation of volcanic rocks and the fluids in the underlying geothermal basement. The entire aquifer is influenced by rising gasses rich in CO_2 and H_2S, present in different ratios

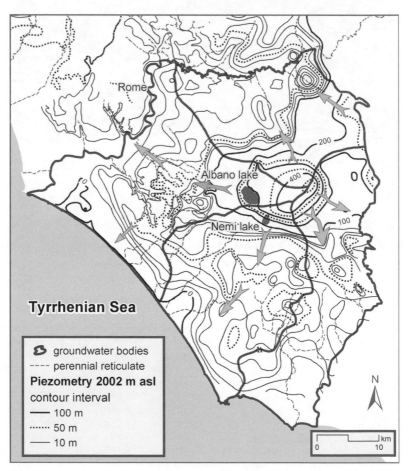

Figure 5.3.2   Piezometric map of the study area with indication of rivers and main groundwater flow directions.

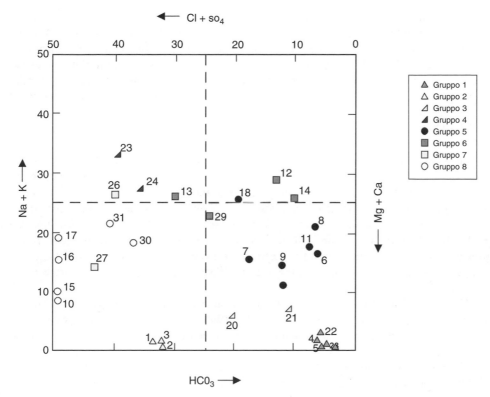

Figure 5.3.3 Different groups of spring water identified in the Albano area. Each group has a different location within the structure.

according to the different parts of the aquifer. At the margins of the volcanic structures mixing phenomena with the liquid component at high temperatures occur.

Consequently, the Colli Albani volcanic aquifers are characterised by a spatial variability of natural background chemistry. This variability depends on the different compositions of the multistrata rocks composing the volcano, the direction and length of the flow paths, the interactions with deep-seated geothermal gasses and fluids.

5.3.3 PRESSURES AND IMPACTS

Groundwater abstraction is the main pressure on the Colli Albani aquifer. There are about 33,000 authorised (or awaiting authorisation) groundwater abstraction points in the area. About two-thirds of these are for private household use, and the other third for productive uses including agriculture, industry and drinking water.

The total volume abstracted from the entire hydrostructure, comprising all four water bodies, amounts to about 345 million m³/year. This is equivalent to 74% of the natural recharge (mean annual effective infiltration) (Table 5.3.1) (Capelli *et al*., 2005).

The pollution sources are determined by analysis of land use characteristics. Most of the diffuse pollution sources originate from agricultural areas (38% of the area) (Autorità di bacino del Fiume Tevere, 2003). There is also a diffuse pollution contribution from the discontinuous urban fabric (19%) in rural areas. In these areas agricultural activities are intermixed with sparse houses, where urban waste water treatment systems are generally lacking. The main source of point source pollution is believed to originate from household and industrial discharges.

The Colli Albani aquifer is an elevated aquifer and its piezometric level reflects, in general, the truncated cone shape of the volcanic structure.

Intense abstraction has not only depressed water level close to the abstraction points (local impacts), but it has also led to a general lowering of the piezometric level across the whole area. Aquifer recharge mainly occurs in the top part of the volcanic structure within the caldera. The lowering of piezometric levels has also led to reduction in the base flow of many of the watercourses in the area. Water levels in lakes connected hydraulically to the aquifer are also decreasing and wetlands in the depressions on the slopes of the volcano are tending to decrease in extent. Near the coast overabstraction causes saline water intrusion from the sea and brackish water intrusion from the River Tevere's delta.

The assessment of quantitative status of groundwater resources provided a focus for monitoring campaigns in the 1970s (Ventriglia, 1990). Since 1997 systematic monitoring of about 2000 wells has allowed changes in piezometric level over the last 30 years to be modelled.

The comparison between the piezometric level in the 1970s and in 2002 shows that a substantial change has occurred, especially in the intensely exploited areas. The main consequences are:

- lowering of the water table;

- lowering of the level of the lakes (about 3 meters in the Albano lake in the last decade);

- 60% reduction of the total base flow to watercourses: the total discharge in the period 1978–1982 was equal to $4.7\,m^3/sec$, the total discharge in the period 1997–1999 was equal to $2.5\,m^3/sec$;

- saline intrusion in the coastal areas.

In 2004 (Autorità di Bacino del Fiume Tevere, 2004) safeguard measures were adopted in order to limit aquifer exploitation in the Colli Albani hydrostructure. The objective was to reduce total abstraction by 15%. Figure 5.3.4 shows the areas regulated by different safeguard measures according to the classification based on abstraction rates.

The main impacts on terrestrial ecosystems dependent on water are variations in the extent of wetlands. Significant variations in macrobenthos associated to a decrease of the flow in the watercourses have not been identified until now, except for the stretches with presence of discharges. However, there has been a general reduction in populations of fish and amphibians.

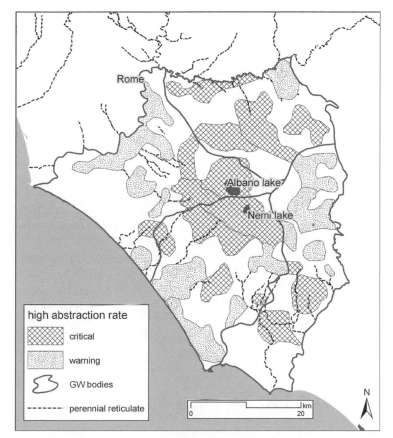

Figure 5.3.4 Identification of areas subject to high withdrawal rates and regulated by specific norms for the authorization of surface and groundwater resource use.

5.3.4 FIRST ASSESSMENT OF GROUNDWATER CHEMICAL STATUS

In 2007 the Tevere River Basin Authority promoted a groundwater quality monitoring campaign (Arpa Lazio, 2007). It consists of about 10–15 monitoring points in each groundwater body, located along the principal groundwater flow paths. The parameters measured were selected according to previous knowledge of groundwater quality and comprise inorganic pollutants, metals and dangerous substances.

Pesticides and other dangerous substances were not detected at any location and so they are not deal with this case study.

The following methodology is adopted for the assessment of GWB status:

• comparison between experimental data and the threshold values at national level in Italian proposal for transposition of the Groundwater Directive;

- analysis of compounds exceeding thresholds for each location;
- in-depth analysis.

Table 5.3.2 shows threshold values for the monitored parameters. For conductivity the threshold value is 2500 µS/cm.

Table 5.3.3 shows threshold values exceedances for each groundwater body. Samples containing high values of conductivity and chloride were those taken in coastal zones where marine intrusion problems exist.

Table 5.3.2   Threshold values for metals and inorganic pollutants (investigated in this case study) set at a national level in Italian proposal for transposition of the Groundwater Directive (2006/118/EC).

Compounds	Threshold values (µg/L)
Metals	
Arsenic	10
Cadmium	5
Total Chromo	50
Mercury	1
Nickel	20
Plumb	10
Copper	1000
Selenium	10
Zinc	3000
Inorganic pollutants	
Fluoride	1500
Nitrate	500
Chloride	250000
Ammonium	500

Table 5.3.3   Exceedances of Threshold Values in groundwater bodies in the study area.

		Conductivity (µS/cm) TV = 2500		Chloride (mg/l) TV = 250		Fluoride (µg/l) TV = 1500		Arsenic (µg/l) TV = 10		Ammonium (mg/l) TV = 10	
gwb	n. samples	max	n. > TV	max	n. > TV	max	n. > TV	max	n. > TV	max	n. > TV
1	8	2260	0	220	0	1950	1	80	1	11.89	1
2	15	1147	0	83	0	5400	4	21	5	0.46	0
3	9	2054	0	54	0	1600	1	10	0	1.12	1
4	12	3593	1	500	1	1700	1	14	2	11.20	4

n – number of samples.
max – maximum concentration/parameter value.
gwb – Groundwater Body.
TV – Threshold Value.

As for the high values of ammonium, these could be explained by:

- to organic deposits in the River Tevere's delta area;

- point source pollution (to be investigated).

Regarding high concentrations of Arsenic and Fluoride in the groundwater, correlation with the gas movement from the basement can be hypothesised because this is a volcanic area. As a matter of fact the map of the values distribution (Figure 5.3.5) shows a high concentration of Arsenic and Fluoride just in the area corresponding to the caldera and other zones aligned with important structural elements.

Due to the natural variability in groundwater quality a more suitable delimitation of groundwater bodies may need to be considered by taking into account different areas with different natural background level. In this way it will be possible to define threshold values higher than the national TV where needed.

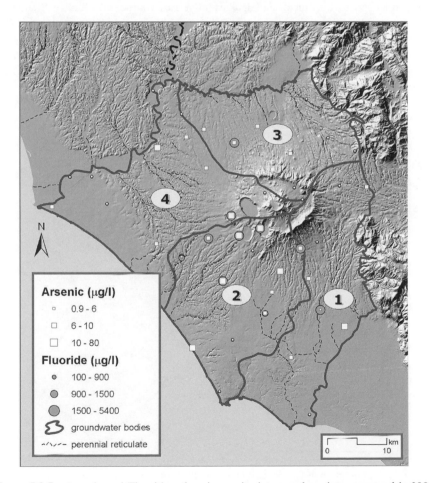

Figure 5.3.5 Arsenic and Fluoride values in monitoring sample points measured in 2007.

Considering the intensive overexploitation of the Colli Albani area it is important to investigate mobilisation of chemicals induced by abstraction.

A first in-depth examination has been complete in the Castelporziano area. This is a National protected area, consisting of a strip of Mediterranean forest that has remained unaltered for many centuries.

Reclamation works carried out in the surrounding marshland during the 1930s and recent groundwater exploitation for agricultural, household and industrial uses modified aquifer recharge in the areas near the protected area, causing progressive infiltration of saline water from the sea and the final stretch of the Tevere river. Today this area is monitored and studied by different scientific groups.

Conductivity and piezometric levels were considered to be significant parameters for measuring the effects of saline intrusion on terrestrial ecosystems since salinity variations may cause adaptive changes in the flora and fauna of the protected area's terrestrial ecosystem.

Methodologies for the identification of salinity levels that may induce these adaptive changes still need to be consolidated. Currently, Environmental Quality Standards for conductivity are being assessed by expert judgement, by considering trend analysis rather than threshold values.

A detailed reconstruction of the phenomenon is shown in the map in Figure 5.3.6, where dedicated internal (Bucci, 2006) and external sampling points have been installed for the protected area (Autorità di Bacino del Fiume Tevere, 2006).

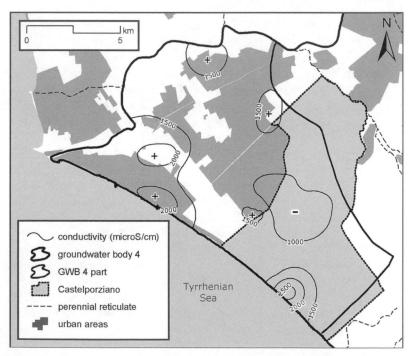

Figure 5.3.6   Delimitation of Conductivity levels in the protected area of Castelporziano measured in 2004.

The map shows that areas characterised by higher conductivity levels are located in the most intensely urbanized areas and areas characterised by the lowest conductivity levels are located at the centre of the protected area.

Historical data on groundwater conductivity levels in this area before saline intrusion are not available Therefore, the lowest salinity values were selected for the purpose of Natural Background Level value identification (about $1000\,\mu S/cm$).

The threshold value was fixed to $2000\,\mu S/cm$ by experts, lower than the threshold value in the Italian proposal for transposition of the Groundwater Directive ($2500\,\mu S/cm$), taking into account the natural characteristics of the area (Di Domenicantonio *et al.*, 2007).

The Castelporziano case study showed that the threshold values for parameters measuring marine water intrusion were exceeded in several coastal areas, showing a general upward trend of the values from 1999 to 2004. For a better management the part of the Groundwater body 4 evidenced in Figure 5.3.6 should be a separate body.

5.3.5   CONCLUSION

The chemical composition of the water in the Colli Albani aquifer system is very variable due to the nature of the rocks and the interaction with gasses and fluids originating from depth.

Under these conditions the identification of threshold values (on the basis of natural background levels) and trend reversal measures is significant only if the geochemical characteristics of groundwater and its interaction with the receptors is known.

A general lowering of the piezometric level is observed due to intense groundwater abstraction. Therefore, quantitative recovery is the first objective of the programme of measures to be adopted in the Water Framework Directive River Basin Management Plan for this area. The Colli Albani aquifer is currently subject to safeguard measures aimed at limiting abstraction permits in critical areas where demand for abstraction is concentrated.

REFERENCES

A.R.P.A. Agenzia per la Protezione dell'Ambiente del Lazio. 2007. *Convenzione per la definizione degli aspetti ecologici nell'ambito della sperimentazione delle linee guida di cui alla direttiva 2000/60/EC dei corpi idrici laziali*. Final report. (Unpublished report). Roma.

Autorità di Bacino del Fiume Tevere, Autorità dei Bacini Regionali del Lazio, Regione Lazio, 2003. *Studi idrogeologici per la definizione degli strumenti necessari alla redazione dei piani stralcio relativi agli acquiferi vulcanici del territorio della Regione Lazio*. (Unpublished report), Roma.

Autorità di Bacino del Fiume Tevere, 2004. *Misure di salvaguardia nel bacino del fiume Tevere tra Castel Giubileo e la foce*. Gazzetta Ufficiale della repubblica Italiana, 89, 14.

Autorità di Bacino del Fiume Tevere, 2006. *Studi idrogeologici per la definizione degli strumenti necessari alla redazione del piano stralcio per l'uso compatibile delle risorse idriche sotterranee nell'ambito degli acquiferi del delta del Tevere*. (Unpublished report), Roma.

Bucci M., 2006. *Stato delle risorse idriche, Il sistema ambientale della tenuta presidenziale di Castelporziano*. Accademia Nazionale delle Scienze detta dei Quaranta, Scritti e documenti XXVII, 327–387.

Capelli G. and Mazza R., 2005. *Water criticality in the Colli Albani (Rome, Italy)*. Giornale di Geologia, **1**, 263–73.

Capelli G., Mazza R. and Gazzetti C., 2005. *Strumenti e strategie per la tutela e l'uso compatibile della risorsa idrica del Lazio. Gli acquiferi vulcanici*. Pitagora editrice, Bologna.

Di Domenicantonio A., Ruisi M. and Traversa P. 2007. *Groundwater natural background levels and threshold definition in the Colli Albani volcanic aquifers in Central Italy* http://www.igme.es/bridge

Ventriglia U. 1990. *Idrogeologia della Provincia di Roma. Regione Vulcanica dei Colli Albani*. Provincia di Roma, Roma, Volume III.

5.4

Monitoring the Environmental Supporting Conditions of Groundwater Dependent Terrestrial Ecosystems in Ireland

Garrett Kilroy[1], Catherine Coxon[2], Donal Daly[3],
Áine O'Connor[4], Fiona Dunne[5], Paul Johnston[6], Jim Ryan[7],
Henning Moe[8] and Matthew Craig[3]

[1,2] *Geology Department, School of Natural Sciences, Trinity College Dublin, Dublin, Ireland*
[3] *Environmental Protection Agency, Richview, Clonskeagh, Dublin 14, Ireland*
[4,7] *National Parks & Wildlife Service, Department of Environment, Heritage & Local Government, Dublin, Ireland*
[5] *Ecological Consultant, Dublin, Ireland*
[6] *Civil, Structural and Environmental Engineering Department, Trinity College Dublin, Dublin, Ireland*
[8] *CDM Ireland Ltd, O'Connell Bridge House, Dublin, Ireland*

Groundwater Monitoring Edited by Philippe Quevauviller, Anne-Marie Fouillac, Johannes Grath and Rob Ward
© 2009 John Wiley & Sons, Ltd

5.4.1   INTRODUCTION

The European Water Framework Directive (WFD) (2000/60/EC) requires at least good status for all groundwater, rivers, lakes, estuarine and coastal waters by 2015. The status of all water bodies are currently being classified in advance of the first River Basin Man-aement Plans (RBMP) due in 2009. The classification of groundwater bodies (GWBs) includes the requirement to assess the 'significant damage' to groundwater dependent terrestrial ecosystems (GWDTEs) caused by anthropogenic pressures on the associated GWB. GWDTEs are defined in this chapter as wetlands that depend on a significant pro-portion of their water supply (quality and quantity) from groundwater. Whilst the WFD does include objectives to protect wetlands and useful guidance has been prepared (EC, 2003) ambiguity still remains on what constitues signficant damage or how they should be assessed. In determining significant damage within GWDTEs, the environmental sup-porting conditions required to maintain dependent plant and animal communities in a favourable state must be defined. Knowledge of these environmental supporting condi-tions provides the basis for determining the sensitivity of GWDTEs to anthropogenic pressures (chemical and abstraction) and improves understanding of the relevant path-ways for water and contaminants to GWDTEs. This knowledge will inform the design of monitoring networks in the associated GWB and any subsequent management mea-sures needed to either protect pristine habitats or rehabilitate damaged ones. In addition, this knowledge will contribute to the achievement of favourable conservation status as required under the European Habitats Directive (92/43/EEC).

This study arises from a project carried out under the auspices of the Irish WFD Groundwater Working Group. The project's main aim is to develop a conceptual frame-work that assists in evaluation of the environmental supporting conditions required for different wetland types (Kilroy *et al.*, 2008). This framework will provide the basis for the assessment of 'significantly damaged' GWDTEs, as required under the WFD for GWB classification.

5.4.2   GROUNDWATER DEPENDENT TERRESTRIAL ECOSYSTEMS IN IRELAND

The ecology of GWDTEs is fundamentally reliant on the supporting geohydrological and hydrochemical conditions. Therefore, significant changes to these conditions due to human activities can cause ecological damage to these wetlands. Understanding the connection between geohydrology and ecology is fundamental to the management of these wetlands under the WFD. There are numerous types of GWDTE in Ireland, but raised bogs, fens and turloughs represent three common but contrasting wetland types which were chosen for this study. These wetland types were selected for study because the supporting geohydrological and hydrochemical conditions for each wetland type are different and each has been subject to several research projects. Therefore their study will help inform future decisions on monitoring and management of wetlands.

In Ireland there are three types of peatland: fens, raised bogs and blanket bogs. There are some $2000\,km^2$ of actively growing peatlands in Ireland (Foss, 1998). The main water source for many fens is mineral-rich (usually calcium-rich) groundwater and sur-face water and so they have a neutral or slightly alkaline pH. Rainfall is the primary source of water for bogs and as a result such peatlands are acidic and poor in minerals (ombotrophic). Most raised bogs in Ireland have developed from fens – as the peat continued to accumulate, it formed a flattened dome, slightly higher than the surrounding area and hence the name. The peat of raised bogs consists largely of water and sphagnum mosses, and they are known to have reached thicknesses of 15 m in Ireland. Peatlands support specialised vegetation comprising of plants that have adapted to living in waterlogged conditions and to maximising nutrient uptake and conservation (Doyle and Ó'Críodáin, 2003). The main pressures on peatlands include peat removal for fuel, afforestation and drainage for land reclamation and agriculture.

Turloughs are depressions in karst that usually become inundated with groundwater during the winter and drain in summer through swallow holes connected to underground water systems (Coxon, 1987). Typically, these ecosystems contain distinctive aquatic and terrestrial plant and animal communities adapted to fluctuating water levels. The variability in plant and invertebrate communities between turloughs is primarily due to different hydrogeomorphological characteristics, but also depends on the range of farm animal grazing practices when the turlough is empty during the summer. Over 300 turloughs have been documented in Ireland and the main anthropogenic pressures include artificial drainage to facilitate agriculture and pollution from nutrient inputs (Sheehy-Skeffington *et al.*, 2006).

5.4.3   A CONCEPTUAL FRAMEWORK FOR ASSESSING THE ENVIRONMENTAL SUPPORTING CONDITIONS OF GROUNDWATER DEPENDENT TERRESTRIAL ECOSYSTEMS

The development of sound conceptual models for different wetland types is an essential starting point for understanding, monitoring and managing wetlands. Subsequent investigation and monitoring will help validate, modify or refine these conceptual models. A source-pathway-receptor approach was used as the context for developing a conceptual framework for GWDTEs. Therefore, an essential starting point is to examine the important pathways of water transport from the contributing aquifer to different types of wetland.

The identification of a habitat's environmental supporting conditions requires a clear understanding of the water supply mechanisms of the particular wetland type. This in turn helps identify where and what to monitor, particularly the key ecological indicators that are used to evaluate the condition of the wetland. There have been several water supply mechanisms typologies proposed for wetlands (e.g. in the UK: Wheeler and Shaw (2000), in the USA: Clairain (2002), and in Canada: Warner and Rubec (1997)). A simple water supply mechanism typology was adopted for this study (based on Lloyd and Tellam, 1995), which can be readily applied to all Irish GWDTEs. This typology focuses on the general geo-hydrological setting within which the wetland is situated. Two basic geo-hydrological settings are proposed: groundwater discharge zone wetlands and groundwater flow-through depression wetlands. In groundwater discharge zone wetlands the wetlands and any associated open surface water represent the main groundwater discharge zone. These wetlands include fens, wet heaths, transition mires and petrifying springs. In groundwater flow through depression wetlands, the wetland is situated in a topographical depression within the regional groundwater flow aquifer system. These wetlands include turloughs and raised bogs.

The groundwater requirements of different wetlands will include quantifiable attributes such as groundwater contribution, groundwater level and chemical flux. However, there are few detailed studies in Ireland that relate these attributes to quantifiable responses in the wetland ecosystems. The few studies that do exist will be essential for guiding the development of conceptual models for the first WFD river basin management cycle.

A conceptual framework for assessing the environmental supporting conditions of GWDTEs under the WFD is presented in Figure 5.4.1. The three axes aim to incorporate the main geohydrological and hydrochemical attributes of GWDTEs. The location of a habitat within this cube can inform the key monitoring and management issues for different GWDTE types by highlighting important environmental supporting conditions. The three axes, which are described in more detail below, represent the main intrinsic groundwater conditions required by different wetlands to maintain their habitats in good condition.

The relative groundwater contribution to the wetland (x-axis) can be calculated by carrying out a water balance and will, for example, be more significant for fens (which are minerotrophic, and are fed by surface and subsurface water) than raised bogs (which are mainly ombrotrophic or dependent on rainfall input). The variability of groundwater level (y-axis) can be measured through changes in water table and can be expressed, for example, as a coefficient of variation. This will often need to be measured both in the

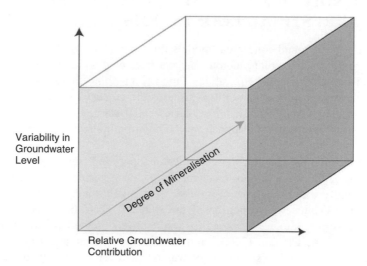

Figure 5.4.1 A conceptual framework encompassing the main environmental supporting conditions of GWDTEs. The location of a particular habitat within the cube highlights its key geohydrological and hydrochemical supporting conditions and thus informs monitoring and management priorities.

wetland itself and in the contributing aquifer. Some wetlands, such as turloughs, can have a dynamic range of groundwater levels as they drain and fill seasonally. The vegetation in such habitats has adapted to alternating periods of inundation and exposure. In contrast, fens require more stable groundwater levels and the vegetation and animals within these habitats require constant wetting. Many groundwater dependent wetlands will require particular chemical conditions (z-axis) to maintain a habitat in a favourable condition, e.g. total hardness, calcium concentration. For example petrifying springs (habitats often found with fens) are reliant on a continuous supply of base-rich groundwater to ensure tufa formation.

Once the key environmental supporting conditions are established for a particular wetland the next step is to establish its sensitivity to particular pressures. One approach is to examine the vegetation communities of GWDTEs in order to characterise their groundwater dependency and sensitivity to certain pressures. Ellenberg (1988) defined a set of indicator values for the vascular plants of central Europe, indicating their 'ecological behaviour' in relation to environmental conditions such as soil moisture (F), soil/water pH (R), nitrogen/soil fertility (N), light (L), salinity (S), temperature (T) and continentality (K). For each variable there is a scale, e.g. Moisture (F) ranging from 1–12, dry to wet or nitrogen/soil fertility (N) ranging from 1 to 9, infertile to richly fertile. Species with wide amplitude tend to fall in the middle of the scale, while more specialised species occur at either end of the scale (i.e. those with a preference for dry conditions or those for aquatic conditions). Therefore, information provided by the occurrence of a plant species can infer particular environmental conditions. Ellenberg values can also be obtained on a whole wetland site basis by averaging the values from all communities within a site. This mean value for all communities on a site indicates the relative groundwater influence between sites, allowing for a comparison between sites. Weightings can be used to prioritise one site over another – for example Tynan *et al.* (2006) calculated the Ellenberg N value for plant communities within turloughs in western Ireland and used the percentage area of oligotrophic communities within each turlough as a weighting to rank each turlough site.

5.4.4 APPLICATION OF THE CONCEPTUAL FRAMEWORK TO CASE STUDIES

Three contrasting wetlands typical of the Irish environment are examined here as a basis for developing a conceptual framework for assessing groundwater dependent wetlands. Figure 5.4.2 presents the locations of the three study sites. These sites are part of a network of GWDTEs identified during the characterisation of Irish river basin districts for Article 5 of the WFD (EPA and RBDs, 2005).

All three wetlands have been subject to previous investigations and research projects, their water supply mechanisms are reasonably well understood and all are Special Areas of Conservation (SAC) containing priority habitats in Annex I of the European Habitats Directive (92/43/EEC).

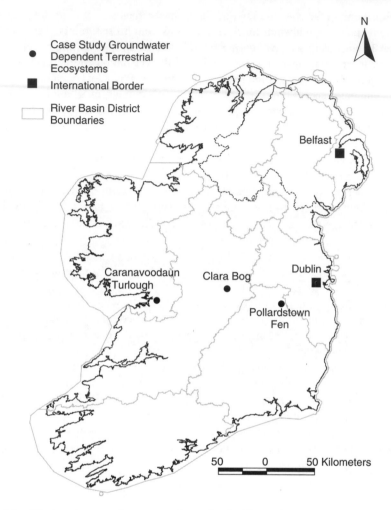

Figure 5.4.2 Site location map for case study wetlands. These sites are groundwater dependent terrestrial ecosystems (GWDTE) and a subset of Ireland's Special Areas of Conservation (SAC).

5.4.4.1   Pollardstown Fen

Geo-hydrological Setting and Dependent Ecosystem Characteristics

Pollardstown Fen is a SAC extending over 2.35 km^2, located on the northern boundary of the Curragh aquifer (a Pleistocene fluvioglacial sand and gravel aquifer) in Co. Kildare. Groundwater from this aquifer largely maintains this wetland via a series of springs and seepage zones (Kuczynska, 2006). It is the largest spring-fed fen in Ireland, with approximately 40 springs providing a continuous supply of calcium rich groundwater to the fen. These springs are located mainly at the margins of the fen, along distinct seepage areas. It is this continuous input of groundwater that maintains the fen and prevents succession to a more ombrotrophic system, ultimately developing into a raised bog.

Pollardstown Fen is designated as a SAC owing to the occurrence of a number of Annex I habitats – alkaline fen, *Cladium* fen* and Petrifying springs* (* denotes priority habitats under the EU Habitats Directive) – and Annex II species (NPWS, 2007). The three species of whorl snail *Vertigo angustior, V. geyeri* and *V. moulinsiana* (all Annex II species) occur in the fen, the only known Irish site where they occur together. The most sensitive receptor is *Vertigo geyeri* which requires a stable set of environmental conditions and is extremely vulnerable to even small changes in groundwater level (±5 cm).

Environmental Supporting Conditions

The main pathway of groundwater flow to the fen is along flow lines through the sand and gravel aquifer from south to north. Some data were examined to explore the y-axis of the conceptual framework in a more quantitative way as it might apply to Pollardstown Fen. Borehole MB 30 was used as it is located on the groundwater divide south of Pollardstown Fen. Therefore changes in level at this borehole will also result in a change in the groundwater contribution being delivered to the wetland. Figure 5.4.3 presents the time series for MB 30.

To the south of MB 30 a motorway has been constructed which involved extensive dewatering during the construction phase. The relative contribution of groundwater to the wetland is significant, with the majority of water supply being provided by the Curragh aquifer. Groundwater that supplies Pollardstown Fen via springs and seepage zones is alkaline and extremely calcareous with mean hardness values up to 500 mg/l $CaCO_3$ (Reynolds, 2003). This supersaturated water results in the deposition of calcium carbonate 'tufa' (petrifying springs, a priority habitat).

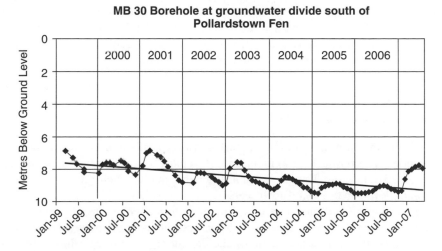

Figure 5.4.3 Temporal series of groundwater levels (metres below ground) at borehole MB 30 located at the groundwater divide south of Pollardstown Fen.

Sensitivity of Receptor and Implications for WFD

It is clear from Figure 5.4.3 that a pressure on the groundwater body is resulting in a lowering of the groundwater levels in the aquifer feeding the fen. Droughts induced by climate change and construction of the motorway south of the fen may be amongst the contributing factors. This is also resulting in a lowering of the groundwater levels at the fen and thus drying out of the seepage zones along the fen margin which are critical for supporting the fen habitats and species. The habitat is highly sensitive to changes in groundwater level, with impacts likely to the most sensitive species where there is a change in groundwater contribution of 5% or greater (Kuczynska, 2006). Monitoring programmes have detected changes in *Vertigo* populations and in the vegetation, as a result of changes in groundwater level variability. For this reason the groundwater body containing Pollardstown Fen would likely be classified as being at poor status under the WFD because there is significant damage to a groundwater dependent terrestrial ecosystem. Consequently, corrective action is required through a programme of measures in the WFD river basin management plans.

5.4.4.2   Caranavoodaun Turlough

Geo-hydrological Setting and Dependent Ecosystem Characteristics

Caranavoodaun Turlough is approximately $0.31\,\mathrm{km^2}$ in size when flooded and is part of the Castletaylor Complex SAC located in Co. Galway and situated in a Carboniferous Limestone karst aquifer. It is surrounded by hazel woodland and limestone heath and has a shallow basin containing areas of exposed limestone, including a swallow hole by which it empties and fills. Caranavoodaun turlough is entirely groundwater-fed and this groundwater input is dominated by shallow epikarst flow in the top $2-5\,\mathrm{m}$ in the karst limestone.

The interest of the site lies partly in the diversity of habitats within a small area. There are five Annex I habitats present within the site, including turlough habitat. The site also shows the transition from wetland to surrounding habitats. Caranavoodaun turlough itself is considered a good example of an extremely oligotrophic and calcareous turlough that is relatively undamaged.

Environmental Supporting Conditions

Depth of flooding from the base of the turlough is typically 1 m but can range from 1 to 1.5 metres. The turlough can empty and fill rapidly via its swallow hole as evident from Figure 5.4.4 below. Caranavoodaun turlough exhibits much greater variability in groundwater levels than Pollardstown Fen.

The inflowing groundwater is a hard limestone water. Following flooding, turlough waters lose carbon dioxide and increase in pH, becoming supersaturated with calcium carbonate and forming a characteristic calcite deposit on the turlough floor. The trophic status of the temporary water bodies in turloughs has received little attention to date but is the subject of a current study on nutrient and algal dynamics in turlough waters. Total

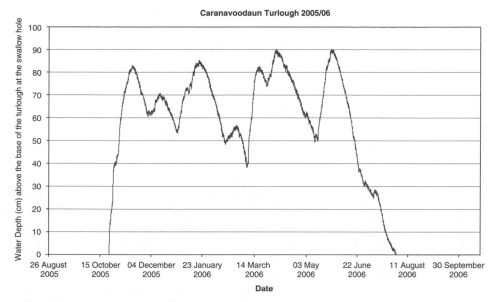

Figure 5.4.4 Temporal series of water depth (cm) above the base of the turlough at the swallow hole, which empties and fills with groundwater from the surrounding karst limestone aquifer.

phosphorus (TP) concentrations in Caranavoodaun Turlough in the 2006–07 flooding season (October to April) ranged from 6 to 12 μ g/l (mean 10 μ g/l), placing it at the upper boundary of the oligotrophic category according to the OECD (1982) criteria for lakes (Pereira, pers. comm.). This mean TP concentration falls in the bottom 23% of TP values in the 22 turloughs surveyed in this year, thus the site is at the more nutrient-sensitive end of the range. Chlorophyll *a* concentrations ranged from 0.7 to 2 μ g/l, falling in the bottom 27% of the 22 sites surveyed (Pereira, pers. comm.)

Sensitivity of Receptor and Implications for WFD

The turlough vegetation is predominantly fen and aquatic plant communities, with drier plant communities making up only a small proportion of the vegetation (Goodwillie, 1992). Using the plant community data from Goodwillie, Tynan *et al*. (2006) in their turlough study calculated the Ellenberg N values (nitrogen/soil fertility) for a number of turloughs including Caranavoodaun. They obtained an average N value for the site of 3.15. Ellenberg N values of less then 4 indicate infertile sites (Hill *et al*., 1999); therefore indicating that Caranavoodaun is highly oligotrophic.

Karst groundwater environments, where turloughs typically occur, have been identified as particularly vulnerable to pollution and can result in contaminants, such as phosphorus, being rapidly transmitted from groundwater to surface water ecosystems via conduit flow and spring discharge (Kilroy and Coxon, 2005). Given that Caranavoodaun turlough has a naturally oligotrophic status, anthropogenic nutrient inputs would have a damaging effect on the turlough ecology. Therefore, whilst not currently damaged, Caranavoodaun turlough has being classified as 'at risk' and will require measures for the contributing GWB under the WFD to prevent any future damage.

5.4.4.3   Clara Bog

Geo-hydrological Setting and Dependent Ecosystem Characteristics

Clara Bog is located in Co. Offaly in the centre of Ireland and is 6.65 km² in size. It is underlain by a poorly productive Carboniferous Limestone aquifer. Above this is a layer of glacial till, of varying thickness within the basin. This till layer is overlain by a layer of lacustrine clay, again of varying thickness. In places, a thin layer of lake marl has been deposited over the clay. Before the development of ombrotrophic peat, the basin went through a fen phase and this layer of minerotrophic fen peat can be found below the ombrotrophic peat mass. The greatest peat depth is 12 m. Raised bogs have two distinct hydrological zones: the acrotelm (the living 'skin' at the top of the bog comprising sphagnum and other bog plants) and the catotelm (located between acrotelm and the mineral subsoil).

The vegetation of Clara bog comprises three main areas – the high bog (acrotelm), the soaks and the lagg zone (Schouten, 2002). Clara Bog has been designated a SAC for the presence of active raised bog and bog woodland habitats, both priority habitats under EU Habitats Directive. Clara bog is regarded as one of the most important raised bogs in the country, with well developed hummock hollow complexes and one of the few remaining soak systems, as well as a number of rare species.

Environmental Supporting Conditions

Hydraulic conductivity in the acrotelm for Clara Bog has been measured as typically >25 m/d and for the catotelm as low as 10^{-5} m/d (Daly *et al.*, 1994). Hydraulic conductivity measurements and potentiometric mapping indicate the bog is relatively independent of the regional groundwater flow system. However, at the fringes of the bog (lagg zone) there is evidence of mixing of upwelling groundwater flow and surface water runoff from the bog. Groundwater plays two distinct roles in Clara bog. Firstly, the regional groundwater has a support function for the whole raised bog, even though it may not directly influence the plant communities. Secondly, the lagg zone is where the vegetation is directly influenced by groundwater upwelling.

Sensitivity of Receptor and Implications for WFD

Clara bog has suffered much damage – a third of the bog has been lost to peat cutting, a road built in the 18th century bisects bog into Clara East and Clara West and an extensive network of drains was installed in 1980s. More recently a programme of research has been undertaken on Clara bog and restoration measures implemented to offset some of the damage. The bog remains threatened, however, by turf cutting on its southern margins, particularly where the groundwater table has been lowered. Subsidence due to drainage has had a major impact on Clara Bog and it is estimated that where the bog surface gradient is more than 0.3 m/100 m the acrotelm becomes damaged and the bog has not been functioning properly either hydrologically or ecologically in these areas (Daly *et al.*, 1994).

While the bog is relatively independent of the regional hydrology because of the low hydraulic conductivity of the deeper peat layers (catotelm), in the longer term compaction of the basal peat layers could occur if the regional groundwater levels drop significantly (Schouten, 2002). The lagg zone depends strongly on groundwater and can be significantly impacted by surrounding hydrological conditions. For example, the lagg zone north of the bog, which is bordered by permeable esker deposits, has been impacted by arterial drainage of the River Brosna to the north. Lowering of the river water level resulted in increased groundwater flow to the river and away from the lagg zone. Under the WFD whilst the surrounding groundwater body may not be contributing significantly to the water supply of the bog, measures may be required to ensure its supporting function is not impaired and remedial measures for the lagg zones will be required. Detailed topographic surveys of the bog will be required to monitor the on-going subsidence issue.

5.4.5 DISCUSSION AND CONCLUSIONS

The proposed conceptual framework for assessing the environmental supporting conditions of GWDTEs is being developed and tested using data from the three study wetlands. Based on the available the data the main environmental supporting conditions for specific habitats of the three study wetlands are presented in Figure 5.4.5.

The location of these habitats within the conceptual framework highlights their key geohydrological and hydrochemical conditions and thus informs monitoring and management priorities. For example, Caranavoodaun turlough requires much greater variability in groundwater levels than the other two habitats, which require stable groundwater level conditions. Pollardstown fen needs a significant groundwater contribution, stable levels and groundwater quality that is base-rich to ensure the alkaline fen characteristics are maintained. The high bog element of Clara Bog has the lowest overall groundwater

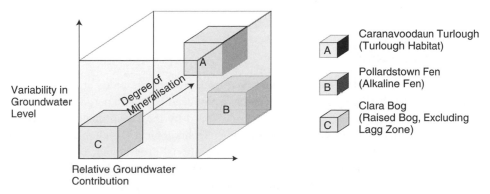

Figure 5.4.5 A proposed conceptual framework indicating the main environmental supporting conditions for specific habitats in the three GWDTE case studies.

contribution; however, the lagg zone element of this habitat will require much greater groundwater inputs.

The environmental supporting conditions required by different GWDTEs will dictate different monitoring and management strategies. The geohydrological monitoring of raised bogs may focus particularly on the lagg zones that fringe the bogs whereas fens are more directly dependent on groundwater and require monitoring along the flow lines to the supply points. The geohydrological monitoring needs of turloughs will vary depending on the degree and nature of karstification, the presence of epikarst and connectivity to surface water sources. Monitoring will need to take into account the complexity of each wetland, thereby providing key indicators of the habitats sensitivity to groundwater inputs.

Recently progress has been made to develop wetland typologies that specifically address the requirements of the WFD; Rodríquez-Rodríquez and Benavente (2007) used a hydro-morphological and water balance approach for wetlands in Southern Spain and Dahl *et al*. (2007) have developed a groundwater-surface water interaction typology for riparian wetlands in Denmark. Different wetland types will require different assessment and monitoring approaches. However, common to all will be the necessity to identify the key environmental supporting conditions required to maintain these habitats in a favourable condition.

Further work is needed to characterise significant damage in different GWDTE types due to different pressures in the contributing groundwater body. In particular it is necessary to identify relevant sensitive species within each wetland type and their relationship with corresponding geohydrological and hydrochemical characteristics. A significant monitoring effort will be required to develop understanding of the relevant pathways for water and contaminants to GWDTEs during the early stages of the WFD river basin planning process. As knowledge gaps are addressed and understanding of groundwater–surface water interactions improves this monitoring effort may be reduced. Progress has been made concerning the integration of wetlands into the WFD implementation process; however, further work is needed to improve understanding of the relationship between the biotic and the geohydrological/hydrochemical characteristics of GWDTEs.

Acknowledgements

A significant amount of this chapter is derived from a paper presented at the 2007 IAH Congress in Lisbon (Kilroy *et al*., 2007) and is reproduced here with permission from the Conference Organisers. This research was part funded by the Environmental Protection Agency (EPA) under the Environmental Research Technological Development and Innovation Programme under the Productive Sector Operational Programme 2000–2006. The authors gratefully acknowledge the provision of groundwater level data at Pollardstown Fen from Michael Browne (Environmental Protection Agency), nutrient data and level data at Caranavoodaun turlough from Helder Pereira and Owen Naughton respectively (both Trinity College Dublin).

REFERENCES

Clairain E. J., 2002. *Hydrogeomorphic Approach to Assessing Wetland Functions: Guidelines for Developing Regional Guidebooks*. U.S. Army Engineer Research and Development Center, Vicksburg, MS.

Coxon C., 1987. The spatial distribution of turloughs. *Irish Geography*, **20**, 24–42.

Dahl M., Nilsson B., Langhoff J. H. and Refsgaard J. C., 2007. Review of classification systems and new multi-scale typology of groundwater–surface water interaction. *Journal of Hydrology*, **344**(1–2), 1–16.

Daly D., Johnston P. and Flynn R., 1994. The hydrodynamics of raised bogs: an issue for conservation. In: T. Keane and E. Daly (eds), The Balance of Water – Present and Future. *Proceedings of AGMET Group (Ireland) and Agricultural Group of the Royal Metereological Society (UK) Conference, Dublin*, pp 105–21. AGMET Group, Meteorological Service, Dublin.

Doyle G. J. and Ó'Críodáin C., 2003. Peatlands – fens and bogs. In: M. Otte (ed.) *Wetlands of Ireland – Distribution, Ecology, Uses and Economic Value*, pp. 79–108. University College Dublin Press, Dublin.

EC (European Communities), 2003. Common Implementation Strategy for the Water Framework Directive (2000/60/EC) Guidance Document No. 12 – The role of wetlands in the Water Framework Directive. Luxembourg, Office for Official Publications of the European Communities.

Ellenberg H., 1988. *Vegetation Ecology of Central Europe*. 4th Edition. Cambridge University Press, Cambridge.

EPA and RBDs (Environmental Protection Agency and River Basin Districts) (2005) The Characterisation and Analysis of Ireland's River Basin Districts – National Summary Report (Ireland) 2005. Available at www.wfdireland.ie.

Foss P. J., 1998. National overview of the peatland resource in Ireland. In: G. O'Leary and F. Gormley (eds), *Towards a Conservation Strategy for the Bogs of Ireland*. Irish Peatland Conservation Council, pp. 3–20.

Goodwilli R., 1992. Turloughs over 10ha. Vegetation survey and evaluation. A report for the National Parks and Wildlife Service, Dublin.

Hill M. O., Mountford J. O., Roy D. B. and Bunce R. G. H., 1999. ECOFACT 2 Technical Annex-Ellenberg's indicator values for British Plants. Centre for Ecology & Hydrology, Monk Woods.

Kilroy G. and Coxon C., 2005. Temporal variability of phosphorus fractions in Irish karst springs. *Environmental Geology*, **47**(3): 421–30.

Kilroy G., Coxon C., Daly D., *et al*., 2007. A geohydrological basis for assessing risk, monitoring requirements and ecological sensitivity for groundwater dependent wetlands for management purposes under the Water Framework Directive. In: Ribeiro, L., Chambel, A. and Condesso de Melo, M.T. (eds), *Conference Proceedings of the 35th Congress of the International Association of Hydrogeologists (IAH), 17-21 September 2007, Lisbon, Portugal*.

Kilroy G., Dunne F., Ryan J., O'Connor A., Daly D., Craig M., Coxon C., Johnston P., Moe H. 2008. A Framework for the Assessment of Groundwater-Dependent Terrestrial Ecosystems under the Water Framework Directive. *Environmental Research Centre Report No. 12. Environmental Protection Agency, Wexford*. Available at www.epa.ie.

Kuczynska A., 2006. Eco-hydrology of Pollardstown Fen, Unpublished PhD Thesis. Trinity College, Dublin.

Lloyd J. W. and Tellam J. H., 1995. Groundwater-fed wetlands in the UK. In: Hughes and Heathwaite (eds), *Hydrology and Hydrochemistry of British Wetlands*. John Wiley & Sons Ltd, Chichester.

NPWS (National Parks and Wildlife Service) (2007) Site synopsis for Pollardstown fen SAC. Available at www.npws.ie.

OECD (Organisation for Economic Cooperation and Development) (1982) Eutrophication of Waters, Monitoring, Assessment and Control. Paris, OECD.

Pereira H. C., pers. comm. Unpublished data on water chemistry of Caranavoodaun turlough, October 2006 to April 2007, from monitoring in progress for research project on Dynamics of Nutrients and Algae in Irish Turloughs (part of research project on Assessing the Conservation Status of Turloughs, funded by the Irish National Parks and Wildlife Service). Data provided 22 June 2007.

Reynolds J., 2003. Fauna of turloughs and other wetlands. In: M. Otte (ed.), *Wetlands of Ireland – Distribution, Ecology, Uses and Economic Value*, pp. 135–59. University College Dublin Press, Dublin.

Rodríquez-Rodríquez M. and Benavente J., 2007. Definitions of wetland typology for hydro-morphological elements within the WFD. A caset study in Southern Spain. Water Resource Management, **22**(7), 797–821.

Schouten M. G. C., 2002. Conservation and restoration of raised bogs; geological, hydrological and ecological studies. Department of the Environment and Local Government/Staatsbosbeheer.

Sheehy-Skeffington M., Moran J., O'Connor A., *et al.*, 2006. Turloughs – Ireland's unique wetland habitat. *Biological Conservation*, **133**, 265–90.

Tynan S., Gill M. and Johnston P., 2006. Development of a methodology for the characterisation of a karstic groundwater body with particular emphasis on the linkage with associated ecosystems such as turlough ecosystems. Final Research Report to the Environmental Protection Agency (Project No. 2002-W-DS-08-M1). EPA, Wexford.

Warner and Rubec, 1997. The Canadian wetland classification system. The Wetlands Research Centre, University of Waterloo, Ontario.

Wheeler B. D. and Shaw S. C., 2000. A wetland framework for impact assessment at statutory sites in eastern England (WETMECS). Report to Environment Agency.

5.5

Use of WETMECs Typology to Aid Understanding of Groundwater-Dependent Terrestrial Ecosystems in England and Wales

Mark Whiteman[1], Bryan Wheeler[2], Sue Shaw[3], Tim Lewis[4], Mark Grout[5] and Kathryn Tanner[6]

[1] *Environment Agency – England and Wales, Leeds, United Kingdom*
[2,3] *University of Sheffield, Wetland Research Group, Department of Animal & Plant Sciences, Sheffield, United Kingdom*
[4] *Entec UK Ltd, Canon Court, Abbey Lawn, Abbey Foregate, Shrewsbury, United Kingdom*
[5] *Environment Agency – England and Wales, Peterborough, United Kingdom*
[6] *Environment Agency – England and Wales, Preston, United Kingdom*

Groundwater Monitoring Edited by Philippe Quevauviller, Anne-Marie Fouillac, Johannes Grath and Rob Ward
© 2009 John Wiley & Sons, Ltd

5.5.1 INTRODUCTION

The Environment Agency of England and Wales has statutory conservation duties under European Directives including the Habitats Directive (92/43 EEC) and Water Framework Directive (2000/60/EC) and domestic UK legislation (Water Act 2003; Countryside and Rights of Way Act 2000; Environment Act 1995), all of which seek to conserve and enhance habitats and species dependent on water. It has statutory duties to further and promote the conservation and enhancement of flora and fauna dependent on the aquatic environment under the Environment Act (1995) and the Countryside and Rights of Way Act (2000).

These duties require the development of conceptual understanding of groundwater-dependent terrestrial ecosystems (GWDE), and the Environment Agency has been working with statutory conservation organisations (Natural England and the Countryside Council for Wales) since 1998 to develop an understanding of the links between ecology and hydrology for wetlands. As part of this initiative, a project called the 'Wetland Framework' has been developed through a partnership between the Wetland Research Group at the University of Sheffield, the Environment Agency, English Nature (now Natural England), and the Countryside Council for Wales.

The importance of water supply mechanisms in determining the distribution and composition of plant communities has been recognised for some time (Wheeler and Shaw, 1995; Wheeler, 1999). The Agency had initiated hydrological investigation and monitoring of wetlands since the late 1980s (Lloyd and Tellam, 1995; Whiteman, 2000), but there remained a critical need to bring together and integrate the data from hydrological and ecological studies of wetlands. The main objectives of the 'Wetland Framework' project were therefore to:

• link wetland vegetation types to environmental conditions;

• link wetland vegetation types to mechanisms of water supply;

• link water supply mechanisms to the hydrogeological, topographical and landscape circumstances that produce these;

• use this understanding to assess and predict the impact of specific environmental changes (water level drawdown, nutrient enrichment);

• identify the main categories ('habitats') of wetlands.

The primary impetus for the project arose from the requirement to undertake detailed assessments of the impacts of a wide range of consented and unconsented activities upon conservation features (habitats and species) of European importance. The project also complemented work by Natural England and the Countryside Council for Wales on conservation objectives for designated sites and their features, by identifying environmental factors that are critical for the maintenance (or enhancement) of conservation features, and distinguishing these from factors that are not so critical. The Wetland Framework also provides a basis for assessing the potential for these objectives to be sustained or enhanced in specific wetland sites.

The main outputs of the project have been the:

- identification of generic **Wet**land Water Supply **Mec**hanisms (WETMECs);

- identification of primary wetland hydrochemical 'habitats' (particularly base-richness and fertility categories);

- assessment of water supply and habitat conditions required by selected vegetation types;

- assessment of how selected wetland sites 'work' ecohydrologically, with particular reference to their conservation interest.

This chapter provides an introduction to the concept and potential use of WETMECs. These are perhaps best seen as conceptual units which take particular account of the impact of 'top-layer effects' in the supply and distribution of water within wetlands and which can form an 'add-on' to wider conceptual hydrogeological models. They can help identify the components of water supply that sustain habitat features of conservation importance in wetlands.

The Wetland Framework is exclusively concerned with *mires* (i.e. bogs, fens and some swamps). Even within this category, there may well be other WETMECs that have not yet been identified within the UK, especially perhaps in parts of Scotland. A series of smaller-scale studies have tried to identify a similar functionally-driven classification in wet woodlands (Barsoum *et al*. 2005), wet heath (Mountford, Rose and Bromley, 2005), wet dunes (Davy *et al*., 2006), swamps and ditches (Mountford in Wheeler *et al*., 2004) and wet grasslands (Gowing *et al*., 2002; Gowing in Wheeler *et al*., 2004). The eco-hydrology of mires, ditches, swamps and wet grasslands have been drawn together and summarised in a user-friendly format in the publication 'Eco-hydrological Guidelines for Lowland Wetland Plant Communities' by Wheeler *et al*., 2004. The practical use of the Guidelines for wetland management are discussed by Whiteman *et al*., 2006.

5.5.2 WETMECs AS CONCEPTUAL MODELS

5.5.2.1 Data and Methods

Ecological and hydrogeological data from over 1500 stand samples from over 200 wetlands throughout England and Wales were analysed in order to identify the main wetland water supply mechanisms (Wheeler *et al.*, 2009). Data available for each vegetation stand included:

- vegetation – species composition and derivatives;

- 'top-layer' stratigraphy – wetland infill (peat etc.) and underlying material;

- hydrochemical variables – base-richness, fertility, redox potential etc.;

- hydrological variables – water tables, flow etc.;

- topographical variables – overall topography (estimated), surface microtopography;

- topographical and spatial information about apparent groundwater and surface water sources, and potential drainage features.

Values were either measured or categorised and estimated using ranked categories.

In addition, information available for each wetland site included estimated annual precipitation and actual evaporation (Low Flows 2000 data) (Young *et al*., 2003), water level data from piezometers and gauge-boards (some sites only) (mainly Environment Agency data) as well as geological data and conceptual hydrological models.

A cluster analysis of all 'water-related' variables was performed using Ward's Method (incremental error sum of squares), resulting in a 36-cluster stage being selected for examination based on a moving average best cut significance test (t-statistic).

5.5.2.2 Data Interpretation

Each of the 36 clusters was interpreted in terms of the water supply mechanisms that it appeared to represent. This resulted in the identification of 20 WETMECs (Table 5.5.1), some of which were subdivided into sub-types. In some cases more than one cluster was allocated to a single WETMEC.

Many of these types are familiar to hydrogeologists and peatland ecologists, including Type 1 (Raised Bogs), Type 10 (Permanent Seepage Slopes) and even Type 12 (Fluctuating Seepage Basins). However, other types are more subtle products of the interactions between water supply, topography and the characteristics of the 'top-layer' (see below), which although on first inspection appear to be variations on the 'groundwater seepage' mechanism, have distinctive and recurrent ecological consequences.

WETMECs are primarily self-standing hydrological categories. It is only when they are interpreted and identified with particular regard to the ecological characteristics of the areas of wetland concerned and added to the other Framework categories (landscape

Table 5.5.1 Wetland Water Supply Mechanisms (WETMECs).

WETMECs	
1: Domed Ombrogenous Surfaces ('Raised Bogs')	11: Intermittent & Part-Drained Seepages
2: Buoyant Ombrogenous Surfaces ('Quag Bogs')	12: Fluctuating Seepage Basins
3: Buoyant Weakly Minerotrophic Surfaces ('Transition Bogs')	13: Seepage Percolation Basins
4: Drained Ombrotrophic Surfaces in Bogs and Fens	14: Seepage Percolation Troughs
5: Summer 'Dry' Floodplains	15: Seepage Flow Tracks
6: Surface Water Percolation Floodplains	16: Groundwater-flushed Bottoms
7: Groundwater Floodplains	17: Groundwater-flushed Slopes
8: Groundwater-fed Bottoms with Aquitard	18: Percolation Troughs
9: Groundwater-fed Bottoms	19: Flow Tracks
10: Permanent Seepage Slopes	20: Percolation Basins

type, base-richness, fertility, management) that we can identify eco-hydrological units (wetland 'habitats'). The Framework approach has been 'bottom-up', in that units and their limits have been derived from an analysis of data collected from individual stands of vegetation. This differs from some existing wetland categorisations, which have been essentially 'top-down' in character – i.e. based on subdivisions derived from intuitive appraisal of the identity of the main units.

5.5.2.3 Top-layer Controls – an Important Conceptual Feature of WETMECs

Definition of the 'Top-layer'

WETMECs are partly based upon the potential importance of 'top-layer' conditions to the water supply of wetlands. The 'top-layer' is the uppermost substratum, including the wetland infill, particularly those deposits which have tended to be excluded from many geological maps, or whose character is often uncertain and variable (e.g. Head), together with various paludogenic deposits (e.g. marl, gyttja and peat). The 'top-layer' effect includes features such as lithological variation within drift deposits coupled with features more localised to the wetland, such as induration layers below the site, organic 'seals' lining the site, and variation in the character of the peat infill, from an effective aquitard to highly transmissive horizons. These have sometimes been disregarded by hydrogeologists, either because they are poorly-characterised, poorly-documented, or else thought unlikely to be important.

'Top layer' control refers to influences upon the hydrodynamics of mires imposed by these superficial deposits, which can be subdivided into three components (Figure 5.5.1), the surface layer, bottom layer, and basal substratum. The disposition and character of the top layers is variable horizontally as well as vertically and specific layers are not necessarily laterally persistent. This creates a patchiness beneath or within the wetland deposit and is one of the reasons for the occurrence of different WETMECs within a single wetland site. The lack of information on the hydraulic properties of these deposits and their heterogeneity can make it difficult to use regional groundwater models to assess hydrological conditions in mire sites. Borehole logs and piezometric data are useful but

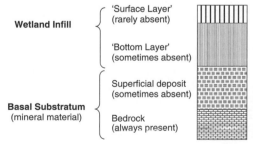

The **Wetland Infill** consists of material deposited by and within the wetland.

The 'surface layer' can correspond to the *acrotelm* of some peatland ecologists, the 'bottom layer' to the *catotelm*. They often differ in their hydraulic characteristics.

The **Basal Substratum** is mineral material, most often a superficial deposit (Till, Head *etc.*) which is not always well-documented on geological maps *etc.* Where this is absent, the wetland infill lies directly on bedrock.

Figure 5.5.1 Top layer characteristics.

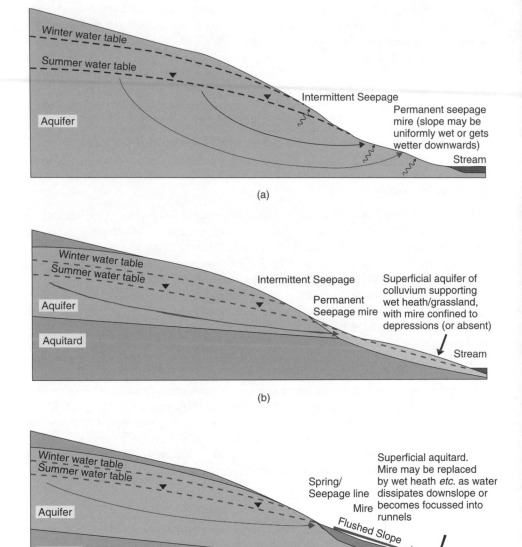

Figure 5.5.2   Schematic sections showing the influence of superficial deposits of varying composition downslope of groundwater outflows upon the gross patterns of water flow and the habitats that develop. (a) Seepage face in permeable bedrock without superficial deposits. (b) Seepage face associated with a superficial aquifer downslope. (c) Seepage line associated with flushing of a superficial aquitard downslope.

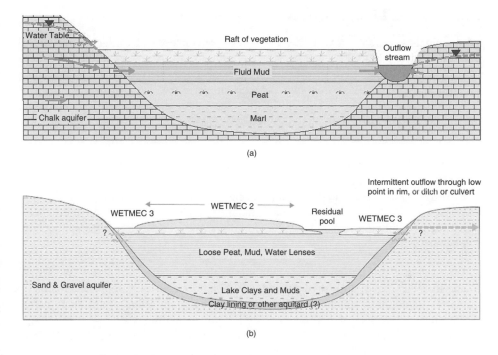

Figure 5.5.3 Topographic basins with contrasting WETMECs.

may be too sparse to provide conclusive information for heterogeneous sites, especially as sometimes boreholes are not located within, or even close to, areas of greatest ecological interest.

Examples of Top-layer Control

Local variation in superficial deposits can act as a major determinant of the characteristics of wetland habitats on slopes, and can influence their response to potentially damaging operations such as groundwater abstraction or ditching. The superficially similar sloping wetlands shown in Figure 5.5.2 have different vulnerabilities, and, for example, the seepage slope (Figure 5.5.2a) and the flushed slope (Figure 5.5.2c) have distinct vegetation communities even though both are supplied by groundwater seepage.

Topographic basins may also support different WETMECs. In the basin depicted in Figure 5.5.3a, there is hydraulic contact between the underlying aquifer and the basin via a layer of fluid mud, and a raft of vegetation floats on the surface. However, if the basin is sealed from the aquifer by a thin clay lining, as shown in Figure 5.5.3b, a quite different vegetation type may result, in this case primarily supported by rainfall inputs.

5.5.3   PRACTICAL USES OF WETMECs

Hydrogeologists are familiar with the development of conceptual hydrogeological models for wetland sites. WETMECs can be seen as 'add-ons' to these which, though generic, extend conceptual models to take better account of the properties of the wetland itself and its infill. A WETMEC-based approach can influence the way in which wetlands are perceived and investigated. We would identify the following uses, both conceptual and practical, as a consequence of 'thinking like a WETMEC'.

1. **Ecohydrological units:** WETMECs, when interpreted, can be regarded as basic ecohydrological units within wetlands. In some respects they can be seen as 'fundamental units', comparable to the species in taxonomy (though they are more variable, and show much more intergradation, than do species). As units, WETMECs have obvious potential value for purposes of description, communication, mapping etc.

2. **Investigative units:** Although, within a wetland site, adjoining WETMECs are likely to have some hydraulic connection, because they represent distinctive water supply mechanisms, WETMECs can provide an appropriate basis for the design and location of hydrological investigations within wetlands. Hydrological monitoring installations could be stratified to ensure that different WETMECs are appropriately represented. WETMECs also emphasise the need for measurements to be made within the wetland itself, as well as within its surroundings (see below).

3. **Conservation units:** WETMECs provide a potential basis for the subdivision of sites for conservation activities, assessments and monitoring. Different WETMECs may have different water supplies, and different water and vegetation management requirements. Moreover, superficially similar WETMECs can sometimes have strikingly different vulnerabilities.

4. **Vegetation-related units:** WETMECs form part of the influences on the distribution of plant communities in wetland sites. In some instances, an individual WETMEC may be co-extensive with an individual plant community, but at some sites a WETMEC may support several communities, because their occurrence is determined also by other factors.

Overall, 'thinking like a WETMEC' means shifting investigative focus somewhat away from, for example, *single* target water levels etc. to a holistic consideration of water supply mechanisms and delivery or protection of wetland regimes including hydrological (groundwater level, groundwater flow, seepage, surface water flow), hydrochemical and site management.

5.5.4 SOME EXAMPLES OF THE USES OF WETMECs

5.5.4.1 WETMECs and Site Investigations

WETMECs may provide an appropriate basis for the design and location of hydrological investigations within wetlands. They highlight the possible need for monitoring installations to be situated *within* wetland sites and, sometimes, within specific WETMECs.

At Wybunbury Moss (Cheshire, UK), for example, piezometers or dip-wells installed within the mineral ground around the mire basin (Figure 5.5.4a) may not adequately reflect the hydrodynamics of the basin itself. The mire basin (a subsidence feature) is undoubtedly fed by groundwater outflow from a sand and gravel aquifer along the northern margin. It appears that along the southern margin the basin may directly adjoin Wilkesley Halite. However the high piezometric head within the Halite close to the mire suggests that groundwater flow from this into the mire basin may be very limited, possibly because 'sealing muds' provide a partially confining layer.

WETMECs have helped to clarify the hydrodynamics of the basin itself (Figure 5.5.4b) and therefore its vulnerability to groundwater quantitative pressures (abstractions) and chemical pressures. Wybunbury Moss receives a mixture of direct precipitation and input from groundwater. The WETMEC 2 surface is fed more or less exclusively by precipitation. Visible groundwater outflow occurs from the mineral aquifer into the lagg water-track (WETMEC 15) around part of the basin (other parts may be fed just by surface run-off). However, ditching and reduction of the surface level of the bog has provided the potential for near-surface ingress of groundwater, particularly along the northern margin, so that parts of the mire have effectively become a Seepage Percolation Basin (WETMEC 13). The presence of this groundwater-driven flow is probably the main reason why enrichment from septic tanks (etc.) has had such a pervasive effect upon parts of this mire, but other factors may also have contributed to this – such as the dereliction of the northern marginal drain, reducing surface flow of enriched water eastwards along the northern margin. It is likely that in a more natural condition much of this inflow would have been intercepted and dispersed by the lagg stream.

5.5.4.2 Wetland Impact Assessment

Knowing how well connected the wetland is with the aquifer is of practical importance to help answer questions faced by regulators and wetland managers (Lerner *et al.*, 2007), such as:

- whether reduced groundwater discharge to a wetland as a result of abstraction has an adverse impact on the ecology; and

- if there is an adverse impact, what options are effective to mitigate or remove the impact to ensure that favourable condition can be met (e.g. can reduced groundwater flow be compensated for by increased surface water supply, without significant damage to the ecosystem?)

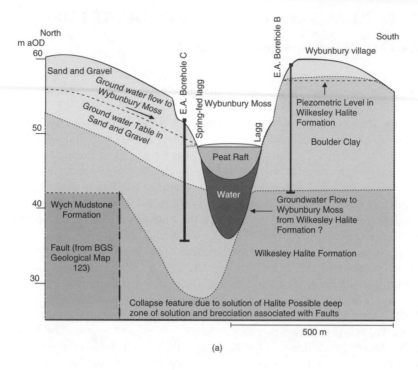

(a)

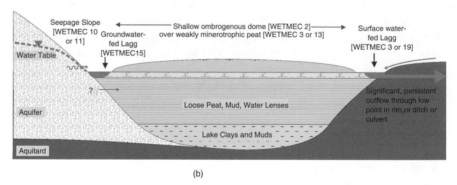

(b)

Figure 5.5.4 Geological cross-section and conceptual model of Wybunbury Moss, Cheshire. (Figure 5.5.4a has been redrawn and slightly modified from Environment Agency, 2003).

The Environment Agency has recently investigated these impacts at conservation sites of international importance in compliance with the EU Habitats Directive Review of Consents. The conceptual understanding of wetlands gained through identification of WETMECs is of high value in these investigations.

At Foulden Common, Norfolk (Eastern England), the following WETMECs were identified:

10a – Localised strong seepage

10b – Diffuse seepage

11a – Permeable partial seepage

12 – Fluctuating seepage basins (Figure 5.5.5)

An eco-hydrological conceptual model for the site was developed to include these water supply mechanisms. This helped guide the specification of local improvements within a regional scale groundwater model including, albeit relatively crudely, representation of 'top-layer' effects. Knowledge of the relevant WETMECs helped build confidence that model behaviour was appropriate for the site, even at locations where no specific hydrometric data were available. Although due regard must be taken of simplifications and assumptions inherent within the model, consideration of WETMECs helped in the identification of several criteria that could be used to assess the relative acceptability of groundwater abstraction scenarios using the model (Entec, 2007):

- continued discharge of groundwater to the site via springs and seepages – assessed by the relative volume of groundwater discharge to streams;

- maintenance of an upward hydraulic gradient from the chalk to the near surface deposits – assessed by the relative elevations of groundwater levels within the chalk and the top active layer in the model;

- maintenance of an upward flow of groundwater from the chalk to the near surface deposits – assessed by the relative volume of flow to the top active layer in the model;

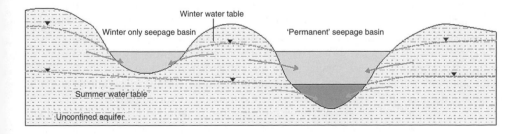

Figure 5.5.5 WETMEC 12 – fluctuating seepage basin.

- impacts on groundwater level in the top active layer of the model as an indicator of abstraction effects on the depth to the water table;

- impacts on soil moisture characteristics, especially with regard to 'ooziness', stress thresholds and water saturation – assessed by consideration of typical soil hydraulic properties to relate modelled changes in water level to probable changes in soil moisture and capillary flux contribution to evaporation, with implications for vegetation 'health'.

The *Schoenus nigricans–Juncus subnodulosus* mire (M13) vegetation community (National Vegetation Classification, Rodwell 1991–1995) represents the ecological feature of the site which is most sensitive to changes in the hydrology (e.g. to near surface groundwater levels and 'flushing'). The criteria were developed for this feature to help determine acceptable levels of abstraction on the basis that, if M13 were adequately protected, less sensitive communities would not be adversely affected.

5.5.5   DISCUSSION AND CONCLUSIONS

We consider that the Wetland Framework has considerable relevance to the study and conservation of wetlands. Given their complexity, a tool which categorises parts of wetlands in a manner which reflects their function and dependence on specific environmental factors is of obvious benefit in aiding assessment of impacts and developing conservation and restoration programmes. These outcomes are of critical importance given the importance of wetlands for wildlife conservation, and their often significant role in flood mitigation and regulation of water quality.

WETMECs are generic conceptual units, and it is likely that at some sites unusual water supply mechanisms may occur that are not easily related to existing WETMECs. Hence WETMECs should be seen as hypotheses which can be tested, modified and developed as further data become available.

One of the most important outcomes of the Wetland Framework analyses is that they highlight the difficulties of making predictions of water regimes within wetlands based on localised piezometric data and broad hydrological models. In particular, the analyses suggest that local variation in 'top-layer' conditions in and near the wetland have a great impact upon water supply and conditions relevant to the conservation interest of the site.

The top-layer effect may be of little consequence to the development of regional groundwater models for the vicinity of the site, but may be critical to the water supply and retention characteristics of the wetland itself. As the top layer can show considerable small-scale variation, initial specification and subsequent analysis of behaviour of relatively coarse scale groundwater models needs careful assessment and interpretation in the context of the known smaller scale information. Detailed stratigraphical surveys and geological borehole logs, potentially supplemented by coring, will often be essential to help understand this 'top-layer' and guide the design of hydrological monitoring. The quantitative importance of the top-layer effect has been little investigated in England and Wales. We suggest that there is a need for a more rigorous assessment of this feature across a representative range of wetlands in England and Wales, with particular respect to the distribution of vegetation types.

Since numerical groundwater models, particularly regional models, can only represent appropriate conceptual detail for the scale at which they are constructed, WETMECs are especially likely to be represented in the models when they occupy relatively large areas. It may be difficult to accommodate the hydraulic detail and patchiness of some more localised WETMECs, which may occupy only some tens of square metres, within the grid sizes typically used for current regional groundwater models. It should, however, be appreciated that very often small patches of WETMECs correspond to particular features of conservation interest (e.g. vegetation-types). Therefore, coarser-scale groundwater models may not be able to represent adequately the spatial variability in the hydrological characteristics of small WETMECs and conservation features. Conversely, we must also consider the impracticality of obtaining sufficient supporting hydrogeological data to warrant construction of finer resolution models. This leads to the conclusion that existing relatively coarse scale models may be used, particularly to assess potential *changes* in hydrogeological behaviour, but that their application needs considerable care. Results from such models must not be used 'blind', and in particular should be interpreted with due regard to small scale variability.

Acknowledgements

The authors would like to acknowledge the contribution of staff of partner organisations, including Natural England and the Countryside Council for Wales, for their financial support of the Wetland Framework project.

Disclaimer

The views expressed in this chapter are those of the authors and do not necessarily represent the formal policy of the Environment Agency for England and Wales.

REFERENCES

Barsoum N., Anderson R., Broadmeadow S., Bishop H. and Nisbet T., 2005. *Eco-hydrological guidelines for wet woodland – Phase I*. English Nature Research Reports Number 619.

Davy A.J., Grootjans A.P., Hiscock K. and Peterson J., 2006. *Eco-hydrological guidelines for dune habitats – Phase 1*. English Nature Research Reports Number 696.

Entec, 2007. *Foulden Common Site Options Plan* – report to Environment Agency, Anglian Region, Peterborough, United Kingdom.

Environment Agency, 2003. A hydrogeological assessment of Wybunbury Moss Environment Agency internal report by John Ingram, Hydrogeology section, Warrington Office.

Gowing D.J.G., Lawson C.S., Youngs E.G., *et al.*, 2002. *The water regime requirements and the response to hydrological change of grassland plant communities*. Final report for DEFRA commissioned project BD1310. Silsoe: Cranfield University.

Lerner D.N., Harris B. and Surridge B., 2007. The role of groundwater in catchment management, *Groundwater and Ecosystems* (Eds Ribeiro. L., Chambel. A. and Condesso de Melo. M.T.). *Proceedings of XXXV IAH Congress, Lisbon*, 17–21 September 2007.

Lloyd J.W. and Tellam J.H., 1995. Groundwater-fed wetlands in the UK, in *Hydrology and Hydrochemistry of British Wetlands*, J.M.R. Hughes and A.L. Heathwaite (eds), 39–61.

Mountford J.O., Rose R.J. and Bromley J., 2005. *Development of eco-hydrological guidelines for wet heaths – Phase 1*. English Nature Research Reports Number 620.

Rodwell J.S. (ed.) (1991–1995) *British Plant Communities*. Volumes 1–5. Cambridge University Press, Cambridge.

Wheeler B.D., 1999. Water and plants in freshwater wetlands. In A. Baird & R.L. Wilby (eds), *Hydroecology: Plants and Water in Terrestrial and Aquatic Ecosystems*, Routledge, London, 127–80.

Wheeler B.D. and Shaw S.C., 1995. Plants as Hydrologists? An assessment of the value of plants as indicators of water conditions in fens. In *Hydrology and Hydrochemistry of British Wetlands*, J.M.R. Hughes and A.L. Heathwaite (eds), 63–93. John Wiley & Sons Ltd, Chichester.

Wheeler B.D. and Shaw S.C., 2000. *A Wetland Framework for Impact Assessment at Statutory sites in Eastern England*. Environment Agency R&D Technical Report W6-068/TR1 and TR2.

Wheeler B.D., Shaw S. and Tanner K., 2009. *A Wetland Framework for Impact Assessment at Statutory sites in England and Wales*. Environment Agency R&D Technical Report.

Wheeler B.D., Gowing D.J.G., Shaw S.C., Mountford J.O. and Money R.P., 2004. *Eco-hydrological Guidelines for Lowland Wetland Plant Communities*, A.W. Brooks, P.V. Jose, and M.I. Whiteman (eds), Environment Agency (Anglian Region).

Whiteman M.I., 2000. Groundwater management near wetlands. Paper presented at the 7th National Hydrology Symposium, Newcastle, United Kingdom, 2000. In: *Proceedings of British Hydrological Society 7th National Hydrology Symposium*, Newcastle, 4-6 September 2000, 5.23 – 5.25.

Whiteman M.I., José P., Grout M.W., Brooks A., Quinn S.A. and Acreman M., 2004. Local impact assessment of wetlands – from hydrological impact to ecological effects. Paper presented at the 2nd British Hydrological Society International Conference, Imperial College, London, 12–16 July 2004. In: *Hydrology: Science & Practice for the 21st Century, Proceedings of 2nd British Hydrological Society International Conference, Imperial College London 12-16 July 2004*, Volume 2, 198–212.

Whiteman M.I., Skinner A., José P.V. and McNish J., 2006. Eco-hydrological Guidelines for Wetland Management. Paper presented at the international conference HydroEco2006 on 'Hydrology and Ecology: The Groundwater/Ecology Connection', Karlovy Vary, Czech Republic, 11-14 September 2006. In: Kovar-Hrkal-Bruthans (eds.) *Proceedings of HydroEco 2006 International Conference – Karlovy Vary, Czech Republic*, 337–40.

Young A.R., Grew R. and Holmes M.G.R., 2003. Low Flows 2000: a national water resources assessment and decision support tool. *Water Science & Technology*, 48, 119–26.

5.6

Groundwater Quality and Quantity Assessment Through a Dedicated Monitoring Network: The Doñana Aquifer Experience (SW Spain)

Marisol Manzano[1], Emilio Custodio[2], Carlos Montes[3] and Carlos Mediavilla[4]

[1] *Technical University of Cartagena (UPCT), Department of Mining, Geological and Topographical Eng., Cartagena, Spain*
[2] *Technical University of Catalonia (UPC), Department of Geotechnics, Barcelona, Spain*
[3] *Autonomous University of Madrid (AUM), Department of Ecology, Madrid, Spain*
[4] *Geological Institute of Spain (IGME), Plaza de España, Sevilla, Spain*

5.6.1   Introduction
5.6.2   Background to Aquifer Geology and Hydrogeology
5.6.3   Needs and Design of a Dedicated Observation Network
5.6.4   Operational Results of the New Monitoring Network: Successes and Failures
5.6.5   Discussion and Conclusions
References

5.6.1   INTRODUCTION

The Doñana aquifer system is located along the SW coast of Spain, between the Guadalquivir and Tinto rivers, not far from the Portuguese border (Figure 5.6.1). Doñana is also the name of a well-known, large natural area in Europe which has been declared a Biosphere Reserve, Ramsar Site and Natural World Heritage Site. Nowadays

Groundwater Monitoring Edited by Philippe Quevauviller, Anne-Marie Fouillac, Johannes Grath and Rob Ward
© 2009 John Wiley & Sons, Ltd

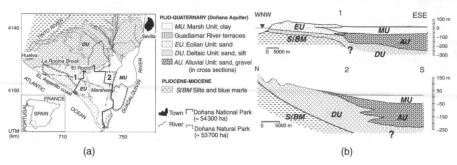

(a)   (b)

Figure 5.6.1 Location, geology and geometry of the Doñana aquifer system. Modified after Manzano, M.; Custodio, E.; Iglesias, M.; Lozano, E. 2008. Groundwater baseline composition & geochemical controls in the Doñana aquifer system (SW Spain). In: Natural Groundwater Quality. W. M. Edmunds & P. Shand (eds.), Blackwell. Chapter 10: 216–232.

more than $1080 \, km^2$ (108,000 ha) of the Doñana area is protected by Spanish law due to its significant environmental value. It contains a great diversity of ecosystems and these include four very different ecodistricts: marshes, aeolian mantles, coastal lines and the Guadalquivir river estuary. It is the habitat of hundreds of plant species, home to threatened animal species like the Imperial Eagle and one of the most endangered mammals in the world, the Iberian lynx. It is also the most important wintering site for waterbirds in the Mediterranean region (Rendón *et al.*, 2008).

Most of the main Doñana area is inhabited. Around the protected zones (Doñana National and Doñana Natural Parks) large areas are under intensive cultivation, and there are also some towns and tourist centres. Human settlement inside the area started in the 1930s and 1940s following reclamation of part of the Guadalquivir marshes for rice cultivation, and later on by introducing pine tree forests. The pine forest plantations failed in the areas with sandy soils and a shallow water-table. Subsequently, extensive planting of eucalyptus trees for timber was successful outside the marshes. These eucalyptus trees replaced the native trees and shrub vegetation in large areas, including some phreatophyte species in the shallow water–table areas.

Late in the 1970s, outside the present protected zones large areas were developed for irrigation by exploiting local groundwater in the sands around the marshes. This was the result of a development plan implemented following studies sponsored by United Nations Food and Agriculture Organization and the Spanish Government (FAO, 1972). Tourism developed as an important economic activity in the 1960s. It is mainly beach-based and is concentrated both seasonally (spring and summer) and spatially (Matalascañas and Mazagón coastal resorts with a capacity of about 300,000 people). Other planned developments were halted in order to protect the area. At the same time, the World Wildlife Fund (WWF) bought and protected for the first time a sector of the territory, which afterwards became the core of the Doñana National Park.

Since 1969 relevant events both for environment protection and for economic development of the area took place in a sequential way (Table 5.6.1). Nowadays the area supports significant agricultural activity and beach–based tourism, whilst at the same time the protected territory has been enlarged and environmental tourism is increasing (Fernández-Delgado, 2006).

Table 5.6.1 Chronology of relevant events for the Doñana environmental conservation (italics) and economic development (normal).

Year	Relevant event
1969	*Creation of the Doñana National Park*
1977	The Spanish Government approved the Almonte-Marismas Agricultural Development Plan (PATAAM). 14,000 ha were planned to be converted into groundwater irrigated land. Some 300 wells were drilled in the 1970s
1979 to 1989	Land parcels are progressively distributed to farmers
1980	*Declaration of Biosphere Reserve (UNESCO-MAB)*
1982	*Declaration of RAMSAR Site (Humid zone of International Importance)*
1988	*Enactment of a Land Plan for the Doñana Region*
1994	*Declaration of UNESCO World Heritage Site*
1990	The PATAAM development is stopped by the Andalusian Government. A study to assess both environment and aquifer sustainability was undertaken. Less than 10,000 ha had been transformed up to this point.
1991	Report to the Government of Andalusia of an 'ad hoc' International Group of Experts for the sustainable development of Doñana

Enactment of the European Water Framework Directive (WFD) in 2000, and of the Groundwater Protection Directive (GWD) in 2006, mean a new paradigm on water cycle understanding and management. Within this paradigm European Member States are driven, among other, by mandatory activities to delineate their water bodies, to analyse the pressures and impacts on them, to monitor their present status and evolution and to protect aquifers, wetlands, rivers and lakes. Groundwater monitoring is a paramount task which implies the construction of dedicated networks to monitor the aquifer flow pattern and quality status, their evolution and the relationships with other water bodies.

As said before, the Doñana aquifer has a national relevance because of its essential role in supporting the ecosystems of the Doñana Nature Reserve. This work compiles the experience gained during the 1990 in the Doñana aquifer, well before the existence of the WFD, after the building up of a dedicated groundwater monitoring network designed to consider the 3-D nature of the flow pattern and the associated groundwater chemical composition.

5.6.2 BACKGROUND TO AQUIFER GEOLOGY AND HYDROGEOLOGY

The climate in the Doñana area is Mediterranean sub-humid, with Atlantic influence: dry summers and wet winters. Mean yearly rainfall is concentrated between October and March and ranges between 500 and 600 mm/year. Inter–annual variability is important and ranges from 250 to 1100 mm. Mean yearly temperature is around 17 °C near the coast and 18 °C in the centre of the area. The mean number of annual sun-hours is close to 3000.

The aquifer system consists of detrital, unconsolidated Plio–Quaternary sediments overlapping impervious Miocene marine marls (Figure 5.6.1). The Pliocene materials are mostly poorly permeable (marls, silts and sandy silts). The Quaternary consist on deltaic and alluvial silts, sands and gravels to the north and southeast, and alluvial and eolian sands to the west. To the southeast the sands and gravel layers are covered by estuarine and marshy clays and silts. Whichever the age, the sediments are dominantly formed by amorphous silica grains, with minor constituents of K- and Na-feldspars, illite, chlorite and kaolinite. Carbonates may be present either as detrital grains or as shell remains, except in the upper sand layers of the western sector, from where they have been dissolved by the locally acidic rainfall recharge.

The Quaternary layers thicken from N to S and from NW to SE. To the SE, the coarse sediments are covered by a thick (50–80 m) sequence of estuarine and marshy clays separated from the ocean by a recent sand spit. The aquifer has a surface area of about 2700 km^2 and variable thickness, from some 20 m inland to >150 m at the coastline. At a regional scale the upper part of the aquifer shows two lithologic domains: a sandy one to the N and W of the marshes, with extensive areas of eolian sands blanket, which roughly behaves as an unconfined aquifer, and a clayey one in the marsh area (1800 km^2 of surface), under which the aquifer is confined.

Recharge occurs mainly by rain infiltration in the sandy areas. Under natural conditions discharge took place as seepage to the ocean, to the streams and the many wetlands on top of the sands, as phreatic evapotranspiration, as lateral and upward flows around

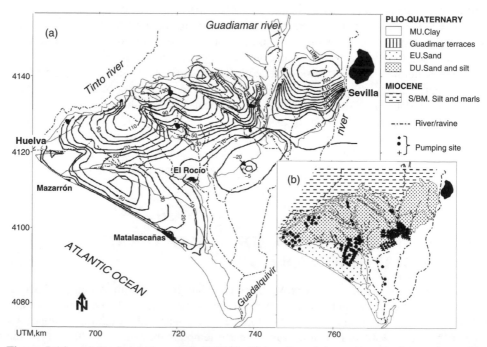

Figure 5.6.2   (a) Regional piezometry in 1997; (b) location of the main groundwater abstraction sites. (Modified after UPC, 1999).

the marshes and through the Quaternary clays. However the increased evapotranspiration by the eucalyptus tree plantations, and the intensive groundwater abstraction since early in the 1980s for crop irrigation and water supply (to the tourist areas), resulted in a piezometric and water-table drawdown (Trick and Custodio, 2004; Custodio and Palancar, 1995). The pumping was concentrated in the unconfined area close to the marshes (Figure 5.6.2) and as a result has partially depleted natural discharge to springs, streams, phreatophyte areas and seepage, and has induced changes in vegetation and in wetland hydrology (Suso and Llamas, 1993; Coleto, 2003; Manzano *et al*., 2005; Custodio, 2000; Custodio *et al*., 2009). The SE sector of the confined area contains almost stagnant, old marine water which has not been flushed out due to the low hydraulic head prevailing since the late Holocene sea level stabilisation, some 6000 years ago (Zazo *et al*., 1996; Manzano *et al*., 2001).

5.6.3   NEEDS AND DESIGN OF A DEDICATED OBSERVATION NETWORK

After the initial hydrogeological studies of the 1970s for the PATAAM (Table 5.6.1) a further set of studies was undertaken late in the 1980s. Several hydrodynamic and groundwater flow modelling studies at regional and local scales have now been carried out to understand the aquifer behaviour (Trick and Custodio, 2004) and its relationship to the wetlands in the water-table areas (Lozano *et al*., 2005; Coleto, 2003; Manzano *et al*., 2007b). The whole aquifer system has been modelled at various times with the last one following the large Aznalcóllar mining accident in the Guadiamar river, a tributary to the area (UPC, 1999). Most of the studies (supported by public research funding) have been carried out by University teams. These include the Complutense University of Madrid (UCM), the Technical University of Catalonia (UPC) and the Technical University of Cartagena (UPCT). Hydrochemical and environmental isotopes studies have also been performed to help develop a conceptual model for the origin and evolution of groundwater chemistry, to get insight into groundwater recharge and transit times, and to understand the role of groundwater in the wetlands (Baonza *et al*., 1984; Poncela *et al*., 1992; Iglesias, 1999; Manzano *et al*., 2001, 2008). Much of the work has concentrated in the westernmost recharge area, roughly coinciding with the outcrop of the Eolian Unit (Figure 5.6.1a), because a large part of the groundwater flowing to the marshes within the National Park originated from this sandy area and also because it holds some of the most active agricultural areas and the most productive water wells.

A piezometric and a nitrate monitoring network designed to observe the possible impact of groundwater abstractions and agricultural activities has been in operation since early in the 1980s. This monitoring has been carried out by the Geological and Mining Institute of Spain (IGME) for the Guadalquivir Water Authority (CHG). However, until 1992 the network consisted mostly of multiscreened boreholes and agricultural wells. Thus, both groundwater samples and piezometric measurements integrated flow lines with different transit times and heads. In many areas groundwater has important vertical flow components, either downward or upward. The information obtained from the monitoring was insufficient to identify vertical gradients, chemical variations with depth and travel times. In the 1980s electrical conductivity, downhole temperature logging

(Custodio *et al.*, 1996) and saline tracing tests confirmed the existence of vertical flows (Figure 5.6.3) and guided the subsequent design and installation of a dedicated discrete level monitoring network in the 1990s.

The design chosen consisted of clusters of boreholes, at up to 65 sites. Each cluster consisted of 2–4 boreholes installed to different depths with single, short well screens (2 m) at the bottom (Figure 5.6.4). The work was financed by the Guadalquivir River Basin Water Authority, which operates it.

In the last decade, IGME has also drilled several clusters of dedicated boreholes, in collaboration with UPC and UPCT. These monitor groundwater levels in areas of special interest and ecological relevance for research studies.

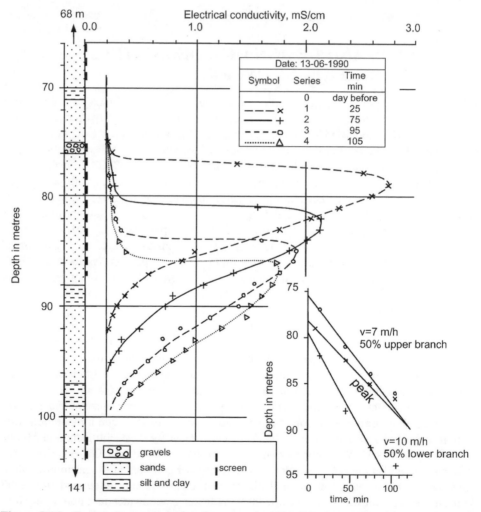

Figure 5.6.3 Downhole tracing with saline water in 1990. Dilution profiles show the existence of vertical downward flows in most of the water table area and supported the design of new observation boreholes with different depths and a single, short screen.

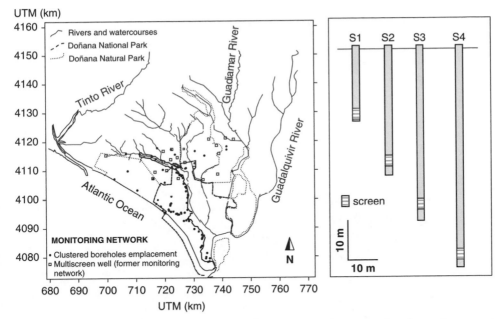

Figure 5.6.4 (Left) Location of the clustered boreholes emplacements that forms the current monitoring network. (Right) Typical emplacement of clustered boreholes with usual dimensions.

Drilling was designed to be carried out without adding water and chemicals that could induce chemical and isotopic changes to groundwater. However, the characteristics of the eolian sands (well sorted and fine) required that drilling fluid additives like bentonite and polymers had to be used in many cases. This has caused some difficulties in obtaining representative groundwater samples.

5.6.4 OPERATIONAL RESULTS OF THE NEW MONITORING NETWORK: SUCCESSES AND FAILURES

In general, the new monitoring network constructed between 1990 and 1996 has been a success. Whereas the studies carried out before 1990 (Baonza *et al.*, 1984; Poncela *et al.*, 1992) had many uncertainties because of the inability to sample along individual flow lines using the available multiscreened wells, the new borehole clusters have allowed the 3–D groundwater flow system to be defined by allowing measurement of vertical head evolution. The new network also granted better definition of flow paths and facilitated the study of chemical evolution and travel times.

A significant number of boreholes were equipped with pressure transducers, and the ones that were not equipped were manually measured on a monthly basis. After only a few months of data the evidence for vertical gradients in different areas was clear (Figure 5.6.5), and after a few years of records the time series has become very useful

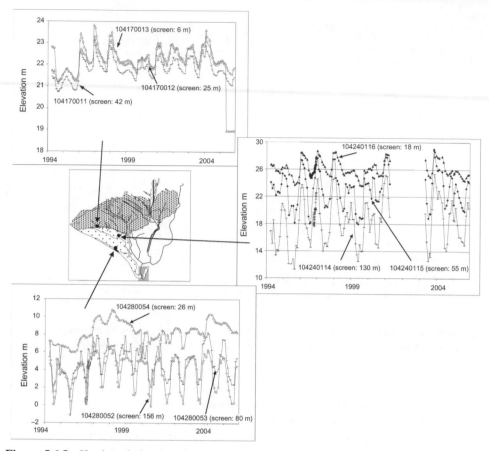

Figure 5.6.5 Head evolution in selected emplacements of clustered boreholes. Screen depth in each borehole is indicated between brackets to show the existence of vertical head gradients.

for recharge calculation and numerical flow model calibration (Trick and Custodio, 2004; UPC, 1999; Lozano, 2004). Currently network monitoring continues for administrative monitoring purposes and for research studies, for example to improve knowledge of wetlands hydrology in the sand area.

Also, the spatial heterogeneity of groundwater–surface water relationships was revealed, especially in relation to wetlands. Most wetlands in the Doñana region are groundwater-dependent, and those located on the sandy recharge area are very sensitive to groundwater flow modifications. Intensive groundwater abstraction for more than two decades in natural discharge areas nearby the marshes has led to a transient period of flow pattern changing which is lasting for decades. Accumulated piezometric drawdowns in the more permeable, deep aquifer layer have induced a water-table lowering, and seepage to wetlands and to phreatophyte areas located in the aquifer's sandy recharge areas has decreased. This resulted in severe hydrological damage to most wetlands by decreasing the inundation and/or the soil saturation frequency and duration (Lozano, 2004; Manzano *et al.*, 2007b).

Groundwater baseline composition was established, as well as the following conceptual model to explain groundwater chemistry origin and evolution along flow lines (Figure 5.6.6) (Iglesias, 1999; Manzano *et al.*, 2001, 2008):

- The unconfined parts of the aquifer contain low mineral content freshwater of the Na–Cl type to the W and of the Ca-HCO$_3$ type to the N, with low hardness and being slightly acidic in the upper part of the western area. Groundwater becomes harder and more alkaline where the terrain still contains remnants of calcite shells. Anthropogenic effects from airborne pollutants and agricultural activities are limited to the upper part (<30 m) of the aquifer. The main processes controlling groundwater chemical evolution are: rain water composition, which is of Na–Cl type; sulphur and other anthropically-derived components from the nearby Huelva industrial site; equilibrium with silica; dissolution of soil CO$_2$; dissolution of Na/K feldspars; dissolution of CaCO$_3$ in the deepest layers of the western unconfined area and everywhere in the northern one. The resulting groundwater mineralization in the unconfined areas ranges from low to moderate (<0.2 to 1 mS.cm^{-1}).

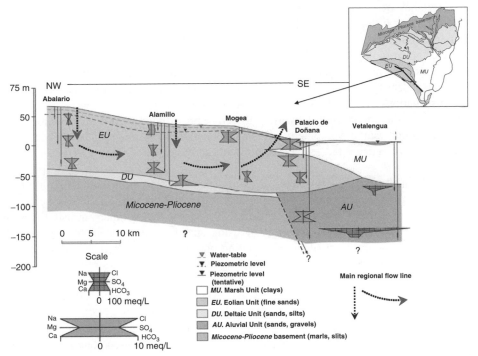

Figure 5.6.6   Vertical evolution of groundwater chemistry in the western water table area as resulted from periodic sampling of five clustered boreholes emplacement. Each diagram is placed at the screened section of the corresponding borehole. Groundwater flow sense is derived from head measurements. Modified after Manzano, M.; Custodio, E.; Iglesias, M.; Lozano, E. 2008. Groundwater baseline composition & geochemical controls in the Doñana aquifer system (SW Spain). In: Natural Groundwater Quality. W. M. Edmunds & P. Shand (eds.), Blackwell. Chapter 10: 216–232. (See Plate 4 for a colour representation).

- Following the main groundwater flow paths from the unconfined to the confined parts of the aquifer (from N to S and from W to E) groundwater becomes increasingly saline due to mixing with old marine water trapped both in the confined permeable layers and the clays. A broad mixing zone develops from NW to SE under the marshes. In the confined sector of the aquifer groundwater composition changes due to: mixing with modified old marine water; equilibrium with calcite; cation exchange (Na/Ca–Mg) in moving fresh–saline waters fronts; sulphate reduction (depletion with respect to conservative mixing with sea water); probable C incorporation from sedimentary organic matter in the clays. The resulting salinities in the confined aquifer range from 1 up to $80\,\text{mS.cm}^{-1}$; the highest values relate to saline water evaporated in old marshes.

Hydrogeochemical and environmental isotope studies currently under way (Manzano *et al.*, 2007a) show that although groundwater composition dominates wetlands chemistry at a regional scale, reactions taking place within the lagoons have a major influence on water salinity, ion composition and isotopic concentration of the nearby shallow groundwater. They also show that the hydrogeochemical processes taking place within some wetlands are changing due to the modifications of wetlands flooding frequency and duration, which in turn is derived from the lowering of the water-table due to intensive groundwater abstraction.

Recent microbiological studies (Velasco *et al.*, 2009) clarify the hydrological connectivity existing between surface water in the lagoons and shallow groundwater. These studies also identify the importance of this connection in controlling the spatial and temporal distributions of microbial communities and their controls on ecological functions.

The use of the observation network is therefore providing new information about the impact of groundwater development and management decisions on groundwater discharge to wetlands by seepage and evapotranspiration, and also on wetland and groundwater chemistry.

An additional benefit of the new monitoring network has been their suitability for carrying out groundwater dating and residence time studies. Tritium studies helped to check the groundwater flow model in different areas and to estimate transit times. A significant advance was the possibility of using specific flow models to interpret ^{3}H data combining the different types of sampling points available, their location either in a recharge or discharge areas, and the sampling methods used (Iglesias, 1999; Manzano *et al.*, 2007a). The work was supported by mixing modelling with the code MULTIS (Richter *et al.*, 1992). This and further works (Manzano *et al.*, (2001, 2008)) provided good knowledge about transit times through the aquifer (Figure 5.6.7).

More recently, the possibility of sampling individual flow lines to characterize (for the first time) microbial communities, bacterial abundance, cell biomass, bacterial biomass and microbial activities of functional groups, is providing insight into the ecological functions played by microbial communities along groundwater flow lines related to wetlands (Velasco *et al.*, 2009).

As well as the successes there are also some failures. Poor sealing of connections between lengths of borehole/well casing and breakages, especially in boreholes drilled through interlayered sands and clays, due to difficulties in introducing the tube, is producing leakage into and out of some of the boreholes. These problems may lead to

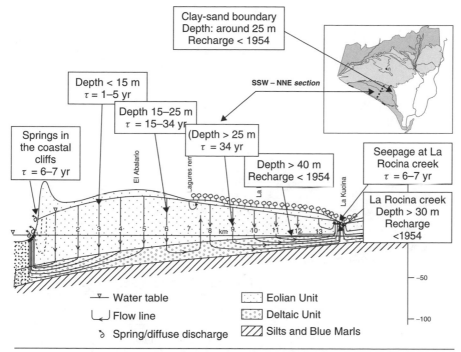

Conceptual flow model in the Littoral Eolian Mantle after ^3H. Estimated transit times (τ) at different depths and locations.

Figure 5.6.7 ^3H based transit times along groundwater flow in the western water table area. Times were quantified by modelling with the MULTIS code (Richter *et al*., 1992). Tritium data belong to clusters of boreholes, springs and short hand-dug wells. Modified alter Manzano, M; Custodio, E.; Higueras, H. 2007a. Groundwater & its functioning at the Doñana RAMSAR site wetlands (SW Spain): role of environmental isotopes to define the flow system. International Symposium in Advances in Isotope Hydrology & its Role in Sustainable Water Resources Management STI/PUB/1310. IAEA, Vienna: 149–160.

poor and inaccurate head measurements. Although this happens rarely, leakage from tube connections also makes it difficult to get representative samples, especially in the deep boreholes (200–300 m) drilled along the contact between the eolian mantle and the marshes (Figure 5.6.8). Layers of high groundwater salinity may produce rapid mixing inside the tube by convection. This requires time consuming purging before sampling, and may produce unreliable data where water quality probes are installed. A combination of these factors has made a few boreholes unusable for hydrochemical and isotopic studies.

The need to use drilling polymers together with the moderate permeability of the sands and the short screen length made it difficult to fully remove the drilling fluid from the aquifer. This may be the cause of sample representativity problems, especially in the short term after drilling. The polymer (drilling fluid additive) remnants are clearly observed where pH values are over 8. Natural pH values are normally 6–7.5. The residual contamination can be removed by purging the tube for long enough periods (3–4 h) before

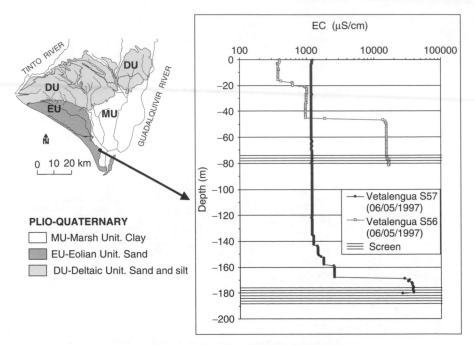

Figure 5.6.8 Electrical conductivity (EC) downhole record in two boreholes (S56 and S57) at Vetalengua emplacement, which belongs to the new monitoring network. The location of the unique screened tranche is shown for both boreholes. The survey was performed a few months after drilling, and the stepped graphs show the entrance of groundwater inside the casing at several depths in both boreholes by this time.

sampling. However this increases the cost of each sampling survey, and limits the number of points that can be surveyed in each field survey.

5.6.5   DISCUSSION AND CONCLUSIONS

Monitoring of groundwater level and quality trends require long-term observations, often at low frequencies and using standardized measurements of groundwater quality (Condesso de Melo *et al.*, 2008). In recent years the European Water Framework (WFD, 2000) and Groundwater (GWD, 2006) Directives have been significant drivers to establish groundwater quality monitoring programmes. To complying with directives requirements means dedicated, objective-oriented designed observation networks are needed. This type of network is still rare across Europe.

In the case of the Doñana aquifer a very effective aquifer monitoring network was established prior to the enactment of the WFD. The aquifer is located in the Guadalquivir River Basin, and has been managed by the Guadalquivir River Basin Water Authority (Confederación Hidrográfica del Guadalquivir, CHG), of the Spanish Ministry of the Environment, until the end of 2008, and currently by the Andalusian Water Agency. The Doñana aquifer sustains a large environmentally protected area of international relevance.

At the same time, surrounding the Doñana National Park are very active agricultural zones irrigated with local groundwater. In the late 1980s the expertise relating to environment protection increased as did agriculture-based economic development. At this time, the CHG recognized the need for a better knowledge of the aquifer in the areas where there is intensive abstraction of groundwater for irrigation close to the protected ecosystems. The international obligations of the Spanish Government on the conservation of the wetland area (Ramsar Convention) played a significant role in establishing the new attitude of the Water Authority.

The design of the new monitoring network was based in a detailed study of the hydrogeological information provided by the previously existing monitoring points, (multiscreened pumping wells) and on geological information derived from a large study supported by FAO and the Spanish Government in the 1970s. Borehole tracer tests was also important for confirming the existence of vertical flows in most of the studied area, and to help design a network of borehole cluster sites covering the most sensitive zones. Despite some operational problems during construction and the difficulty in using some boreholes for water quality monitoring, the data obtained from the new network – time series data on aquifer head, and chemical and isotopic data – has provided a much better understanding of aquifer characteristics. These include recharge and flow rates, groundwater flow patterns, origin and evolution of the natural and acquired groundwater composition, hydrogeochemical processes taking place in different areas of the aquifer, transit times, water age, and much more. More recently, the possibility of sampling individual flow lines for microorganism studies is providing insight into the ecological functions played by microbial communities in shallow groundwater related to wetlands.

Effective groundwater management in this area needs the involvement of stakeholders. They need to accept a share of the responsibility and co-operate in monitoring, surveillance and decision-making. In particular there needs to be adequate participation by nature conservationists. Efforts by public Spanish and Andalusian organizations to incorporate stakeholders have so far been unsuccessful, but this is a goal to be pursued because it is the only effective way of managing water in such a complex system and in which groundwater plays a key role (Custodio *et al.*, 2009). The large number of stakeholders, the rivalry between the different municipal territories and the shortage of funds is the reason for the slow progress. Notwithstanding this, public awareness has clearly advanced since 1992, when the International Experts Report was issued.

REFERENCES

Baonza E., Plata A. and Silgado A., 1984. Hidrología isotópica de las aguas subterráneas del Parque Nacional de Doñana y zona de influencia [Groundwater isotopic hydrology of the Doñana National Park and its area of influence]. Centro de Estudios y Experimentación de Obras Públicas, Madrid. Cuadernos de Investigación, C7: 1–139.

Coleto I., 2003. Funciones hidrológicas y biogeoquímicas de las formaciones palustres hipogénicas de los mantos eólicos de El Abalario-Doñana (Huelva) [Hydrological and biochemical functions of hypogenic lagoon formations of the El Abalario–Doñana (Huelva) eolian mantles]. Doctoral Thesis. Autonomous University of Madrid, Faculty of Biology, Madrid, Spain.

Condesso de Melo M. T., Custodio E., Edmunds W. M. and Loosli H. H., 2008. Monitoring and characterisation of natural groundwater quality. In: *Natural Groundwater Quality*. W. M. Edmunds and P. Shand (eds), Blackwell, Oxford. Chapter 7: 155–77.

Custodio E., 2000. Groundwater–dependent wetlands. *Acta Geologica Hungarica*, **43** (2): 173–202.

Custodio E. and Palancar M., 1995. Las aguas subterráneas en Doñana [Groundwater in Doñana]. *Revista de Obras Públicas*, Madrid, **142** (3340): 31–53.

Custodio E., Manzano M. and Iglesias M. 1996. Análisis térmico preliminar de los acuíferos de Doñana [Preliminary thermal analysis of the Doñana aquifers]. IV Simposio sobre el Agua en Andalucía (SIAGA–96). Almería II: 57–87.

Custodio E., Montes C. and Manzano M. 2009. Las aguas subterráneas en el Área de Doñana: implicaciones ecológicas y sociales [Groundwater in Doñana Area: ecological and social implications]. Agencia Andaluza del Agua. Sevilla.

GWD, 2006. Directive 2006/118/EC of the European Parliament and of the Council of 12 December 2006 on the protection of groundwater against pollution and deterioration (Groundwater Directive). *Official Journal* **372**, 27.12. 2006.

Fernández-Delgado C., 2006. Conservation management of a European natural area: Doñana National Park, Spain. In: Groom, M. J., Meffe, G. K., Carroll, C. R. (eds), *Principles of Conservation Biology*. Third ed. Sinauer Associates. Sunderland: 536–43.

Iglesias M., 1999. Caracterización hidrogeoquímica del flujo del agua subterránea en El Abalario, Doñana, Huelva [Hydrochemical characterization of groundwater flow in El Abalario, Doñana, Huelva]. Doctoral Thesis. Civil Engineering School, Technical University of Catalonia, Barcelona, Spain.

Konikov L. F. and Rodríguez-Arévalo F. J., 1993. Advection and diffusion in a variable-salinity confining layer. *Water Resources Research*, **29** (8): 2747–61.

Lozano, E., 2004. Las aguas subterráneas en los Cotos de Doñana y su influencia en las lagunas. [Groundwater in the Cotos of Doñana and its influence on lagoons]. Doctoral Thesis. Civil Engineering School, Technical University of Catalonia, Barcelona, Spain.

Lozano E.; Delgado F., Manzano M., Custodio E. and Coleto C., 2005. Hydrochemical characterisation of ground and surface waters in 'the Cotos' area, Doñana National Park, southwestern Spain. In: E. M. Bocanegra; M. A. Hernández y E. Usunoff (eds), *Groundwater and Human Development*. International Association of Hydrogeologists, Selected Papers 6: 217–31.

Manzano M., Custodio E., Loosli H. H., Cabrera M. C., Riera X. and Custodio J., 2001. Palaeowater in coastal aquifers of Spain. In: Palaeowaters in Coastal Europe: Evolution of Groundwater since the late Pleistocene. (Edmunds, W. M. and Milne, C. J., eds.). Geological Society London, Sp. Publ. 189: 107–38.

Manzano M., Custodio E., Mediavilla C. and Montes C., 2005. Effects of localised intensive aquifer exploitation on the Doñana wetlands (SW Spain). Groundwater Intensive Use. Intern. Assoc. Hydrogeologists, Selected Papers, 7, Balkema, Leiden: 295–306.

Manzano M., Custodio E. and Higueras H., 2007a. Groundwater and its functioning at the Doñana RAMSAR site wetlands (SW Spain): role of environmental isotopes to define the flow system. International Symposium in Advances in Isotope Hydrology and its Role in Sustainable Water Resources Management STI/PUB/1310. IAEA, Vienna: 149–60.

Manzano M., Custodio E., Iglesias M., Lozano E. and Higueras H., 2007b. Relationships between wetlands and the Doñana coastal aquifer (SW Spain). In: L. Ribeiro, A. Chambel & M.T. Condesso de Melo (eds), *Proc. of the XXXV Intern. Assoc. Hydrogeologists Congress: Groundwater and Ecosystems. Lisbon*. In CD, ISBN: 978-989-95297-3-1.

Manzano M., Custodio E., Iglesias M. and Lozano E., 2008. Groundwater baseline composition and geochemical controls in the Doñana aquifer system (SW Spain). In: Edmunds, W. M. and Shand, P. (eds.), *Natural Groundwater Quality*. (Edmunds, W. M. and Shand, P., eds.) Blackwell, Oxford. Chapter 10: 216–32.

Poncela R., Manzano M. and Custodio, E., 1992. Medidas anómalas de tritio en el área de Doñana [Anomalous tritium measurements in the Doñana area]. Hidrogeología y Recursos Hidráulicos, Madrid. XVII: 351–65

Rendón M. A., Green A. J., Aguilera E. and Almaraz, P., 2008. Status, distribution and long-term changes in the waterbird community wintering in Doñana, South-West Spain. *Biological Conservation*. **141**: 1371–88.

Richter J., Szymczak P. and Jordan H., 1992. A computer program for the interpretation of isotope hydrogeologic data. Tracer Hydrology. *Proceedings of the 6th Int. Symp. on Water Tracing, Karlsuhe*: 461–462.

Suso J. and Llamas M. R., 1993. Influence of groundwater development of the Doñana National Park ecosystem (Spain). *Journal of Hydrology*, **141** (1–4): 239–70.

Trick Th. and Custodio E., 2004. Hydrodynamic characteristics of the western Doñana Region (area of El Abalario), Huelva, Spain. *Hydrogeology Journal*, **12**: 321–35.

UPC, 1999. Modelo regional de flujo subterráneo del sistema acuífero Almonte–Marismas y su entorno [Regional groundwater flow model of the Almonte–Marismas aquifer and its surroundings]. Grup d'Hidrologia Subterrània/Dept. Enginyeria del Terreny/Universitat Politècnica de Catalunya. Barcelona. 1–144 + annexs. (internal).

Velasco S., Acebes P., López-Archilla A. I., Montes C. and Guerrero M. C. 2009. Environmental factors controlling the spatiotemporal distribution of microbial communities in a coastal, sandy aquifer system (Doñana, SW Spain). *Hydrogeology Journal*, **17**: 766–780.

WFD, 2000. Directive of 23 October establishing a framework for Community action in the field of water policy (Water Framework Directive). Official Journal L **327**, 22.12.2000.

Zazo C., Goy J. L., Lario J. and Silva P. G., 1996. Littoral zone and rapid climatic changes during the last 20,000 years: the Iberian study case. Z. Geomorph. N.F.; Berlin–Stuttgart, Suppl. **102**: 119–34.

5.7

Llobregat Delta Aquifer

Josep Mª Niñerola[1], Enric Queralt[2] and Emilio Custodio[3]

[1] Agència Catalana de l'Aigua (ACA), Àrea de Planificació Provença, Barcelona, Spain
[2] Comunitat d'Usuaris d'Aigua Subterrània del Baix Llobregat (CUADLL), Barcelona, Spain
[3] Technical University of Catalonia (UPC), Department of Geotechnics, Barcelona, Spain

5.7.1   INTRODUCTION

The Llobregat river lower valley and delta are located at the southwest of the Barcelona's Metropolitan Area (Figure 5.7.1). They consist of about $30\,km^2$ of alluvial valley, up to 1 km wide, and a delta of $80\,km^2$. The system contains very transmissive aquifers fed by rainfall infiltration and small creeks draining the mountainous boundary, and especially and dominantly by river water infiltration in the bed and in the irrigated agricultural areas along the valley, which use river water brought by two canals. The importance of the aquifer system stems from being a key groundwater source and reserve for this densely populated area (ca. 4 million), the associated industrial settlements, and the remaining irrigated agricultural fields.

Until the midst of the nineteenth century, the area was mostly rural, although some factories were using hydraulic power and water from the river, mostly for the textile

Groundwater Monitoring Edited by Philippe Quevauviller, Anne-Marie Fouillac, Johannes Grath and Rob Ward
© 2009 John Wiley & Sons, Ltd

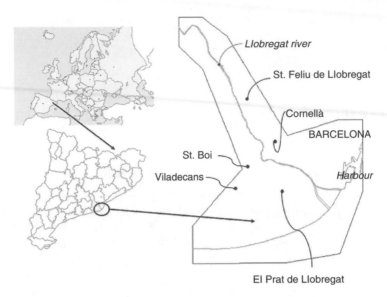

Figure 5.7.1   Llobregat's lower valley and delta aquifers situation map.

industry. Groundwater abstraction development was not significant. In the last third of the nineteenth century, flowing deep artesian wells in the delta were successfully drilled, and abstracted water greatly improved local health and irrigation of crops, as well as attracted factories. Also in the same period, the Barcelona's Water Supply Company built up several deep, large diameter wells at the valley mouth to pump water by using steam engines and piston pumps, and to distribute it to Barcelona after being impelled to water storage reservoirs in the surrounding mountains. Different authors described the situation at the time, such as Santa María and Marín (1910). When the flowing wells ceased to flow due to increasing abstraction rate they were substituted by drilled wells with pumps, but it was not until the 1940s that turbine pumps became easily available and this prompted an exponential growth of groundwater abstraction, especially in the 1950s and early 1960s. The aquifer system, easily recharged, was a key factor for town supply and for industrial settlements demanding large quantities of water (textiles, paper, chemical, machinery, cars).

The consequences of this intensive development were a deep piezometric drawdown cone in the confined delta aquifer and in the water-table aquifer in the lower valley, with conspicuous seasonal fluctuations (reduced storage) and problems in dry years. Also a penetrating seawater intrusion body developed in the deep aquifer, that was already clearly noticed in mid the 1960s. A series of wells were salinized, although they continued in operation, mostly for cooling purposes. This helped in delaying the spread of the saline groundwater body.

With the exception of a pioneer study of the early twentieth century (Santa Maria and Marin, 1910), the first detailed hydrogeological studies started in the mid-1960s (MOP, 1966; Llamas and Molist, 1967; Custodio, 1967, 2007; Gámez, 2007), and a piezometric monitoring network consisting in about 40 points was installed, with several tubes of different depth in each emplacement. The idea of combined use of surface and groundwater was born at that time (Llamas, 1969).

The Barcelona's Water Supply Company was concerned by the deteriorating situation and the possible depletion of groundwater reserves in summer, especially in dry years. Already in the 1940s it started enhanced recharge in the upper part of the lower valley by careful scrapping of river bed in favourable moments and in a tract where alluvium was free of fine materials layers. Late in the 1960s excess treated river water in a recently constructed treatment plant was recharged in dedicated drilled wells and in dual purpose wells, that is, wells that could be used for recharge and for abstraction (Custodio *et al.*, 1977).

Also the different factories in the area were concerned and after a technical meeting with hydrogeologists of the Water Authority (CAPO at that time) and the Technical University of Catalonia (UPC) in 1970 they decided to invest in reducing pumping to fight the increasing pumping costs due to piezometric level depletion and the already well perceived risk of seawater intrusion. As a consequence a Groundwater Users Community was created (CUADLL) to introduce self–management by local factories and suppliers, including farmers, and to collaborate and obtain effective control measures from the Water Authority. This was a great accomplishment that lasted up to the present, and around it a series of activities have been carried out or are under project. Current groundwater abstraction is only a fraction of that in the early 1970s, and groundwater use has changed objectives (Figure 5.7.2). Currently the aquifer system is a highly valuable and key element for local supply in periods of droughts and emergencies, and functions close to a large surface reservoir, but with a series of advantages and also drawbacks.

Groundwater management is quite complex due to the progressive urbanization of the area and the large number of transport infrastructures (roads, railroads, airport, large harbour and its service areas ...), which means a reduction of recharge and water quality

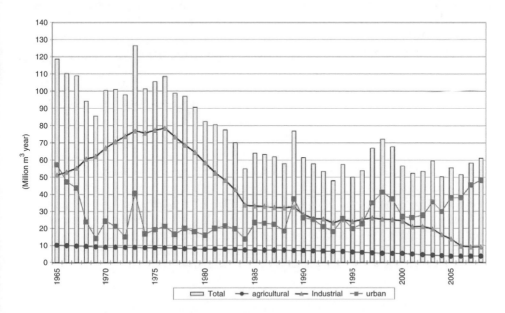

Figure 5.7.2 Llobregat's lower valley and delta groundwater abstractions evolution, 1965–2008.

risks. Artificial recharge is to be increased by means of basins. An injection saline water barrier started its operation in 2007 and is being enlarged. Recharge water comes from advanced tertiary treated waste water with final reverse osmosis salinity reduction.

5.7.2   AQUIFER DESCRIPTION

The delta is formed by a sedimentary package ranging from Pliocene to Quaternary in age, even though its current size is the result of the delta's progradation during the last Holocene post–glacial marine high stand. Figure 5.7.3 shows the schematic geologic profile of the Llobregat's low Valley and Delta. In the low Valley there is an unconfined aquifer and in the delta there are one upper and unconfined aquifer and one deep and confined aquifer. These two aquifers partially merge at the delta boundaries. A body of silt, sandy clays and clays separate both aquifers, the thickness increasing towards the sea, at the time it produces a more effective isolation of the two superimposed aquifers (Manzano *et al.*, 1986; Marqués, 1975; Gámez, 2007). The most important aquifer for human water supply is the one of the low valley and deep delta's aquifer, this last being also economically important for factories. Further descriptions can be found in Custodio (1981, 2007).

The piezometric levels of the low Valley and deep delta aquifers are principally controlled by groundwater abstraction at Cornellà and El Prat (Figure 5.7.4). These levels have been below mean sea level for more than fifty years, which led to the existing deep saline intrusion (Bayó *et al.*, 1977; Custodio, 1981, 1987). Also the excavation of docks of the Barcelona harbour inshore, in the south-eastern part of the delta, where both the upper and the deep aquifers merge, is a serious cause of saline water inflow. Seawater intrusion was early detected in the 1960s studies and has been numerically modelled (Abarca *et al.*, 2006; Iribar *et al.*, 1997; Vázquez–Suñe *et al.*, 2006). The intermediate aquitard in the area close to the coast also contains old marine water trapped during its formation, which is mostly immobile (Manzano *et al.*, 1992).

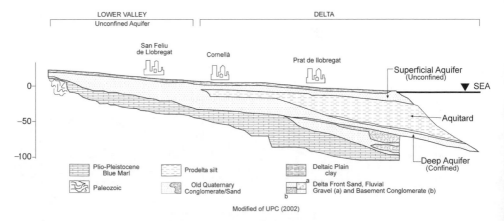

Figure 5.7.3   Schematic longitudinal cross-section of the Llobregat lower valley and delta alluvial formations.

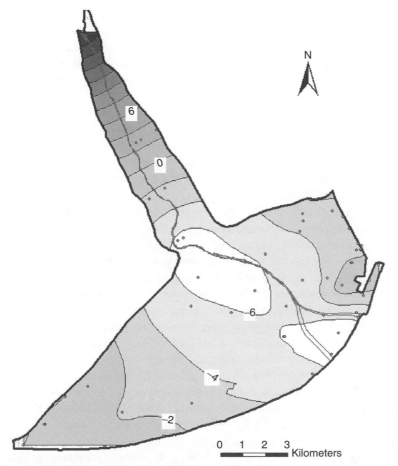

Figure 5.7.4 Piezometric level map of the Llobregat's lower valley and deep delta aquifers (average values 1999–2009) and the monitoring network (points).

5.7.3 AQUIFER STATUS

Intensive groundwater abstraction has produced quantity and quality problems, which prompted the Water Authority to carry out studies and modelling. This was behind the creation of the Groundwater Users Community (CUADLL) aiming to improve the situation and halt the disposal of contaminants in the territory, especially as filling materials of former gravel pits. Now, after the 1985 Water Act, and especially after the enactment of the European Water Framework Directive in 2000, the aquifer status has to be monitored, deterioration trends reversed and action is needed to move towards a good chemical (and also quantitative) status. Current situation has been already assessed (Niñerola and Ortuño, 2008; Niñerola, 2008).

The current Water Authority, the Catalan Water Agency (ACA), has a quantity monitoring network that comprises fifty piezometers in the lower valley and the deep delta aquifer (0.4 points/km^2) (Figure 5.7.4) and about thirty more in the unconfined aquifer

(0.3 points/km^2). Some of this control points are used since the 1960's when the Water Authority of the Eastern Pyrenees Basins (CAPO), together with the Public Works Geological Service (SGOP) started the groundwater's surveillance programme (MOP, 1966). Afterwards the Study for the Hydrologic Plan of the Eastern Pyrenees basins of 1985 enlarged this monitoring network, which nowadays remains almost unchanged. Current trend is towards automatic piezometric level measurement.

There is also a groundwater quality monitoring network based on chemical analysis of operating water wells. This chemical network has two origins, firstly created by the groundwater users and secondly by the Water Authority. Prat's Council, first, and then through Aigües del Prat S.A. (a public sector water supply company) monitored industries and wells from country houses, as well as the water supply of Viladecans, El Prat, and Sant Boi towns.

Since the 1970s, there is a noticeable, progressive salinization of the delta confined aquifer, the most important contamination that suffers this area. This is the reason why a salinity monitoring network was created. Later on, in order to know how the chemical status of the water impacts on agricultural production, the Baix Llobregat Consortium of the Agrarian Park, a public sector entity, started sampling waters from the confined and the unconfined aquifer in 2000. In 2005, the different monitoring networks (the one created by Aigües del Prat, the one by the Agrarian Parc and the wells of the Barcelona's Water Supply Company), were integrated to form the monitoring network of the Groundwater Users Community (CUADLL) (Solà, 2004).

The chemical parameters usually analyzed are chosen according to historical experience. Electrical conductivity, temperature, pH and Eh are measured in the field. Otherwise, in the lower valley and confined delta area aquifers, the laboratory analyzes physicochemical values and major ions (electrical conductivity, pH, Na, K, Ca, Mg, Cl, HCO_3, SO_4, NO_3, NO_2, NH_4, oxydability, anionic tensioactive agents and TOC), heavy metals (Fe, Mn, Cu, Zn, Cr–VI, As, Hg, Pb), halogenated volatile organic compounds (like TCE and PCE) and non–halogenated VOC's, as well as microbiological parameters. Pesticides are analyzed in the unconfined delta aquifer instead of VOC's.

The Water Authority (now the Catalan Water Agency, ACA), after a period of irregular controls in the unified monitoring network, established a periodic surveillance for each point in 1994. Then, since 2006, the Water Authority and the Groundwater Users Community have unified criteria avoiding repetitions and optimizing efforts, achieving a network's density of 1 point/km^2 (Figure 5.7.5).

Specific monitoring is designed whereby a new contamination appears or when the public administration projects a new large infrastructure.

So Llobregat's aquifers are quantitatively and chemically monitored and this allows evaluating the groundwater status at a large scale. Besides, many construction works and big infrastructures have been constructed recently in the Llobregat's delta area, so that each construction has an associated specific monitoring plan in order to control the possible impacts on the aquifers and on the deltaic environment. A flow and transport numerical model was prepared in 2004 by the Groundwater Hydrology Group of the Geoengineering Department, Technical University of Catalonia (UPC), for the ACA, and modified by the Foundation International Centre for Groundwater Hydrology (FCIHS), also for the ACA, and recalibrated by the CUADLL. It is used to assess these impacts,

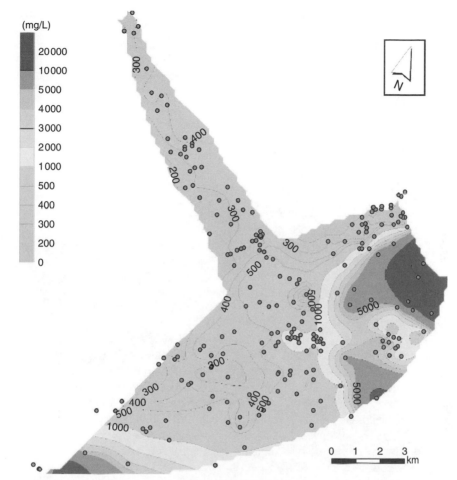

Figure 5.7.5 Isochloride (mg/L Cl) maps of the Llobregat's lower valley and deep delta aquifers in 2007. Points correspond the monitoring sampling network.

and then define the correction and compensatory measures and the specific monitoring for each case.

The knowledge of aquifer status should allow evaluating the impact of any large infrastructure and considering synergies among different ones. The monitoring network allows improving the hydrogeological conceptual model and as a consequence the numerical model.

5.7.4 ISSUES RELATED TO THE EUROPEAN WATER FRAMEWORK DIRECTIVE

During the process of characterization of water bodies (carried out to fulfil requirements of articles 5, 6 and 7 of the 2000/60/CE European Water Framework Directive), it

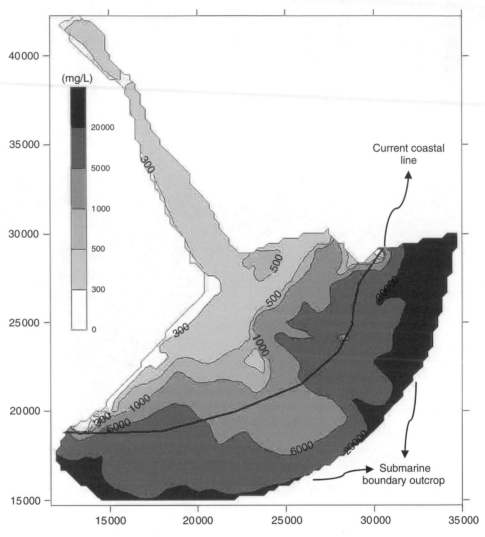

Figure 5.7.6 Chloride (mg/L Cl) content foreseen for 2045 of the Llobregat's lower valley and deep delta aquifers, simulated by the CUADLL with the UPC-ACA flow and transport numerical model, when no corrective measures are undertaken. Elaborated by CUADLL.

was considered that the Llobregat's river lower valley and delta were under risk of not attaining the environmental objectives by 2015. This is partly due to (over)abstraction exceeding recharge, which induces sea water intrusion, mainly in the delta deep confined aquifer. This is also linked to the intensive industrial activity in the whole Llobregat's watershed, to intensive agriculture in the delta, to urban expansion and the establishment of new industrial zones, and to the construction of large infrastructures affecting mainly the upper delta aquifer. The most significant ones are the deviation of the final tract of the Llobregat river, the airport and harbour enlargements, the waste water tertiary

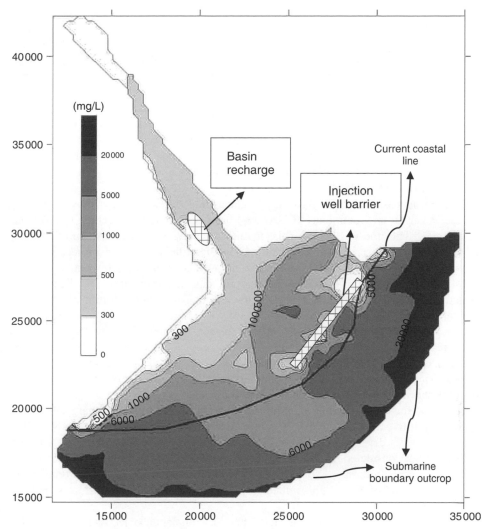

Figure 5.7.7 Chloride (mg/L Cl) content for 2005 of the Llobregat's lower valley and deep delta aquifers, simulated by the CUADLL with the UPC-ACA flow and transport numerical model, considering artificial recharge by infiltration basins (lower valley) and deep injection and pumping barriers (delta). Elaborated by CUADLL.

treatment plant near the coast, which share space with the close to be in operation reverse osmosis desalinization plant, the construction of two 'metro' lines, and the high speed train platform. (Figures 5.7.6 and 5.7.7)

All these stresses negatively affect aquifer recharge and groundwater quality, since they increase surface runoff, modify groundwater flow, discharge contaminants or reactivate old pollution sources, and interconnect different aquifer layers which have diverse quality waters.

In order to ensure a sustainable exploitation from an environmental and water resource availability point of view, and at the same time recover the good quantitative and chemical status, a set of measures and actions has been defined. They are included in a programme of actions to be carried out in 2009, which will be included in the River Basin Management Plan to be operational in 2012. These actions comprise regulations, management criteria and infrastructure built up.

Criteria about exploitation and technical regulations to authorize works have been established, as well as for the follow up of infrastructure construction. This is carried out through the METALL (technical board for the Llobregat's aquifers), which establishes preventive as well as corrective measures addressed to the public operators of the different infrastructures, in order to minimize impacts during the works and then during their exploitation (Ribera, 2008).

Relationships between the Agencia Catalana de l'Aigua (ACA, Water Authority) and the Water Users Community (CUADLL) have been reinforced. The CUADLL has been charged with the follow up of the monitoring network, the follow up of grouting out-of-use wells, the preparation and follow up of artificial recharge projects, and the preparation of periodical reports on the evolution of aquifers with respect to different infrastructures being constructed. A process for citizen participation has been implemented, which will continue in a new framework of participative forum called 'Water Debate'.

Improvements have been achieved through the construction of new water urban waste water treatment infrastructures (tertiary treatments, reversible electrodialysis and reverse osmosis), enlargement of a collector for brines produced in the potash salts mining area in the Llobregat's mid basin and by some industries, and other actions to manage storm waters in the watershed municipalities.

In order to increase available water resources by means of artificial recharge, the installation of a positive hydraulic barrier to control and reduce seawater intrusion, a 60 million hm^3/year capacity desalination plant has been constructed. Also the reuse of treated waste water has been considered, including tertiary treatment, reverse osmosis, reversible electrodialysis, microfiltration, ... (depending on artificial recharge quality requirements, on delta wetlands and on river improvements), for agricultural and industrial uses, and also for recharge.

With this set of actions, it is considered that environmental objectives of article 4 of the WFD are reachable. With reference to groundwater quantitative status, the objectives are to balance abstraction and recharge, while for chemical status, objectives are striving to the reduction of contaminants inflow and the reversal of increasing and sustained contamination increasing trends existing until present.

5.7.5   ACTION ACCOMPLISHED AND UNDER WAY

Actions presented in section 5.7.4 imply an investment of about 200 M€. Some are already operative and all of them will be operational in 2011. With their implementation, the integrated water management will be possible by combining treated waste water reuse for agricultural, industrial and environmental uses, desalinized sea water for urban supply, regulated surface water by means of surface reservoirs and groundwater. Thus, the water supply problems to the Barcelona's Metropolitan area should be solved with guarantee until 2025, while re-establishing the good status of the Llobregat's strategic aquifers.

The actions with the greatest influence on groundwater were as follows:

- Treated waste water reuse, to maintain flow in the Llobregat's lower tract ($2\,m^3/s$), wetland maintenance ($0.3\,m^3/s$), substitution of irrigation water ($0.75\,m^3/s$) and supply the hydraulic barrier ($0.1\,m^3/s$). Besides this, changes in water use in the industrial sector are foreseen, with an estimated preliminary reduction of $0.2\,m^3/s$.

- Increase of aquifer artificial recharge by means of infiltration basins. Since 1969, artificial recharge to the main aquifer have been practised, enhancing flowing water infiltration by means of river bed scrapping. In years such as 1974, 1981, 1982 and 1983 up to 10 to 14 million m^3 were infiltrated. To date, a set of infiltration basins are under construction with the goal of reaching a recharge rate of about 15 million m^3.

The construction of a hydraulic barrier to avoid seawater intrusion, to inject $15,000\,m^3$/day to the Llobregat's deep delta aquifer, is now under advanced construction. Injection water, from the nearby urban waste water treatment plant, follows a tertiary treatment and microfiltration, and then half the injection flow, is subject to inverse osmosis treatment. The barrier will consist of 14 wells along a 5.3 km long line, parallel to the coastline, at an average distance of 1 to 1.5 km. Currently 6 wells are already constructed as well as part of the monitoring tubes. In tests carried out since 2007, about 0.7 million m^3 have been injected. Results may be considered as highly positive (Figure 5.7.8).

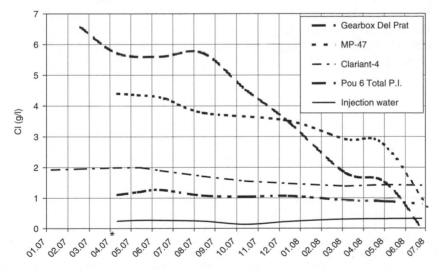

Figure 5.7.8 Effect of the hydraulic injection coastal line in the Llobregat's delta deep aquifer, after the results in some of the 'Polìgon Pretenc' wells, about 1 to 2 km to the north, in the inner boundary of the seawater intrusion wedge. The chloride evolution during the first injection phase is shown.

The set of artificial recharge infrastructures are aimed at the injection of $15\,\text{Mm}^3/\text{y}$. This will be achieved with the help of a numerical model and the evaluation of users needs will allow planning of groundwater resources taking into account all interested parties.

5.7.6   CONCLUSIONS

Actions undertaken and planned in the Llobregat's lower valley and delta should allow complying with article 4b of the WFD environmental objectives by 2015. Besides, they aim to solve the Barcelona's Metropolitan Area supply until 2025 by providing improved quality water in sufficient quantity. A positive injection coastal barrier is being implemented to reduce a large part of seawater intrusion, although numerical simulation under diverse water injection scenarios show that seawater inflow is not fully stopped in the harbour docks excavated in the eastern area of the delta.

Sustainable aquifer exploitation and recovery of good quantitative and chemical status need an agreed integrated water management which has to consider the sensitivity of different social groups, starting from a participation process and a Water Debate forum. In this management, the Water Users Associations will have an active participation, as well as public supply responsible entities, in the framework of the Water Authority of Catalonia.

Acknowledgements

Preliminary ideas and concepts contributed by Drs J. Carrera and E. Vázquez (CSIC) are very much acknowledged. The collaboration of M. Rull, V. Solà, J. Massana and P. Balart, from the CUADLL, and F. Ortuño, V. Colomer and G. Borrás, from the ACA is gratefully acknowledged as regard the preparation of figures. Text processing has been carried out by J. Sánchez–Vila, of the UPC.

REFERENCES

Abarca E., Vázquez–Suñé E., Carrera J., Capino B., Gámez D. and Batlle F., 2006. Optimal design of measures to correct seawater intrusion. *Water Resources Research*, **42**: doi: 10.1029/2005WR004524.

Bayó A., Batista E. and Custodio E., 1977. Sea water encroachment in Catalonia coastal aquifers. General Assembly Intern. Assoc. Hydrogologists. Birmingham UK, XIII.I: F.1–14.

Custodio E., 1967. Études hydrogéochimiques dans le delta du Llobregat, Barcelona (Espagne). Gen. Assem. Bern, *Intern. Assoc. Sci. Hydrol. Publ.* **62**: 135–55.

Custodio E., 1981. Sea water encroachment in the Llobregat delta and Besós areas, near Barcelona (Catalonia, Spain). Sea Water Intrusion Meeting: Intruded and Fosil Groundwater of Marine Origin. Rapp. och Meddelanden, 27. *Sveriges Geologiska Undersökning*, Uppsala: 120–52.

Custodio E., 1987. Sea–water intrusion in the Llobregat delta, near Barcelona (Catalonia, Spain). Groundwater Problems in Coastal Areas, UNESCO Studies and Reports in Hydrology no. 45. UNESCO, París: 436–63.

Custodio E., 2007. Acuíferos detríticos costeros del litoral mediterráneo peninsular: valle bajo y delta del Llobregat. Enseñanza de las Ciencias de la Tierra: Las Aguas Subterráneas. Rev. Asoc. Esp. Enseñanza de las Ciencias de la Tierra. Madrid, **15**(3): 295–304.

Custodio E., Suárez M., Isamat F. J. and Miralles J. M., 1977. Combined use of surface and groundwater in Barcelona Metropolitan Area (Spain). Intern. Assoc. Hydrogeologists, Gen. Assem. Birmingham, XIII.1: 14–27.

Gámez D., 2007. Sequence stratigraphy as a tool for water resources management in alluvial coastal aquifers: application to the Llobregat delta (Barcelona, Spain). Tesis doctoral. Dep. Enginyeria del Terreny, Cartogràfica i Geofísica, Universidad Politécnica de Cataluña. Barcelona: 1–177 + An.

Iribar V., Carrera J., Custodio E. and Medina A., 1997. Inverse modelling of sea water intrusion in the Llobregat delta deep aquifer. *Journal of Hydrology*, **198**(1–4): 226–44.

Llamas M. R., 1969. Combined use of surface and ground water for the water supply to Barcelona (Spain). *Bull. Intern. Assoc. Sci Hydrol*., **14**(3): 119–36.

Llamas M. R. and Molist J., 1967. Hidrología de los deltas de los ríos Besós y Llobregat. *Rev. Agua, Barcelona*, **2**: 139–54.

Manzano M., Peláez M. D. and Serra J., 1986. Sedimentos prodeltaicos en el delta emergido del Llobregat. *Acta Geológica Hispánica* 21–22 (1986–1987): 205–11.

Manzano M., Custodio E. and Carrera J., 1992. Fresh and salt water in the Llobregat delta aquifer: application of ion chromatography to the field data. Study and Modelling of Saltwater Intrusion Into Aquifers. Intern. Center for Numerical Methods in Engineering (CIMNE), Barcelona: 207–28.

Marqués M. A., 1975. Las formaciones cuaternarias del delta del Llobregat. *Acta Geológica Hispánica*. X: 21–38.

Massana J. and Queralt E., 2007. La participación de los usuarios en la ordenación de los acuíferos costeros del Llobregat mediante modelos de flujo y transporte. Los Acuíferos Costeros: Retos y Soluciones, Almería. Instituto Geológico y Minero de España, serie Hidrogeología y aguas subterráneas no. 23, I: 947–56.

MOP, 1966. Estudio de los recursos hidráulicos totales de las cuencas de los ríos Besós y Bajo Llobregat. Comisaría de Aguas del Pirineo Oriental y Servicio Geológico de Obras Públicas. Ministerio de Obras Públicas. Barcelona. 4 vols.

Niñerola J. M., 2008. Calidad y contaminación de las aguas subterráneas: caso de estudio en Cataluña. Las Aguas Subterráneas en España ante las Directivas Europeas, Retos y Perspectivas. Instituto Geológico y Minero de España, Madrid: 93–108.

Niñerola J. M. and Ortuño F., 2008. The quantitative and chemical status of groundwater bodies in Catalonia: state of the art on implementation of the Water Framework Directive and Groundwater Directive. EU Groundwater Policy Development, UNESCO, Paris.

Ribera F., 2008. Las medidas preventivas y las medidas correctoras de las afecciones hidrogeológicas generadas por grandes infraestructuras. Jorn. Agua y las Infraestructuras en el Medio Subterráneo. Barcelona. Assoc. Intern. Hidrogeólogos–Grupo Español (in press).

Santa María L. and Marín A., 1910. Estudios hidrológicos en la Cuenca del río Llobregat. Bol. Com. Mapa Geológico de España, Madrid: 31–52.

Solà V., 2004. La red de control químico de la CUADLL. VIII Simposio de Hidrogeología. Zaragoza. *Asociación Española de Hidrogeológos*: 597–606.

Vázquez–Suñé E., Abarca E., Carrera J., *et al*., 2006. Groundwater modelling as a tool for the European Water Framework Directive (WFD) application: The Llobregat case. *Physics and Chemistry of the Earth*, **31**(17): 1015–29.

5.8

Determination of Natural Background Levels and Threshold Values in the Neogene Aquifer (Flanders)

Marleen Coetsiers and Kristine Walraevens

Now at ERM-Belgium; Ghent University, Laboratory for Applied Geology and Hydrogeology, Gent, Belgium

5.8.1   INTRODUCTION

In the frame of the European BRIDGE project of the EU 6th Framework Program a methodology was developed for the determination of threshold values for pollutants in groundwater bodies in support of the status provisions of the Water Framework Directive (WFD) (2000/60/EC) and the Groundwater Daughter Directive (GWDD) (2006/118/EC). Following the GWDD good chemical status must be based on compliance to existing

Groundwater Monitoring Edited by Philippe Quevauviller, Anne-Marie Fouillac, Johannes Grath and Rob Ward
© 2009 John Wiley & Sons, Ltd

quality standards and on threshold values (TV) to be established by Member States. Threshold values are quality standards for polluting substances in groundwater that need to be formulated by the Member States for status assessment of groundwater bodies. These threshold values represent the concentration of a pollutant that can not be exceedd in order to protect the environment and human health. The determination of natural background levels is of major interest given the GWDD that recognises that groundwater with natural high concentrations of a certain substance can not be defined as poor status for this reason alone. In the BRIDGE project a tiered approach was suggested to determine the status of a groundwater body (Müller *et al.*, 2006). Each tier involves increasingly sophisticated levels of data assembling and analysis. This developed methodology was applied to 14 groundwater bodies across Europe, among which the CKS_0200_GWL_1 groundwater body in Flanders. This groundwater body is part of the Central Campine System, which occurs in the northeast of Flanders and contains the Neogene Aquifer.

The Central Campine System is located in the northeast of Flanders, in the provinces of Antwerp and Limburg, at the border with the Netherlands. The Neogene Aquifer and the Pleistocene Campine Complex Aquifer constitute the Central Campine System. Four groundwater bodies are distinguished in the region of which the CKS_0200_GWL_1 is the largest one. This groundwater body contains the unconfined part of the Neogene Aquifer. This Neogene Aquifer contains large volumes of drinking water resources and is extensively pumped for public water-supply. Furthermore groundwater abstraction for agricultural, industrial and private purposes results in large pumped volumes of ground-water. Drinking water companies in the provinces of Antwerp, Limburg and Brabant extract about 90 million m^3 per year. Additionally, industries, farms and households account for more than 200 million m^3 per year. A large number of groundwater depen-dent terrestrial ecosystems is present in the discharge areas of the CKS_0200_GWL_1 groundwater body. In the north of the region coalmines and metallurgic industries were operating in the past and have caused pollution with heavy metals.

5.8.2   CHARACTERIZATION OF THE GROUNDWATER BODY

5.8.2.1   Physical and Hydrogeological Description

The Central Campine System occurs in the northeast of Flanders where it covers an area of $4210\,km^2$ and is subdivided into four groundwater bodies (Figure 5.8.1). The Central Campine System is bounded by the occurrence of the Boom Clay in the south and south-west. In the east the boundary is formed by the groundwater divide between the Scheldt and Meuse river basins. Towards the north the Campine Basin continues over the border with the Netherlands. The largest groundwater body in the Central Campine System is built up by the unconfined Neogene deposits: CKS_200_GWL_1. In the south an erosion gully is present in the Boom Clay and a different GWB is delineated: CKS_250_GWL_1. Here, the Neogene Aquifer is in direct contact with the underlying Oligocene Aquifer. To the north the semi-confined part of the Neogene Aquifer, where it is overlain by the Campine Clay, is delineated as the CKS_0200_GWL_2 groundwater body. Finally,

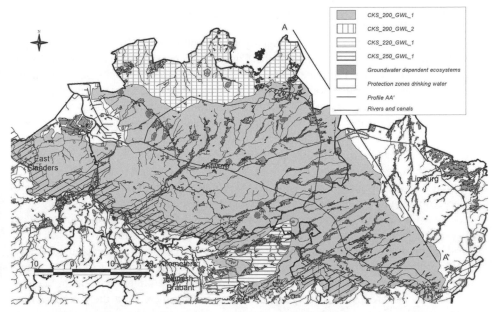

Figure 5.8.1 Location of the groundwater bodies in the Central Campine System, groundwater dependent terrestrial ecosystems and protection zones of drinking water wells. (See Plate 5 for a colour representation).

the CKS_0220_GWL_1 is formed by the Pleistocene deposits on top of the Campine Clay.

The topography in the area is mainly dominated by the occurrence of two cuestas and the Campine Plateau. In the south the occurrence of the Boom Clay has formed a cuesta in the landscape because clay is less affected by erosion. In the north the occurrence of the Campine Clay is also expressed in the landscape as a cuesta, which forms the northern watershed between the Scheldt and Meuse basins. In the southeast and east the Campine Plateau behaves as this watershed. In between these higher elements a saddle-shaped basin is formed. The ground elevation ranges from +100 m a.s.l. on the highest points of the Campine Plateau to less than +5 m a.s.l. in the basin. The main land use in the Campine area is agriculture with 51.2% of the area, followed by urban use (27.8%), woodland (12.3%) and other uses (8.7%). Groundwater dependent ecosystems are mainly present on the flanks of the saddle-shaped basin (Fig. 5.8.1), where groundwater discharge occurs by means of rivulets and small rivers.

The Campine Basin was filled from Late-Oligocene to Pliocene times with marine to continental sediments. The deposits dip gently towards the north-northeast and are in the east disturbed by different faults. The base of the aquifer is formed by the occurrence of the Boom Clay. The groundwater body CKS_0200_GWL_1 consists of a succession of Miocene to Quaternary sands, alternating with more or less important clay layers. Locally, a very high glauconite content is observed and also mica, phosphate boulders, shells, limonitic lenses and lignite layers are encountered. The groundwater body has a phreatic nature and belongs to the 'Sands and gravels' type and the 'marine deposits' subtype, according to the European Aquifer Typology (Wendland *et al.*, 2008; Pauwels *et al*., 2006). The thickness reaches up to 433 m and the transmissivity ranges from 336 to 878 m^2/d (CIW, 2004).

Groundwater recharge takes place mainly in the topographically elevated areas like the Campine Plateau and the cuesta of the Campine Complex. Groundwater infiltrating on the west side of the Campine Plateau will flow towards the Scheldt basin in the west, while water infiltrating on the east side of the plateau flows towards the Meuse basin. In the north of the study area groundwater infiltrates mainly on the cuesta of the Campine Complex. Water infiltrating on the south flank will flow to the Scheldt basin and water infiltrating on the north side will flow to the Meuse basin. In the south of the study area, smaller local infiltration areas exist where remnant hills form topographical elevations. In between the small rivers local recharge areas are present resulting in small flow cycles that are superposed on the regional groundwater flow cycle. Groundwater discharge occurs mainly by outflow to rivers and rivulets in the lower lying areas.

Coetsiers (2007) and Coetsiers & Walraevens (2006a) studied the natural groundwater quality in the Neogene Aquifer. Shallow groundwater in CKS_0200_GWL_1 has a low mineralisation degree and is acid because the sediments are strongly decalcified. To the west and in the deeper parts the deposits still contain calcite, resulting in calcite dissolution reactions (cross-hatched parts in Figure 5.8.2). In the largest part of the aquifer, silicate dissolution will influence groundwater quality by increasing the cation and silicate content. Redox reactions play an important role in the chemistry of groundwater

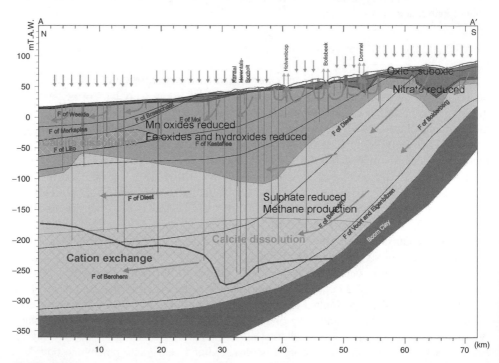

Figure 5.8.2 Characterization of groundwater hydrochemical evolution along profile showing redox boundaries and recharge-throughflow-discharge zones (cross-hatching indicates presence of calcite).

quality in CKS_0200_GWL_1. A succession of oxic, suboxic, nitric, Mn, Fe, sulphidic and methanic zones is observed. In Figure 5.8.2 the redox zones are indicated in the profile AA'. In the deepest parts of the groundwater body the marine depositional environment is still slightly present and cation exchange occurs, leading to the exchange of Ca^{2+} from the groundwater for Na^+, K^+ and Mg^{2+} adsorbed to the clay minerals. Pyrite oxidation in the unsaturated zone causes an increase in sulphate content. Elevated arsenic concentrations are encountered and are related to the redox conditions. The Fe^{2+} originating from pyrite oxidation is immediately oxidized to Fe^{3+} and precipitates as oxides or hydroxides. The arsenic and other heavy metals released from the oxidation of pyrite are incorporated in these iron oxides and hydroxides. In more strongly reducing conditions the arsenic enters into solution by reduction of the iron (hydr)oxides. Phosphate concentrations in groundwater can locally be high due to the dissolution of the phosphate mineral vivianite ($Fe_3(PO_4)_2.8H_2O$).

5.8.2.2 Identification of Pressures on the Groundwater Body

The Neogene Aquifer is a very important resource for drinking water abstraction. In the Neogene Aquifer more than 300 M m^3/year is abstracted for drinking water, industrial uses, agricultural uses and private households. CIW (2004) executed a preliminary quantitative evaluation of the groundwater body by looking at time series of piezometric measurements of more than 10 years long. The CKS_0200_GWL_1 has a stable trend with seasonal variations and has been given a 'good quantitative status' although the number of measuring points in the groundwater body is low. A preliminary qualitative evaluation of the groundwater bodies in Flanders is based on the nitrate concentration in the phreatic monitoring network. This phreatic monitoring network contains ca. 5300 monitoring points in Flanders. The Flemish Environmental Administration defines a good qualitative status if at least 95% of monitoring points in the GWB is not exceeding the standard for nitrate (50 mg/l). By doing so, all phreatic groundwater bodies, like CKS_0200_GWL_1 have a poor status and are at risk.

Diffuse pollution from former metallurgic industries and mining activities is responsible for elevated Zn, Ni, Cd, Pb and Cu concentrations in the CKS_0200_GWL_1 (Walraevens *et al.*, 2003). CIW (2004) identified three point sources of pollution in CKS_0200_GWL_1: the nonferrous pollution in Balen and Olen and the chloride pollution at the river Grote Laak. These point sources are linked to heavy polluting industry in the past. The groundwater pollution with heavy metals in Balen is caused by different point and line sources. The largest part of the pollution is due to indirect draining and leaching out. A large amount of heavy metals has fallen in the neighbourhood of the industrial sites by means of atmospheric deposition. The residues (cinders) of these nonferrous activities were used for asphalting of roads and heightening of land.

5.8.2.3 Review of Monitoring and Impacts

The primary monitoring network in Flanders consists of monitoring wells with one or more screens in the major aquifers and contains a total of 376 screens. Piezometric measurements are executed monthly but groundwater chemistry is up to now not measured on a regular base. During 2005–2006 about 130 deep monitoring wells with screens in the major aquifers have been installed in Flanders to enlarge the primary monitoring network. 42 screens of the primary network at 24 locations are installed in the CKS_0200_GWL_1 groundwater body.

The phreatic monitoring network contains 2113 wells with ca. 5300 monitoring points and is operational since the beginning of 2004 (Eppinger, 2005). Nearly each well is equipped with three screens at different depths. Twice to four times a year the groundwater quality is analysed and piezometry is measured. 475 screens of the phreatic monitoring network at 254 locations are installed in the CKS_0200_GWL_1 groundwater body.

The surface water monitoring network in Flanders contains 3250 measuring points in brooks, rivers, canals, lakes and the sea. These points are not all sampled in one year. There is a subdivision into different monitoring networks: the physico-chemical, Nutrient Management Plan, biological, bacteriological and water soils monitoring networks.

Groundwater depression cones are present close to large pumping wells. Close to large pumping activities the groundwater flow pattern will be disturbed although the overall groundwater flow still reflects the natural flow pattern. Lowering of the water table in the neighborhood of pumping wells can cause changes in the redox environment in the aquifer. A larger part of the aquifer becomes oxidized leading to the oxidation of pyrite, which was formerly present in the reduced zone. Pyrite oxidation can lead to an increase of the arsenic content in the groundwater.

At Grobbendonk an artificial recharge pilot project is running since 1975. Water from the Albert Canal is infiltrating in an open basin in order to compensate the lowering of the groundwater table caused by drinking water extraction. The canal water has undergone purification before it is infiltrated. The total infiltration capacity is $150\,m^3$ per hour in the sand layers of the Formation of Diest and Berchem. There is however no chemical data available.

As anthropogenic pollution mainly affects the phreatic zone of groundwater systems, the qualitative evaluation in the Flemish region is limited to phreatic or shallow groundwater bodies. The preliminary evaluation of groundwater quality was performed for each groundwater body by means of the results of the phreatic monitoring network, focusing on the nitrate concentration. The phreatic groundwater bodies CKS_0200_GWL_1, CKS_0250_GWL_1 and CKS_0220_GWL_1 have a poor qualitative status while the semi-confined groundwater body CKS_0200_GWL_2 has a good status.

5.8.3   GROUNDWATER STATUS EVALUATION BY THRESHOLD VALUES

5.8.3.1   Assessing the Natural Background Level (NBL)

There is no national approach to derive natural background levels in groundwater in Flanders but a large amount of monitoring data is available. For this reason the natural background levels are derived based on the simplified pre-selection approach as described in Müller *et al.* (2006). A database with chemical analyses of 540 groundwater samples is available. Groundwater samples with $NO_3 > 10$ mg/l were excluded from the database. Furthermore the database should fulfill some minimum requirements:

- samples with an error $>10\%$ on the ionic balance should be removed;

- samples of unknown depth should be removed;

- monitoring data not attachable to aquifer typologies should be removed;

- data from hydrothermal aquifers should be removed;

- data from salty aquifers should be removed;

- time series should be eliminated by median averaging.

After pre-selection and data processing, analyses of 453 groundwater samples remained. For statistical purposes half of the detection limit was used for samples with analytical results below detection limit. The natural background level (NBL) is determined as the concentration at the 90-percentile or the 97.7-percentile. In case of pH the 10-percentile and 2.3-percentile are used to define the lower boundary for the NBL.

5.8.3.2 Selection of the Reference Quality Standard

Threshold values (TV) are established with reference to natural background levels (NBL) and a chosen reference standard (REF). There are two possible cases for calculating the TV (Müller *et al.*, 2006):

- Case 1: if NBL < REF: TV = (REF + NBL)/2

- Case 2: if NBL $\geqslant$ REF : TV = NBL

The receptors for groundwater of the CKS_0200_GWL_1 groundwater body are groundwater itself, groundwater dependent terrestrial ecosystems, aquatic ecosystems and drinking water. On Plate 5 the groundwater protection zones for public drinking water companies are indicated in red, while the groundwater dependent terrestrial ecosystems are indicated in green. The dependent aquatic and terrestrial ecosystems and protected areas are mainly rivers and wetlands, which are fed by groundwater. These zones coincide with areas where groundwater outflow occurs. The areas with a nature function that coincide with humid and wet areas are delineated as the groundwater dependent terrestrial ecosystems. In Flanders authorities are legally bound by VLAREM II (Decision of the Flemish Government implying general and sectoral provisions towards environmental protection (VLAREM II) (June 1, 1995; BS July 31, 1995)) to handle the defined environmental quality standards for the planning and the execution of the policy. The standards are guiding the evaluation of requests for environmental licenses. All surface waters in Flanders have to comply with the basic water quality standards. Selected surface waters in Flanders have been designated a specific destination (drinking water, swimming water, fish water and shellfish water) according to the Decision of the Flemish Executive of

21/10/1987 'B.S. 6/1/1988). Specific, more stringent standards have been set for specific destinations. In the study area, some surface waters have been designated drinking water and fish water destinations.

The REF values used for the different receptors are:

- 'groundwater itself': the Environmental Quality Standards for groundwater as defined in the Vlarem II legislation (Annex 2.41 of VLAREM II); these standards however are largely based on drinking water standards;

- 'groundwater dependent terrestrial ecosystems': the Environmental Quality Standards (EQS) for List 1 and List 2 dangerous substances and for Hardness Related List 2 dangerous substances; EC Dangerous Substances Directive (76/464/EEC);

- 'aquatic ecosystems': EQS for surface water as a function of the destination defined in VLAREM II (Annex 2.3.1 to 2.3.5 of VLAREM II);

- 'drinking water': Drinking Water Standards (WHO, 2006).

The parameters that are handled are: pH, EC, Na, K, Ca, Mg, Fe, Mn, NH_4, SO_4, NO_3, NO_2, PO_4, Al, Zn, As, Cd, Cr, Cu, Hg, Ni, Pb, Sb, F and B.

5.8.3.3   Results and Compliance Testing

The threshold values are calculated for the receptors groundwater itself, groundwater dependent terrestrial ecosystems, aquatic ecosystems and drinking water supply (Coetsiers & Walraevens, 2006b). The results of the calculation of TVs for the receptor drinking water supply are given in Table 5.8.1. In case of pH the 10-percentile and 2.3-percentile are used to define the lower boundary for the NBL.

The calculated TV for the receptor 'drinking water' is larger than the REF (case 2) for the parameters:

90-percentile: pH (5 < 5.5), K, Fe, Mn, NH_4, Al, As

97.7-percentile: pH (4 < 5.5), PO_4, K, Fe, Mn, NH_4, SO_4, Al, As

For the receptor groundwater itself the same parameters have TV higher than REF, with the exception of pH and PO_4 for which no REF values for groundwater are defined in Vlarem II. In the case of TV determination for the receptor 'groundwater dependent terrestrial ecosystems' the TV is larger than the REF for pH, Fe and As both at the 90 percentile and at the 97.7 percentile used for NBL. For the hardness related dangerous substances the TV is higher than REF for Cr, Cu, Pb and Zn for different hardness bands in the case of the 90-percentile used as NBL and for Cr, Cu, Pb, Ni and Zn in the case of the 97.7-percentile used as NBL. The calculated TV for the receptor 'aquatic ecosystems' in the case of basic surface water quality is above REF for pH, Fe and Mn for the 90 percentile used as NBL and for O_2, pH, SO_4, Zn, As, Fe and Mn for the 97.7 percentile used as NBL. In the case of drinking water destination of surface water the TV is larger than REF for pH, Fe and Mn both when the 90 percentile or 97.7 percentile is used as

Table 5.8.1 Determination of the Natural Background level (NBL) based on the 90 and 97.7 percentiles and of the Threshold Values (TV) with drinking water standards used as REF.

Parameter	Unit	REF	NBL 90-percentile	TV 90-percentile	Case	NBL 97.7-percentile	TV 97.7-percentile	Case
pH (high)		9.5	7.2	8.4	1	8.1	8.8	1
pH (low)		5.5	5.0	5.0	2	4.0	4.0	2
EC	μS/cm	1500	718	1109	1	976	1238	1
Ca	ppm	250	113	181	1	182	216	1
Mg	ppm	50	12	31.0	1	22	36	1
Na	ppm	150	40	60.0	1	65	108	1
K	ppm	12	12	12.0	2	22.1	22.1	2
PO_4	ppm	2.2	1.0	1.6	1	2.5	2.5	2
Fe	ppm	0.2	44.2	44.2	2	87	87	2
Mn	ppm	0.05	1.1	1.1	2	2.6	2.6	2
NO_3	ppm	50	2.0	26.0	1	6.6	28.3	1
NO_2	ppm	0.1	0.03	0.07	1	0.07	0.09	1
NH_4	ppm	0.5	1.0	1.0	2	4.1	4.1	2
Cl	ppm	400	65	233	1	138	269	1
SO_4	ppm	250	183	217	1	311	311	2
Al	ppb	200	1554	1554	2	2934	2934	2
As	ppb	10	30	30	2	69	69	2
Zn	ppm	5	0.13	2.6	1	0.4	2.7	1
Cd	ppb	5	0.5	2.8	1	1.1	3.1	1
Cr	ppb	50	4.0	27.0	1	12.2	31.1	1
Cu	ppb	3000	8	1504	1	12.3	1506	1
Hg	ppb	1	0.2	0.6	1	0.3	0.7	1
Ni	ppb	50	13.6	31.8	1	42.2	46.1	1
Pb	ppb	50	5.0	27.5	1	9.4	29.7	1
Sb	ppb	10	1.0	5.5	1	1.7	5.9	1
F	ppm	1.5	0.4	1.2	1	0.8	1.2	1
B	ppb	2	0.3	1.2	1	0.7	1.4	1

NBL. In the case of fish water destination of surface water the TV is above REF for pH when the 90 percentile is used as NBL and for pH, NH_4 and NO_2 when the 97.7 percentile is used as NBL.

The obtained NBL and TV for the receptor 'drinking water' are compared to the standards (Background Standards BS and Sanitation Standards SS) for groundwater sanitation defined in the Flemish Soil Sanitation Legislation (VLAREBO), shown in Table 5.8.2. The TV is higher than the groundwater sanitation standard for As, Cu, Pb and Zn when both the 90 and 97.7 percentile are used as NBL and additionally for Ni in case of the 97.7 percentile used as NBL. These exceedings of the sanitation standard are due to high determined NBL in the Neogene Aquifer in the case of As and Ni, and due to high REF values in the case of Cu, Pb and Zn. For high As concentrations a natural origin exists in the Neogene Aquifer indicating that the sanitation standard for As is too stringent for the Neogene Aquifer. Walraevens *et al*. (2003) already remarked that there are strong indications that the BS is too low compared to the natural values observed in Flanders

Table 5.8.2 Comparison of TV at 90- and 97.7-percentile for the receptor 'drinking water' and standards for groundwater sanitation in the Flemish Soil Sanitation Legislation (VLAREBO).

Heavy metals and metalloids	Sanitation standard (μg/l)	Background standard (μg/l)	TV 90-percentile	TV 97.7-percentile
As	20	5	40	69
Cd	5	1	2.8	3.1
Cr	50	10	27.0	31.1
Cu	100	20	1504	1506
Hg	1	0.05	0.6	0.7
Pb	20	5	27.5	29.7
Ni	40	10	31.8	46.1
Zn	500	60	2600	2700

for Zn and As. For the other parameters they found no convincing ground to reconsider the BS.

5.8.4 CONCLUSIONS

Groundwater quality threshold values for the CKS_0200_GWL_1 groundwater body are calculated for 4 different receptors: groundwater itself, groundwater dependent terrestrial ecosystems, aquatic ecosystems and drinking water. In a first step the natural background level (NBL) is calculated in case of the 90- and 97.7-percentiles. The TVs are calculated based on the NBL and a REF value derived from environmental quality standards. The REF values vary between the different receptors. Generally it is seen that for the 97.7-percentile calculations, TV is above REF for more parameters. Since these higher TVs for K, Fe, Mn, NH_4, SO_4, PO_4, Al and As have a natural origin, it is recommended to use the 97.7-percentile in preference to the 90-percentile for this well-characterized groundwater body. The CKS-0200_GWL_1 is characterized by low pH and high levels of Fe, Mn, As, NH_4, PO_4, Al, SO_4 and Zn. The exceedings of the REF at the 97.7-percentile are thus of a natural origin and it is recommended to use the 97.7-percentile for the NBL rather than the 90-percentile since geochemistry of the groundwater body is well known. The methodology for TV derivation duly takes the possibility of high natural background into account.

REFERENCES

CIW, 2004. Characterization of the Flemish part of the international stream district of the Scheldt. Coordination Commission Integral Water Management.

Coetsiers M., 2007. Investigation of the hydrogeological and hydrochemical state of the Neogene Aquifer in Flanders with the use of modelling and isotope hydrochemistry. PhD, Ghent University, Belgium [in Dutch].

Coetsiers M. and Walraevens K., 2006a. Chemical characterization of the Neogene Aquifer, Belgium. *Hydrogeology Journal* 14: 1556–8.

Coetsiers M. and Walraevens K., 2006b. Groundwater natural background levels and threshold definition in the CKS_0200_GWL_1 groundwater body of the Central Campine System (Flanders, Belgium). Report to the EU project BRIDGE (Contract N° SSPI-2004-006538). Ghent, Ghent University.

Coetsiers M., Van Camp M. and Walraevens K., 2003. Reference Aquifer: The Neogene Aquifer of Flanders, Belgium. In: Edmunds WM, Shand P (eds.) *Natural Baseline Quality in European Aquifers: A Basis for Aquifer Management*. EC-project EVK1-CT-1999-00006. Final Report. British Geological Survey, Wallingford.

Eppinger R., 2005. The phreatic groundwater monitoring network – a new view on the quality evolution of shallow groundwater in Flanders with respect to the occurrence of nitrate (in Dutch).

Hart A., Müller D., Blum A., *et al.*, 2006. Preliminary methodology to derive environmental threshold values. Specific targeted EU-research project BRIDGE (contract N° SSPI-2004-006538)-report D15. www.wfd-bridge.net

Müller D., Blum A., Hookey J., Kunkel R., Scheidleder A., Tomlin C. and Wendland F., 2006. Final proposal of a methodology to set up groundwater threshold values in Europe. Specific targeted EU-research project BRIDGE (contract no. SSPI-2004-006538)-report D18. www.wfd-bridge.net

Pauwels H., *et al.*, 2006. Impact of hydrogeological conditions on pollution behaviour in groundwater and related ecosystems. BRGM. BRIDGE (contract N° SSPI-2004-006538) Deliverable D9.

Walraevens K., Mahauden M. and Coetsiers M., 2003. Natural background concentrations of trace elements in aquifers of the Flemish Region, as a reference for the governmental sanitation policy. *8th Int. FZK/TNO Conference on Contaminated Soil (ConSoil), Ghent 2003, Proc.*, 215–24.

Wendland F., Blum A., Coetsiers M., *et al.*, 2008. European aquifer typology: a practical framework for an overview of major groundwater composition at European scale. *Environmental Geology*, 55, 77–85. DOI 10.1007/s00254-007-0966-5

WHO, 2006. *Guidelines for Drinking-Water Quality*. World Health Organisation.

Part 6
Groundwater Measurements

6.1

Metrological Principles Applied to Groundwater Assessment and Monitoring

Philippe Quevauviller[1], Ariane Blum[2] and Stéphane Roy[2]

[1] *Vrije Universiteit Brussel (VUB), Department of Water Engineering, Brussels, Belgium*
[2] *BRGM, Orleans cedex, France*

6.1.1   INTRODUCTION

Groundwater analyses are currently performed in routine and research laboratories for the control of quality of water used for human consumption and of the chemical contamination levels according to EC legislation (see Chapter 1.1 of this book). A good accuracy is

Groundwater Monitoring Edited by Philippe Quevauviller, Anne-Marie Fouillac, Johannes Grath and Rob Ward
© 2009 John Wiley & Sons, Ltd

a prerequisite for analysis for achieving a good comparability of data and hence allowing sound decisions to be taken by authorities. The lack of analytical quality control renders useless many of the results obtained in monitoring campaigns and it is obvious that many studies performed in the past have not been performed with a sufficient quality control mechanism and that many of their conclusions were biased. Poor performance by analytical laboratories and consequent incomparability of data create economic losses due to extra-analyses, court actions etc. The costs in terms of quality of life can hardly be estimated but is certainly very high.

Similarly to general environmental monitoring strategies, sound groundwater monitoring calls for measurement systems capable of producing chemical (and biological in the case of e.g. microbiological measurements) data of demonstrated quality. In this chapter, the focus will be placed on chemical analyses which are at the core of the implementation of the Groundwater Directive 2006/118/EC (European Commission, 2006). In this context, monitoring groundwater chemical status relies on the analysis of a wide variety of chemical substances and physico-chemical parameters (e.g. conductivity) at various levels of concentrations. The last decade has seen an increasing awareness for metrological principles to be followed for ensuring comparable environmental measurements, including for groundwater, which has been reflected by the development of a number of guidelines and documented standards, e.g. for managerial aspects and technical operations (e.g. sampling, method validation etc.), and tools, e.g. reference materials, proficiency testing schemes (Quevauviller, 1995; Subramanian, 1995; Quevauviller & Maier, 1999; Barceló, 2000).

Like other types of freshwaters (e.g. surface waters), groundwaters have benefited from the progress of metrology. But the reasons why recent developments are so important for groundwaters lay on the generally low levels of pollutants. Generally, concentrations of organic or mineral micropollutants in EU contaminated groundwaters are just above the level of quality standards. As an example in France, pesticides concentrations exceed $2\,\mu g/L$ in less than 1% of the groundwater surveillance control monitoring stations (Ifen, 2004). Most polluted stations show concentrations around the Groundwater directive 2006/118/CE standard, i.e. $0.1\,\mu g/L$. It should be noted that within the last two decades, with the decrease of limits of quantification and uncertainties, chemical analysis of groundwater samples became more reliable.

For this reason, analytical performance of European laboratories is a key issue of the implementation of the Water Framework Directive. By setting minimum technical specifications for chemical analysis, the future Commission Directive (European Commission, 2009) will limit uncertainties and ensure the use of analytical methods for which limits of quantification are in accordance with the standards.

One of the key metrological principles is the achievement of traceability, which is a heritage of metrology – the science of measurements – as conceived for physical measurements (e.g. mass, length, time, temperature etc.) more than one century ago. In this context, the application of metrology concepts to environmental analysis has been examined (Quevauviller, 2001). Discussions within the metrological community generally point out that the direct application of theoretical metrology concepts to chemical and biological measurements is not possible, because of major differences between chemical/biological and physical measurement processes (Valcárcel *et al.*, 1998). For example, chemical analysis results are often strongly dependent upon the nature of samples (whereas physical measurements are less or not affected). A wide variety of analytical problems are encountered in relation to different parameters and matrices (preventing standardised general procedures to be used for all cases), preliminary steps are necessary (e.g. sampling, sample pre-treatment) that may have an effect on the final result etc.

When dealing with groundwater monitoring, theoretical discussions seem to be very distant from real-life situations, and the practice is, in most cases, very far from the metrological theory. Even though the situation has drastically improved within the last few years, the warning made at the beginning of the 1990s (Griepink, 1990; Horwitz, 1992) is still relevant: many environmental chemists still do not pay sufficient attention to the reliability of analytical results and confuse trueness and precision. With respect to traceability, the situation is even worse and this concept is prone to many misunderstandings when it is applied to water quality measurements. This chapter discusses traceability in the context of groundwater monitoring, in particular the various stated references to which chemical environmental data may be traceable.

6.1.2 MEANING OF TRACEABILITY REGARDING GROUNDWATER MEASUREMENTS

JCGM defines traceability as 'the property of a measurement result whereby the result can be related to a reference through a documented unbroken chain of calibrations, each contributing to the measurement uncertainties' (JCGM 200, 2008). In this definition, three key elements may be distinguished, which have been extensively discussed with respect to their applicability to chemical measurements: (1) the link to stated references, (2) the unbroken chain of comparisons and (3) the stated uncertainties. Detailed discussions have already investigated how these elements apply to chemical measurements (Valcárcel & Ríos, 1999; Walsh, 1999; Quevauviller, 2000), let us examine now how they may be understood in the larger context of groundwater monitoring.

In the definition, the *stated references* may be reference methods, reference materials or SI units (kg or mole for chemical measurements) (King, 1997). In theory, all chemical measurements should aim at being traceable to SI units. In practice, measurements correspond to approximations via comparisons of amounts, of instrumental response generated by a number of particles etc., and establishing SI traceability nowadays implies to demonstrate to what extent these approximations are clearly related to the stated references (Valcárcel *et al.*, 1998; Valcárcel & Ríos, 1999). As discussed below, most of the chemical measurements performed in the context of groundwater monitoring are actually traceable to either a reference material (pure substances or matrix certified reference materials) or to a reference method (e.g. standardised method).

The *unbroken chain of calibrations* basically means that there is no loss of information during the analytical procedure (e.g. incomplete recovery, contamination). Achieving this requirement is more or less difficult according to the analytical problem considered. It will be more critical for measurements involving successive analytical steps (e.g. extraction, separation, detection), e.g. for the determination of organic compounds, and less acute for direct measurement procedures (e.g. sensors) which may, however, be faced to other difficulties (e.g. lack of sensitivity or selectivity). Groundwater monitoring brings an additional difficulty, i.e. sample collection (which includes representativeness of the collection), transport and storage. These three steps form an integral part of the traceability chain, which is too often forgotten.

The third key element, the *measurement uncertainties*, is also a critical feature that many analysts still overlook. The theory implies that the uncertainty of a measurement is based on the traceability and the uncertainty of all the stated references that contribute to this measurement. In other words, uncertainty components should be estimated at each step of an analytical process, i.e. the smaller the chain of comparison the better the uncertainty of the final result. Here again theory is confronted to practice when dealing with complex measurements as those performed in the framework of environmental monitoring. This chapter will not discuss uncertainty matters, which were widely discussed in the literature (Maroto *et al.*, 1999; Valcárcel and Ríos, 1999). Discussions will rather focus on the first two elements, which are more likely to be prone to misunderstandings.

Before starting these discussions, it is useful to remind the reader that traceability should not be confused with accuracy. The latter covers the terms *trueness* (closeness of agreement between the 'true value' and the measured value) and *precision* (closeness of agreement between indications or measured quantity values obtained by replicate measurements on the same or similar objects under specified conditions. General aspects of quality assurance in groundwater monitoring (including considerations on accuracy) have been recently discussed in the literature (Quevauviller, 2005; Quevauviller & Roy, 2008) and will not be repeated here. Let us underline the fact that a method that is traceable to a given stated reference is not necessarily accurate (i.e. the stated reference is not necessarily corresponding to the 'true value'), whereas an accurate method is always traceable to what is considered to be the best approximation of the true value (defined as 'a value, which would be obtained by measurement, if the quantity could be completely defined and if all measurement imperfections could be eliminated' (Quevauviller, 2004). At another level, precision and uncertainty are also often confused and considered to be similar concepts, which is not correct since the uncertainty includes both random and systematic errors (while precision is solely linked to random errors) (Valcárcel & Ríos, 1999).

Measurements performed for groundwater monitoring are based on a succession of actions, namely (i) sampling, storage and preservation of representative samples, (ii) pre-treatment of a sample portion for quantitation, (iii) calibration, (iv) final determination and (v) calculation and presentation of results associated with calculated uncertainties. Considering this, we may now consider in further details what types of stated references are used in groundwater monitoring as discussed in a recent paper (Quevauviller, 2005).

6.1.3   RELEVANT METROLOGICAL FEATURES LINKED TO GROUNDWATER MONITORING

6.1.3.1   Groundwater Monitoring and Related Quality Control Needs

Monitoring of groundwater is currently performed by a large number of control laboratories to evaluate the level of pollution by, among others, major or trace elements. A proper evaluation of environmental risks and the necessary comparability of data require that analytical data are of demonstrated accuracy. In this context, external and internal quality control procedures, respectively based on interlaboratory studies and use of reference materials, are essential. Groundwater poses additional analytical difficulties in comparison with e.g. surface water; mainly due to the relatively low concentrations of trace elements, and the occurrence of dissolved organic matter (DOM) and major elements such as chloride, sulphate and iron which are usually in much higher contents than in other types of water. Therefore, the need to control the quality of measurements calls for the availability of Certified Reference Materials (CRMs) of typical groundwater samples. This is discussed below.

The accuracy and uncertainty of a measurement are two basic parameters that should be considered when discussing results of any analysis. Accuracy is of primary importance but if the uncertainty in a result is too high, it is difficult to reach conclusions about the outcome of any experiment or to judge the quality of the water analysed. In water analysis, all basic principles of calibration, elimination of sources of contamination and losses, correction for interferences etc. should be followed (Prichard, 1995; Quevauviller, 2002). Although calibration seems to be so obvious to many analytical chemists, experiences gained in interlaboratory programmes have shown that many systematic errors could be related to calibration errors, showing that insufficient attention had been being paid to this part of the analytical process (Quevauviller, 2002). A typical example concerned estuarine water analysis for which the need to use standard addition procedures was highlighted due to its complex nature, i.e. differences of up to 30% could be observed between results obtained by external calibration and standard additions (Quevauviller *et al.*, 1994).

The increasing awareness for quality assurance has led to the establishment of series of guidelines, the most comprehensive one being the ISO 17025 standard (now adopted as European Norm) which describes the way in which the laboratory should work, its organisation and the way to produce valid results, i.e. involving managerial aspects, quality assessment (inter- and intra-laboratory programmes, use of reference materials), statistical quality control (control charts), maintenance of apparatus, chemicals and reagents, sampling and storage, laboratory analysis, documentation and reporting, and archiving. The following sections examine some of the key features which are included in general metrological principles relevant to groundwater measurements.

6.1.3.2   SI Units

Units of the 'Système International' (SI) correspond to internationally recognised fundamental units that are used in metrology. They establish units of length (metre), mass (kilogram), time (second), temperature (Kelvin) etc. The unit that underpins chemical measurements is the unit of amount of substance, the mol. In principle, all chemical measurement data should be traceable to the mole (King *et al.*, 1999). In practice, contrarily e.g. to the mass standard, there is no '^{12}C mole' standard, and the kg is needed to define the mol (JCGM 200, 2008; ISO, 1993). Therefore, chemical measurements are actually traceable to the mass unit, the kg. Environmental measurements are based on the determination of amount of substance per mass of matrix. One should not confuse this traceability to mass units with the traceability to the 'true value' of the substance in the matrix. This is discussed below.

6.1.3.3   Documented Standards

Standardisation is an important aspect of routine groundwater monitoring. Documented standards (written norms) related to measurement procedures are designed to establish minimum quality requirements and to improve the comparability of analytical results. They also often represent the first step of the introduction of techniques/methods into regulations. In this case, the reference is closely related to the documented protocol, representing one of the main links of the traceability chain. This aspect will be particularly acute when dealing with operationally defined parameters (i.e. parameters determined following a strict analytical protocol) since the traceability chain may be broken if the protocol is not strictly followed. For what concerns groundwater monitoring, standardised procedures (documented standards) have been developed for sampling strategies (ISO 5667 Standards).

   Detailed (documented) guidelines are difficult to set up for sample collection and stor- age, which however remain the primary source of error in environmental monitoring (and hence one of the weakest links of the traceability chain). Recommendations are available in the scientific literature (Quevauviller, 1995; Barceló, 2000), but there are very few examples of documented standards that can formally be used as stated references in the framework of groundwater monitoring. Statistical sampling tools exist (Garfield, 1991; Gy, 1991) but they are often neglected and hardly applicable to practical cases. The nature of the sample and of the substance to be monitored actually dictates the choice of the sam- pling, which is hence adapted case by case. A similar situation is encountered for sample storage for which recommendations are given with respect to protection of the sam- ples from light and elevated temperatures. This situation is obviously unsatisfactory with respect to the comparability of data since no clear stated references may be presently used.

6.1.3.4   Reference Methods

Analytical methods differ in the link between the signal produced by a given deter- mined substance and the signal obtained from the calibration material. In the case of the majority of environmental measurements, including groundwater monitoring, this link is usually related to an amount of substance of established purity and stoichiometry. Most environmental measurements of e.g. organic micropollutants are based on successive analytical steps involving e.g. extraction, separation, and detection steps. This stepwise analysis multiplies the risk that the traceability chain is broken owing to a lack of proper tools (e.g. reference materials, see Section 6.1.3.5). Reference methods are directly linked to the strength of the traceability links. In this respect, so-called primary methods are certainly representing a cornerstone for achieving traceability: these methods have the highest metrological qualities, their uncertainty can be established in terms of SI unit and the analytical results can be accepted without reference to an external calibrating material; they have few random errors and are supposed to be exempt of systematic errors. Using primary methods guarantees, in principle, that measurements will be traceable to SI units, i.e. traceability links will be established to the 'true value' of amount of substance. Therefore, different analytical methods may be referred to as 'reference methods', providing they are linked to primary methods.

This being said, examples of analytical techniques used for groundwater inorganic analysis are given below (as used in the certification of groundwater reference materials, see Section 6.1.3.5 as described by Quevauviller & Roy, 2008). Note that the qualification of 'reference' to these methods in the present case is directly related to their use in a certification round. In other words, not all these methods are defined as 'reference methods' in routine laboratory works. The list below is of course not exhaustive with respect to groundwater monitoring:

- Metallic pollutants, e.g. aluminium, arsenic, cadmium, copper, lead, etc. are currently determined by Electrothermal Atomic Absorption Spectrometry (ETAAS). Hydride Generation Atomic Absorption Spectrometry (HGAAS) was used for the determination of arsenic with volumetric dilution and addition of $NaBH_4$. Inductively Coupled Plasma Atomic Emission Spectrometry (ICP-AES) and Inductively Coupled Plasma Mass Spectrometry (ICP-MS) are also currently used for a range of metallic polluting substances. Another method (now less in use) is Differential Pulse Anodic Stripping Voltammetry (DPASV) for the determination of e.g. cadmium, copper and lead. Pre-treatment methods involve dilution (either volumetric or gravimetric), addition of nitric acid, addition of Pd matrix modifier, or addition of complexing reagent.

- Nutrients (nitrates, phosphate), sulphate, ammonia may be determined by Inductively Coupled Plasma Atomic Emission Spectrometry (ICP-AES). These elements, as well as chloride and bromide, were also determined by Ion Chromatography (IC) with conductivity detection. Potentiometric determination by titration was also applied to the determination of chloride in groundwater, using a specific electrode. Examples of pre-treatment methods included gravimetric dilution, addition of nitric acid and filtration on $0.45\,\mu m$ or $0.2\,\mu m$ membrane. Chloride and nutrients (e.g. nitrate, phosphate, sulphate) were also determined by spectrophotometry. Pre-treatment methods were based on the addition of derivatives for SPEC determination of complexes (e.g. diazo-compound for nitrate, Mo-blue for phosphate, etc.).

Let us note that, besides the above methods, Isotope Dilution Mass Spectrometry (IDMS) is also used for the determination of some trace elements (e.g. Cu, Pb). This method, responding to the definition of a 'primary method', thus firmly anchored other methods used in reference material certification programmes (see Section 6.1.3.5) to the demonstration of their traceability to SI units. This means in turn that the certified values may unambiguously be considered as 'true values' for the elements which were determined by IDMS, i.e. there are hardly any operational analytical steps (e.g. extraction for which recovery would need to be evaluated) that could represent a risk that the traceability chain be broken in this present case (Quevauviller & Roy, 2008).

6.1.3.5   Reference Materials

The role of reference materials is in principle well known. Certified Reference Materials (CRMs) may be calibration materials (pure substances or solutions, or materials of known composition for techniques requiring matrix-matched calibrants, e.g. XRF) or matrix materials representing as far as possible 'real matrices' for the verification of measurement process. Laboratory Reference Materials (also known as Quality Control Materials) have the same basic requirements of representativeness, homogeneity and stability, but these materials are not certified and generally produced at a much smaller scale e.g. for interlaboratory studies or internal quality control (control charts), i.e. to monitor the performance of analytical methods with time (reproducibility) through the establishment of control charts (Hartley, 1990). In this view, control charts and related RMs may be considered as long-term stated references for analytical measurements (see Section 6.1.3.6). It has been stressed that the 'reference' represented by a RM may not always be reliable since, in many cases, the RM does not have the 'same' matrix as the unknown sample (de Bièvre, 1996). This is discussed below.

Official certification bodies attempt, wherever possible, to produce reference materials estimating the true values as closely as possible. In the case of matrix environmental CRMs, this is mainly achieved by employing a variety of methods with different measurement principles in the material certification study (see discussion in Section 6.1.3.3 for what concerns groundwater); if these methods are in good agreement, one may assume (but not firmly demonstrate) that no systematic error has been left undetected, and the reference (certified) values are the closest estimate of the true value. This approach possibly includes definitive methods (e.g. IDMS, see above), which seldom exist for analyses involving an extraction or derivatisation step. In many instances, *consensus* values are accepted as *true* values reflecting the state-of-the-art (hence ensuring data comparability). Discussions are on-going on the fact that many matrix CRMs do not guarantee a full verification of accuracy owing to possible remaining bias (e.g. all extraction methodologies, although being in good agreement, could be biased to a certain degree, with no means to demonstrate it at the present stage). This is a point of discussion below.

As mentioned above, some (certified) RMs are intended for calibration purposes; in these cases, the uncertainty of the certified value is of prime importance since it will affect the final uncertainty of the measured value in the unknown sample. In the case of certified pure substances or calibrating solutions, the uncertainties of certified values are usually negligible in comparison with the method uncertainty. This is not the case of matrix CRMs that are in principle reserved to the validation of methods.

With respect to groundwater, the preparation and certification of groundwater matrix-matched CRMs as carried out by the European Commission (BCR, now based at the Institute for Reference Materials and Measurements, IRMM, in Geel, Belgium) has been achieved by the following steps: (1) preparation of a batch of material representative of the measurement(s) of concern (feasibility study), (2) homogeneity testing, (3) stability testing, (4) interlaboratory studies to detect possible sources of systematic error(s) in the measuring methods, and (5) certification campaign. The first step (feasibility study) is of utmost importance since the results obtained will condition the entire project. As described elsewhere (Quevauviller & Maier, 1999), a feasibility study for the preparation of water CRMs involved investigations on the homogenisation system and sample containers (tests on sorption or volatilisation risks for various types of containers), optimisation of the cleaning procedures to avoid contamination risks, choice of the best vial for storing the samples (bottles or ampoules), stabilisation (using chemical additives or by γ-irradiation) and stability testing (at different temperatures). Additional details on the type of materials successfully used for water CRMs and various preparation procedures are given elsewhere (Quevauviller & Maier, 1999) and are available in the various papers and reports related to BCR-CRMs. CRMs produced for the quality control of groundwater analyses concerned a range of major elements in artificial groundwater considered to be the best compromise as regard to their stabilisation (Benoliel *et al.*, 1997) and trace elements and bromide in natural groundwater materials stored in ampoules (Quevauviller *et al.*, 1998, 1999). These materials are described in specific papers as well as summary publications (Quevauviller, 2005; Quevauviller & Roy, 2008).

There arc a number of suppliers of water CRMs, some of them being specialized in a particular field of interest (e.g. the National Research Council of Canada for marine monitoring). Two main bodies, NIST (USA, formerly NBS) and the European Commission (Institute for Reference Materials and Measurements, BCR) cover several fields and ensure a long term availability of CRMs. Surveys are regularly undertaken by international organisations. It should be noted that CRMs are products of very high value. Typically, the production of a water CRM costs some hundred thousand of euros. Therefore, these materials should be reserved for the final verification of analytical procedures and not as materials for routine measurements (e.g. in control charts). In Europe, CRMs are available at the Institute for Reference Materials and Measurements (IRMM), Retieseweg, B-2440 Geel (Belgium). They are delivered along with certification reports containing a full description of the material preparation, the homogeneity and stability studies, the techniques used in the certification, the technical and statistical evaluation of the results, and all the individual results provided by the certifying laboratories. Direct information can also be obtained by e-mail to BCR.SALES@irmm.jrc.be and through the IRMM Internet Worldwide Web Site at http://www.irmm.jrc.be/mrm.html

6.1.3.6 Statistical Control

When a laboratory works at a constant level of high quality, fluctuations in the results become random and can be predicted statistically (Taylor, 1985). This implies that the limits of determination and detection should be constant and well known. Rules for rounding-off final results should be based on the performance of the method. Furthermore, in the absence of such systematic fluctuations, normal statistics should be applied to study the results wherever necessary (e.g. regression analysis, t- and F-tests, analysis of variance etc.) (Vogelsang, 1987).

Control checks should be carried out as soon as the method is in control in the laboratory. These can be done by setting up control charts which provide a graphical way of interpreting the method's output in time in order to evaluate the reproducibility of the results over a period of time. To do so, one reference material of good homogeneity and stability should be analysed with e.g. 10–20 unknown samples. In order to be able to detect non-random fluctuations in the real analysis, the RM used should pose the same or similar problems to the analytical chemist as the unknown samples do (analytical similarity) and their composition should be homogeneous and stable in time. A detailed description of the different types of control charts (e.g. Shewhart or cusum control charts) is given elsewhere (Vogelsang, 1987).

6.1.3.7 Comparison of Methods

The use of control charts enables to detect whether or not a method is still under control. However, it does not allow detecting a systematic error which could be present from the moment of the introduction of the method in the laboratory. To overcome this risk of undetected bias, results should be verified by using other methods.

All analytical methods have their own particular source of error, e.g. different interferences occurring in spectrometric or voltametric techniques. Independent methods displaying different sources of error may therefore be used to verify the validity of results of chemical analysis. If results of two independent methods are in good agreement, it may be concluded that the risk of error of systematic nature is unlikely; this conclusion is strongest when the two methods differ widely. If the methods have similarities, such as digestion step in the case of voltammetry and spectrometry, a comparison of the results will only allows conclusions to be drawn on the accuracy of the final determination and not the whole analytical procedures.

In some cases, the method of comparison is a technique of which the sources of error are well known and controlled and which is relatively insensitive with respect to human failures, i.e. as 'reference method' as referred to in Section 6.1.3.4 (in particular e.g. isotope dilution mass spectrometry). When possible, the application of such a reference method is used in a good quality control system. However, it must be stressed that good independent or reference methods do not always exist (e.g. there are hardly reference methods for organic analysis). Furthermore, these techniques require a good experience of the technician who might not have a sufficient skill to produce reliable results even with a so-called reference method. Further details are given in the literature (Prichard, 1995; Quevauviller, 2002).

6.1.3.8 Interlaboratory Studies

Interlaboratory studies are useful tools to detect systematic errors linked to a specific method or incorrect application of a method in a laboratory. In general, besides the sampling error, three sources of error can be detected in all analyses: (i) sample pre-treatment (e.g. digestion, extraction, separation, pre-concentration), (ii) final measurement (e.g. calibration errors, spectral interferences, peak overlap), (iii) insufficient experience (e.g. lack of training, care applied to the work, awareness of pitfalls, management, clean bench facilities). A relatively large laboratory where more techniques are applied by different experienced technicians may be in position to eliminate most errors related to (i) and (ii) in a continuously and carefully applied quality control scheme. Smaller laboratories or laboratories having one or two techniques only are not always able to assess the quality of their measurements with respect to (i) and (ii) in the absence of other tools (e.g. use of CRMs as described below). For both types of laboratories, however, interlaboratory studies to evaluate the step (iii) are necessary.

Interlaboratory studies can be conducted for several aims:

1. to detect the pitfalls of analytical techniques and ascertain their performance;

2. to measure the quality of a laboratory or part of a laboratory (e.g. audits for accredited laboratories);

3. to improve the quality of a laboratory in collaborative work in a mutual learning process;

4. to certify reference materials.

When laboratories are working under control (of a recognised QC scheme), interlaboratory studies of types 2 and 4 are conducted only. However, studies of types 1 and 3 also play an important role. The organisation of interlaboratory studies, the types of samples, laboratories invited, management of the trials etc. have been described in detail elsewhere (Quevauviller & Maier, 1999; Quevauviller, 2002).

Interlaboratory studies are a useful and important means to assess the quality of the work done in the laboratory, to motivate laboratory workers and to demonstrate the quality of a laboratory's result to a 'customer'. Experiences obtained in a programme concerning groundwater analysis have been described in the literature (Benoliel *et al.*, 1997; Quevauviller *et al.*, 1998, 1999; Quevauviller & Roy, 2008). In this context, participants of different EC Member States discussed thoroughly the reasons for discrepancy in collaborative trials and were able to improve their methods in the light of these discussions. Many exercises followed a step-by-step approach, i.e. series of studies with samples of increasing difficulty (e.g. synthetic samples, natural samples spiked with the analyte(s) of concern, natural samples or matrix-matched synthetic samples).

Proficiency testing (type 2) enables in principle laboratories to establish 'external' references for evaluating the performance of their methods. As described above, one or more reference materials are distributed by a central organisation to several laboratories for the determination of given substances. Comparing laboratory results (based on different methods) allows detecting possible sources of errors linked to a specific procedure or related to the way a method has been applied by a given laboratory. When the testing focuses on a single method, this enables performance criteria (e.g. precision) to be checked. The references (establishing the traceability link) are again based on reference materials which have to usual requirements. However, contrarily to reference materials used for internal quality control, proficiency testing may involve samples with a limited shelf life which may be distributed to laboratories for analysis of particular parameters that could not be evaluated using stabilised RMs. Examples are 'fresh' materials with a short-term preservation period, e.g. groundwater samples containing unstable substances.

The interlaboratory-based certification (type 4) responds to the requirements of proficiency testing schemes. An example of such scheme as applied to groundwater analyses is described elsewhere (Charlet & Marschal, 2004). As discussed above for reference materials, the measurement values obtained in the framework of interlaboratory studies (using different techniques) may be considered as an anchorage point representing the analytical state-of-the-art; this offers laboratories a mean to achieve comparability (i.e. traceability) of their results to a recognised reference, which is in this case a consensus value (generally the mean of laboratory means). It represents a useful method for achieving comparability of environmental measurements using an external quality control scheme (Benoliel *et al.*, 1997).

6.1.3.9   Uncertainty Assessment

Uncertainty associated with results of chemical analyses is a key element in terms of traceability and reliability of monitoring data. The necessity to consider uncertainties is now explicitly stated in the international normative reference ISO/CEI 17025. Uncertainties on measurement are therefore intrinsically linked to every single analytical result. The determination takes into account each aspect of laboratory measurements and analytical procedures, from sample collection to laboratory analyses. These calculations, however, do not usually take into account the complete analytical chain. In particular, uncertainties related to sampling stages, including sampling itself, packaging, sample conditioning and transportation, etc., are often ignored or poorly quantified. In relation to groundwater analyses, this aspect has been studied recently by Roy and Fouillac (2004) on the basis of the study of two industrial sites exhibiting groundwater contaminated with chemical compounds. The study enabled to estimate the representativeness of groundwater sampling and to calculate the associated sampling uncertainties in order to establish an overall evaluation of the impact of field activities compared to analytical measurements. As an example, it was possible to calculate that for Pb concentrations in the area of 1 µg/l, sampling uncertainty does not exceed 18% of the measured value, if a refined sampling strategy and procedure is applied. In contrast, if sampling is limited to a single sample collected at random ('blind') in the water column, the uncertainty at such low concentration approaches 100%. The two most important effects (accounting for 73% of the variability) were found to be natural inter-borehole variability, i.e. spatial variability, and intra-borehole variability, with different flow patterns depending on the lithological and pedological features of the ground medium (Roy & Fouillac, 2004). Two other minor effects (accounting for 12% and 6% of total variability) were identified. One is related to bacterial activity, an effect directly observed in terms of temporal variability. The other represented an analytical effect that included the entire analytical chain, from the sampling stage to the analysis stage.

6.1.4   CONCLUSIONS

This chapter aims to reconcile groundwater monitoring policy requirements with key metrological principles. It highlights needs for integration of analytical knowledge and expertise at various levels for a proper implementation of groundwater policy. Starting from the key requirement to ensure comparability of chemical data to properly implement policies, i.e. to take appropriate decisions on groundwater management on the basis of monitoring data, one may understand the importance to consider carefully all the features that enable this comparability to be achieved ('traceability'), namely references to SI units, reference methods, reference materials, internal and external quality control procedures etc. At this stage, the most recent advances in analytical knowledge may allow to establish those references for inorganic groundwater analyses. The basic question is to ask whether this knowledge is accessible to routine control laboratories, i.e. whether groundwater monitoring analyses carried out in the framework of large-scale policy-based monitoring programmes are performed in the most 'traceable way' possible.

As a final word, it is recognised that, while the state-of-the-art of groundwater analyses may be considered to be sufficient for monitoring groundwater in support of policy requirements, efforts are still needed to streamline information from the most recent advances in analytical research to policy decision-makers. In this respect, recent European Research projects like EAQC-WISE (Held *et al*., 2008) and activities of experts forums led by the European Commission, e.g. the 'Chemical Monitoring Activity' (Quevauviller, 2008) clearly highlight efforts from all actors to improve and share their knowledge with the ultimate objective of increasing chemical analysis reliability including groundwater ones.

REFERENCES

Barceló D., ed., 2000. *Sample Handling and Trace Analysis of Pollutants*, 2nd Edition, Elsevier, Amsterdam, The Netherlands.

Benoliel M.J., Quevauviller Ph., Rodrigues E., Andrade M.E., Cavaco M.A. and Cortez L., 1997. *Fresenius' J. Anal. Chem.*, **358**, 574.

de Bièvre P., 1996. In: *Accreditation and Quality Assurance in Analytical Chemistry*, H. Günzler ed., Springer, Berlin, Germany.

Charlet Ph. and Marschal A., 2004. *Trends Anal. Chem.*, **23**, 178.

European Commission, 2006a. Directive 2006/118/EC of the European Parliament and of the Council of 12 December 2006 on the protection of groundwater against pollution and deterioration, *Official Journal of the European Communities* L372, 27.12.2006, p. 19.

European Commission, 2009. *Commission Directive laying down, pursuant to Directive 2000/60/EC of the European Parliament and of the Council, technical specifications for chemical analysis and monitoring of water status*, draft version, May 2008.

Garfield F.M., 1991. *Quality Assurance Principles for Analytical Laboratories*, AOAC International, Arlington, USA

Griepink B., 1990. *Fresenius J. Anal. Chem.*, **338**, 360.

Gy P.M., 1991. *Mikrochim. Acta*, **II**, 457.

Hartley T.H., 1990. *Computerized Quality Control: Programs for the Analytical Laboratory*, Ellis Horwood, Chichester, 2nd edn.

Held A., Emons H., Taylor P., 2008. *EAQC-WISE (European Analytical Quality Control in support of the Water Framework Directive via the Water Information System for Europe), Final report of the project: The blue print*, deliverable D25, December 2008, 58 pp.

Horwitz W., 1992. *AOC Intern*. 368.

Ifen, 2004. Pesticides in water. Sixth annual report. Data 2002, Ed. Institut Française de l'Environnement, collection Etudes et travaux n°42.

ISO 5667-11, 1993 – *Water quality – Sampling – Part 11: Guidance on sampling of ground waters*.

JCGM 200, 2008. *International Vocabulary of Metrology – Basic and General Concepts and associated terms (VIM)*.

King B., 1997. *Analyst*, 112, 197.

King B., Walsh M., Carneiro K., *et al*., 1999. *Metrology in Chemistry – Current Activities and Future Requirements in Europe*, EUR Report, EUR 19074 EN, European Commission, Brussels.

Maroto A., Boqué R. and Rius F.X., 1999. *Trends Anal. Chem.*, **18**, 577.

NF EN ISO 5667-1, 2006 – *Water quality – Sampling – Part 1: Guidance on the design of sampling programmes and sampling techniques*.

Prichard E., 1995. *Quality in the Analytical Laboratory*, John Wiley & Sons Ltd., Chichester.

Quevauviller Ph., ed., 1995. *Quality Assurance in Environmental Monitoring – Sampling and Sample Pre-treatment*, VCH, Weinhem, Germany.

Quevauviller Ph., 2000. *J. Environ. Monitor.*, **2**, 292.

Quevauviller Ph., 2001. *Métrologie en chimie de l'environnement*, Tec&Doc Editions, Paris, France.

Quevauviller Ph., 2002. *Quality Assurance for Water Analysis*, John Wiley & Sons Ltd., Chichester.

Quevauviller Ph., 2004. *Trends Anal. Chem.*, **23**, 171–8.

Quevauviller Ph., 2005. *Ecol. Chem. Engin.*, **12**(5-6), 485–501.

Quevauviller Ph., 2006. *J. Soil & Sediments.*, **6**(1), 2–3.

Quevauviller Ph. and Maier E.A., 1999. *Certified Reference Materials and Interlaboratory Studies for Environmental Analysis – The BCR approach*, Elsevier, Amsterdam, The Netherlands.

Quevauviller Ph., Kramer K. and Vinhas T., 1994. *Mar. Pollut. Bull.*, **28**, 506.

Quevauviller Ph., Andersen K., Merry J. and van der Jagt H., 1998. *Analyst*, **123**, 955.

Quevauviller Ph., Benoliel M.J., Andersen K. and J. Merry, 1999. *Trends Anal. Chem.* **18**, 376.

Quevauviller Ph. and Roy S., 2008. Quality assurance for groundwater monitoring, in: *Groundwater Science and Policy – An international overview*, Quevauviller Ph., ed., Chapter 6.3, The Royal Society of Chemistry, Cambridge, 378–404.

Roy S. and Fouillac A-M., 2004. *Trends Anal. Chem.*, **23**, 185.

Subramanian G., ed., 1995. *Quality Assurance in Environmental Monitoring – Instrumental methods*, VCH, Weinhem, Germany.

Taylor J.K., 1985. *Handbook for SRM-Users*, NBS Special Publication 260–100.

Valcárcel M. and Ríos A., 1999. *Anal. Chem.*, **65**, 78A.

Valcárcel M., Ríos A., Maier E., *et al.*, 1998. *Metrology in Chemistry and Biology – A Practical Approach*, EUR Report, EUR 18405 EN, European Commission, Brussels.

Vogelsang J., 1987. *Fresenius' Z. Anal. Chem.*, **328**, 213.

Walsh M.C., 1999. *Trends Anal. Chem.*, **18**, 616.

6.2

Use of Isotopes for Groundwater Characterization and Monitoring

Philippe Négrel, Emmanuelle Petelet-Giraud and Agnés Brenot

BRGM, Orléans cédex, France

Groundwater Monitoring Edited by Philippe Quevauviller, Anne Marie Fouillac, Johannes Grath and Rob Ward
© 2009 John Wiley & Sons, Ltd

6.2.1   INTRODUCTION

One of the main profound changes in the past decades in earth and solar system sciences is the addition of rigorous analytical tools (e.g. the isotopes) in complement to classical descriptive ones. This is now considered as the isotope geology frame (Faure, 1986; Bowen, 1988; Allègre, 2005) in which isotope hydrology and isotope hydrogeology are great challenges (Fritz and Fontes, 1980; Clark and Fritz, 1997; Aggarwal *et al*., 2005). Isotope geology began with dating geologic events through various isotope systematics (Faure, 1986), progressed by the competition of the lunar exploration programs and the associated opportunity to analyse rocks from the moon and culminated with the complete description of the earth's system that led to the Crafoord prize in Geosciences shared by G. Wasserburg and C.J. Allègre in 1986 (Allègre, 1987). On the other hand, isotope hydrology and isotope hydrogeology evolved mainly through the environmental isotopes that included principally the naturally occurring isotopes of elements found in abundance in the earth environment (e.g. H, C, N, O and S). Large developments of isotope hydrogeology were done under the auspice of the International Atomic Energy Agency (IAEA, www.iaea.org) with the joint cooperation between IAEA and the International Hydrological Program of UNESCO (Mook, 2001; Aggarwal *et al*., 2005). Environmental isotopes are now used largely to trace water provenance but also recharge processes, geochemical reactions and reaction rates. The family of environmental isotopes is now growing as new methods allow routine analysis of additional isotopes (Kloppmann, 2003).

With regards to policies and water management, the Water Framework Directive (WFD, 2000/60/EC) main objective's is to prevent further deterioration and protect and enhance the status of aquatic ecosystems. The success of the Directive, in achieving this purpose and its related objectives, will be mainly measured by the status of water bodies (e.g. the units that will be used for reporting and assessing compliance with the Directive's principal environmental objectives). The WFD/Annex II stated, among others, that the characterization of water bodies shall include (1) information on the stratification characteristics of the groundwater within the groundwater body, (2) on the estimates of the directions and rates of exchange of water between the groundwater body and associated surface systems, (3) provide further characterization of the chemical composition of the groundwater, including specification of the contributions from human activity. For that purpose, combined geochemical analysis (major and trace elements), and isotopes ($\delta^{18}O$ and δ^2H, $\delta^{34}S_{SO4}$ and $\delta^{18}O_{SO4}$, strontium, boron, lithium, and uranium are applied on one demonstrative water body in the SW France in the CARISMEAU project (http://carismeau.brgm.fr/). The aim of this chapter is to present some of the main results derived from this project.

6.2.2   ISOTOPES: A WAY TO CHARACTERIZE AND MONITOR GROUNDWATER

6.2.2.1   Isotope Definition

Atoms consist of a nucleus surrounded by electrons. The nucleus is composed of neutrons (with no electric charge) and protons (positively charged). The number of protons

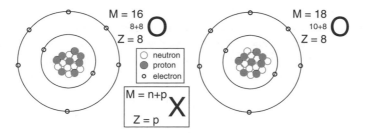

Figure 6.2.1 Atom structure. Example of two isotopes of oxygen - ^{16}O and ^{18}O.

(Z = atomic number), is equal to the number of electrons (negatively charged) surrounding the nucleus, so that the atom as a whole is neutral (Figure 6.2.1).

The sum of protons and neutrons is the nuclear mass number (M). The atomic nuclei are kept together by extremely strong forces between the nucleons (protons and neutrons) with a very small range. Repulsive electrical forces (Coulomb) exist between the protons, the nucleus is stabilised by presence of neutrons. In the most abundant nuclides of the light elements, the numbers of protons and neutrons are equal. Nuclei such as $^{2}_{1}H_1$, $^{4}_{2}He_2$, $^{12}_{6}C_6$, $^{14}_{7}N_7$ or $^{16}_{8}O_8$ are stable. For the heavy elements the number of neutrons far exceeds the number of protons: ^{238}U contains only 92 protons, whereas the largest stable nuclide, the lead isotope ^{208}Pb has an atomic number of 82. Instabilities are caused by an excess of protons or neutrons, for instance $^{3}_{1}H_2$ *and* $^{14}_{6}C_8$ are unstable (or radioactive). Nevertheless, a slight excess of neutrons for light elements does not necessarily result in unstable nuclei.

Isotopes of a chemical element present the same atomic number (same number of protons), they have the same name and are located at the same place in the periodic table of the elements, but they differ by their atomic mass (p + n).

Radioactive isotopes have variable proportions through time in the geological systems, whereas stable isotopes present constant proportions. In the water cycle, proportions of the stable isotopes can be modified by isotope fractionation related to physical/chemical/biological processes. Such isotope fractionations are mainly measured for the light elements such as H, S, N, C, O but there is recent evidence for heavier elements (Fe, Zn...). These modifications can be evidenced by isotopic ratio measurements, pointed out the involved process and then help to reconstruct the 'water story'. At the opposite, radiogenic isotopes of heavy elements are not fractionated by these processes. The link is thus direct between the water and the mineral that release the chemical element in solution through weathering processes (Sr for $^{87}Sr/^{86}Sr$, Nd for $^{144}Nd/^{143}Nd$, Pb for $^{206,\ 207,\ 208}Pb/^{204}Pb$), giving information on the water-rock interactions both in surface and within the aquifers. Details on isotope principals can be found in the IAEA synthesis from Mook (2001).

Among the various isotope systematics used to understand the water cycle and especially the groundwater, in terms of groundwater origin, recharge, dating, pathways, etc., this chapter will only focus on 4 isotopic tools: (1) oxygen and hydrogen of the water molecule to characterize the water origin, the recharge zone, and various processes affecting the isotopic signature (evaporation, mixing, ...); (2) strontium isotopes to evidence water–rock interaction, water circulation, connexion between aquifers, ...;

(3) sulfur and oxygen of sulfates to decipher the origin of dissolved sulfates in water (atmosphere, evaporites, sulfur minerals, anthropogenic sources) and also biological processes that induce isotopic fractionations; and (4) boron isotopes as complementary tool to highlight the source of the salinity and trace the heterogeneities in aquifers.

6.2.2.2 Oxygen and Hydrogen of the Water Molecule

Stable isotopes of the water molecule include O and H. Oxygen isotopes are ^{16}O (99.762%), ^{18}O (0.200%) and ^{17}O (0.038%). Hydrogen isotopes are ^{1}H (99.985%) and ^{2}H (0.015%). Stable isotope ratios for O and H are expressed using the δ notation in part per mil (‰) according to $\delta = [(R_{echantillon}/R_{standard}) - 1] \times 1000$. In this notation, R is the isotope ratio (e.g. $^{18}O/^{16}O$) and the standard is the SMOW (Standard Mean Ocean Water) which value is 0‰.

Without any evaporation and exchange with dissolved gases or rocks, the stable isotopes can be considered as conservative and thus reflect the mixture of the different recharge at the origin of the groundwater.

The hydro-climatic story of an aquifer can thus be reconstituted by looking at the abundance of the stable isotopes of the water molecule. These abundance being driven by environmental condition (T°C), recharge altitude, leading to a specific signature in the groundwater (for synthesis, see Mook, 2001).

6.2.2.3 Strontium Isotopes

Strontium has four isotopes of which the abundance are issued from the nucleosynthesis processes (^{84}Sr, ^{86}Sr, ^{88}Sr) and the last isotope (the radiogenic isotope ^{87}Sr) is produced by the radioactive decay of the ^{87}Rb isotope. The abundance of ^{87}Sr (expressed as the $^{87}Sr/^{86}Sr$ ratio) in a given geological material is a function of the age and the Rb/Sr ratio of this material (as summarized by Négrel, 2006). The $^{87}Sr/^{86}Sr$ ratio variations within an hydrosystem can provide information about the sources of Sr and the different mixing processes involved. Generally, a large range of signatures (through the $^{87}Sr/^{86}Sr$ ratio) exists in waters and, because of the high precision level (20.10^{-6}) of measurement by mass spectrometry, small variations are measurable and can be interpreted (Blum *et al.*, 1994; Millot *et al.*, 2007). Weathering causes rocks with different chemical characteristics and ages to release strontium into water (Pennisi *et al.*, 2000; Négrel, 2006). During interaction between one type of rock and water, the $^{87}Sr/^{86}Sr$ ratio varies according to the Rb/Sr ratios and the age of the weathered material. Since no natural processes fractionate Sr isotopes, the measured differences in the $^{87}Sr/^{86}Sr$ ratios are due to the mixing of Sr derived from various sources with different isotopic compositions (Négrel and Petelet-Giraud, 2005). Furthermore, given the short time scale of the processes studied, the Sr isotope variations are mainly due to the mixture of Sr derived from different sources with different Sr isotope compositions (Kloppmann *et al*., 2001; Millot *et al*., 2007).

6.2.2.4 Sulfur and Oxygen Isotopes and Sulfates

Sulfur displays four stable isotopes: ^{32}S, ^{33}S, ^{34}S and ^{36}S whose mean natural abundances are respectively of 95%, 0.76%, 4.22% and 0.014%. Sulfur is involved in different biochemical processes and because of its various oxidation levels, sulfur is under different form. Study of the isotopic ratio ($^{34}S/^{32}S$) and ($^{18}O/^{16}O$) of dissolved sulfates (SO_4) provides information on the origin of dissolved sulfates and on process involved such as sulfur oxidation and bacterial reduction of dissolved sulfates. As for stable isotopes of the water molecule, the δ notation in part per mil (‰) according to $\delta = [(R_{echantillon}/R_{standard}) - 1] \times 1000$ is used. The standard is the SMOW (Standard Mean Ocean Water) for the $^{18}O/^{16}O$ ratio and the CDT (Cañon Diablo Troïlite) for the $^{34}S/^{32}S$ ratio.

Dissolution of sulfate minerals and oxidation of sulfide minerals, during water/rock interactions, do not induce significant fractionation of S isotopes. Whereas, bacterial reduction of dissolved sulfates, occurring essentially for groundwater, induces an huge fractionation of S isotopes of dissolved sulfates. For O isotopes, only the dissolution of sulfate minerals conserves the initial O isotopic signature of the source materials. The resulting O isotopic composition of dissolved sulfates induced by sulfide oxidation is depending on the part of O incorporated from ambient water and incorporated from the atmosphere and the fractionation induces during this incorporation. As for S isotopes, bacterial reduction of dissolved sulfates also induces an isotopic fractionation of O isotopes. Thus isotopic compositions of dissolved sulfates are both capable to trace source of dissolved sulfates and to trace processes involved in hydrosystems.

6.2.2.5 Boron Isotopes

Boron can be used as a tracer in aquatic systems because of its high solubility in aqueous solutions, measurable abundance in natural waters, and the lack of isotopic effects induced by evaporation, volatization, and oxidation–reduction reactions. Due to the large relative mass difference between ^{10}B and ^{11}B and the high chemical reactivity of boron, significant isotope fractionation produces large variations in the $^{11}B/^{10}B$ ratios in natural samples from different geological environments (up to 90‰ expressed in δ unit, like O and H isotopes; Barth, 1993). Consequently, the boron isotope compositions are sensitive to mixing of pristine and contaminated waters and to water–rock interactions as well as to the origin of salinity (Vengosh *et al*., 1994; Vengosh *et al*., 1995; Barth, 2000a,b; Kloppmann *et al*., 2001; Casanova *et al*., 2002; Pennisi *et al*., 2006a,b; Lemarchand and Gaillardet, 2006). Moreover, during adsorption processes the light isotope, ^{10}B, tends to adsorb preferentially onto clay minerals and the residual fluids is enriched in ^{11}B.

6.2.2.6   Mixture of Two Components

Mixing of two solutions (a, b) with different isotope compositions will led to a mixed solution (i.e. 'mix') that will have a concentration of:

$$X_{mix} = \alpha \, X_a + (1 - \alpha)X_b \qquad (6.2.1)$$

For element X like Sr, B...
   Considering boron, the mixed solution will have a $\delta^{11}B$ of:

$$\delta^{11}B_{mix} = [(B_a \times \delta^{11}B_a \times \alpha) + (B_b \times \delta^{11}B_b \times (1 - \alpha)]/B_{mix} \qquad (6.2.2)$$

where B_a, B_b and B_{mix} are the boron concentration, $\delta^{11}B_a$, $\delta^{11}B_b$ and $\delta^{11}B_{mix}$ are the boron isotope compositions, α is the mixing proportion. A similar mixing equation can be written for Sr following:

$$^{87}Sr/^{86}Sr_m \times [Sr]_m = \alpha(^{87}Sr/^{86}Sr_1 \times [Sr]_1) + (1 - \alpha)(^{87}Sr/^{86}Sr_2 \times [Sr]_2) \qquad (6.2.3)$$

where $(^{87}Sr/^{86}Sr)_m$ is the measured isotopic ratio in the mixture, $(^{87}Sr/^{86}Sr)_1$ and $(^{87}Sr/^{86}Sr)_2$ are the isotope ratios of the first and second end-members, respectively, and $[Sr]_m$, $[Sr]_1$, and $[Sr]_2$ are the Sr contents of the mixture, and the first and second end-members, respectively.

6.2.3   ISOTOPES TOOLS TO CHARACTERIZE THE GROUNDWATER BODIES: EXAMPLE OF THE CARISMEAU RESEARCH PROJECT (SW FRANCE)

6.2.3.1   CARISMEAU Objectives

Within the framework of the directive, the CARISMEAU research project lead by BRGM and the French Water Agency Adour-Garonne, concerns the application of multi isotopic and geochemical characterization of water bodies in SW France. The CARISMEAU research project is focused on the Adour-Garonne district, which covers $116\,000\,km^2$ represents 1/5 of the territory. It is limited by the Massif Central to the east, by the Armorican Massif to the north, by the Pyrenees to the south and by the Atlantic Ocean to the west. The district fully covers the Aquitaine and Midi-Pyrénées regions and partly the Auvergne, Languedoc-Roussillon, Limousin and Poitou-Charente regions. It represents 6900 councils for a total of around 6.7 millions inhabitants.

Based on the WFD/Annex II, aims of CARISMEAU are to provide further character-ization of groundwater bodies or groups of bodies which have been identified as being of primary importance and/or at risk in the district.

For that purpose, combined geochemical analysis (major and trace elements), common isotopic methods with $\delta^{18}O$ and δ^2H of the water molecule and $\delta^{34}S_{SO4}$ and $\delta^{18}O_{SO4}$; innovative isotopic method with strontium isotopes ($^{87}Sr/^{86}Sr$) and potential isotopic methods with boron ($\delta^{11}B$) and lithium (δ^7Li) isotopes as well as U series will be applied on one demonstrative water body: The Eocene sands aquifer water body. This aquifer constitutes a major aquifer used for drinking water, agriculture irrigation, gas storage and thermo-mineral water resource. It extends over the Adour-Garonne district, being an artesian system to the west of the district and confined with piezometric levels around 250 m to the east of the district (Figure 6.2.2).

The Eocene sands aquifer water body was investigated through 42 samples from the different layers of the aquifer (Palaeocene to late Eocene; Figure 6.2.2). Two sampling surveys were carried out during the year 2006.

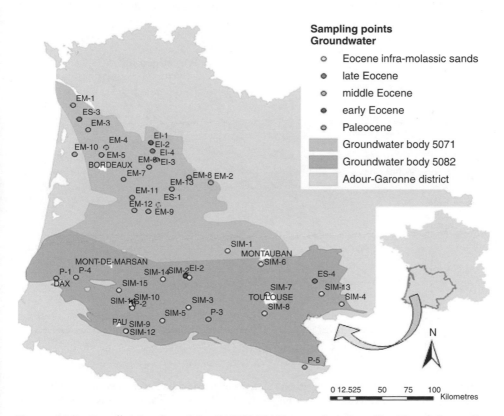

Figure 6.2.2 Sampling location of the CARISMEAU research project. (See Plate 6 for a colour representation).

6.2.3.2   Oxygen and Hydrogen of the Water Molecule: From Recharge to Residence Time Assessment

The $\delta^{18}O$ values of the groundwaters fall in the range -5.6 to $-10.6‰$ vs. SMOW, with the δ^2H values varying between -34.3 and $-72.3‰$ vs. SMOW. There is no relationship between $\delta^{18}O$ and δ^2H values and the salinity with R (correlation coefficient) close to 0.30 and 0.45 between TDS (Total Dissolved Salts) and $\delta^{18}O$–δ^2H values (respectively). All the analysed waters (Figure 6.2.3) plot on or near the global meteoric water line (GMWL $\delta^2H = 8\delta^{18}O + 10$, Craig, 1961) and are well correlated as the correlation coefficient R is close to 0.98.

The groundwater data present a wide range of stable isotopic composition (δ^2H, $\delta^{18}O$) both between the different aquifers and inside a single aquifer (Figure 6.2.4, example of the IMS aquifer). Comparing independently each aquifer level in May and October, it appears:

- *Palaeocene aquifer* collected NW–SE along the Pyrenees border presents a large variation in the $\delta^{18}O$ and δ^2H signature, agreeing with that observed in the district. Sampling points P-1 and P-4 collected in the SW part of the district display the most

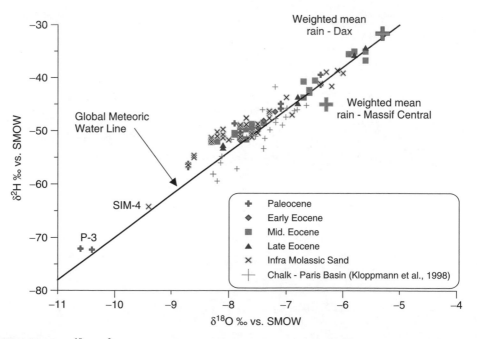

Figure 6.2.3 $\delta^{18}O$ - δ^2H plot for the ground waters collected in the Eocene sand aquifers. GMWL corresponds to the global meteoric water line ($\delta^2H = 8\delta^{18}O + 10$). Chalk – Paris Basin data from Kloppmann *et al.* (1998).

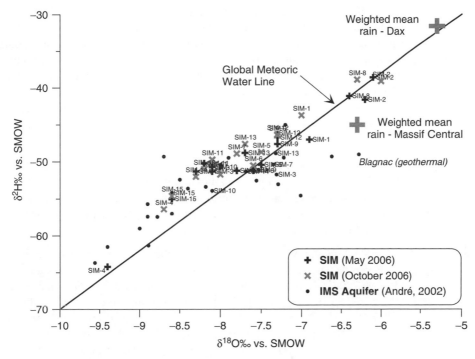

Figure 6.2.4 $\delta^{18}O$ - δ^2H plot for the ground waters collected in the Infra Molassic sand aquifer, comparison between the two surveys and data from André (2002).

enriched $\delta^{18}O$ and δ^2H values, P-5, collected SE of the district has more depleted $\delta^{18}O$ and δ^2H values and P-3 has the most depleted isotopic signature. Such heterogeneity in the $\delta^{18}O$ and δ^2H signatures for the Palaeocene aquifer reflects a variable recharge, either in space and time. The most depleted value correspond to a water recharged with a colder climate than the actual one (>10 000 y) but the likeness in the $\delta^{18}O$ and δ^2H signatures between the two surveys confirms a homogeneous aquifer system without any water input with a different signature between the two periods. This means that the aquifer is confined with no significant recharge.

- *Lower Eocene aquifer* was essentially collected in the northern part of the district plus the point EI-2. The $\delta^{18}O$ and δ^2H signatures are less variable than those of the Paleocene aquifer and fall in the middle range observed as a whole. The values are consistent between the two surveys and are more depleted compared to the actual rainwater input in the Massif Central, reflecting an older recharge which could have occurred under colder climate. This means that the aquifer is semi-confined.

- *Middle Eocene aquifer* has been largely sampled in the northern part of the district and presents a large range in the $\delta^{18}O$ and δ^2H signatures, moreover with the most enriched values (EM-1 and EM-2). These points display a signature very close to that of the rain inputs observed in Dax, reflecting a modern recharge. Most of the samples (EM-5, 6, 7, 8, 10, 11, 12 and 13) shows depleted values that reflect a recharge under colder climate. In between these two groups, the rest of the points (EM-3, EM-9 and EM-4) with intermediate values corresponds either to a mixture between the two previous groups or to close pockets (e.g. small confined part of the aquifer). The large variation of the point EM4 between the two surveys surely reflects hydrological conditions that differ from one survey to the other. This point in located in a relatively unknown area where more investigation, either in the Eocene aquifer or in other ones is required.

- *Upper Eocene aquifer* was sampled in 4 locations in the northern part of the district along a NW–SE profile. Although having similar signature between the two surveys, they display either enriched or depleted values. ES-3 is enriched while ES4 is depleted (recharge under colder climate) and ES-2 displays a signature close to present day rain water in the Massif Central.

- *Infra Molassic Sand aquifer* was largely collected in the southern part of the district along a W–E profile and displays a large range in the $\delta^{18}O$ and δ^2H signatures. SIM-2 and SIM-8 have a signature close to those of rainwater in Dax and Massif Central. SIM-2 is a 58 m depth bore well and a modern recharge is compatible with the observed values. On the contrary, SIM-8 is a 1400 m depth bore well screened between 1030 and 1040 m and the water is the most saline of the district (TDS up to $2.5\,g\,L^{-1}$). The $\delta^{18}O$ and δ^2H signature, suggesting a modern recharge, may results in a rapid circulation of the groundwater in the system. As for the remaining points, they are depleted in ^{18}O and D compared to rainwater, reflecting a recharge under colder climate and a semi confined status for the aquifer.

Taken as a whole, the enriched samples clearly correlate with the present-day recharge when compared to the mean weighted rain in Dax and in the Massif Central. The most enriched ones (EM1, EM2, ES3) originate from the north of the area, in the vicinity of the Gironde estuary and present an isotopic signature quite similar to that of present-day coastal precipitations (mean weighted rain in Dax). Samples SIM2 and 8 also show an enriched signature that can be related to the modern recharge. At the opposite, the most depleted sample (P3) originates from the Paleocene aquifer (860 m depth), and may reflect an old recharge as its signature is clearly lower than that of present-day precipitations from the Massif Central or Pyrenees. The groundwater presents a wide range of variation along the global meteoric water line which excludes significant evaporation of infiltrating waters and any continental effect on the stable isotope composition. These variations cannot be easily correlated with the data spatial location, and are probably mostly due to the period and location of the recharge of the aquifer. The most depleted sample of IMS (SIM4) is located in the eastern border of the study basin and originates from 177 m depth, it may represent an old recharge (as the estimated age of some groundwaters are close to 16–35 ky using ^{14}C, André *et al.*, 2005). Indeed, similar shift was revealed by several studies of paleo-waters in European aquifers and such significant isotopic depletion of groundwaters may be due to a lower recharge temperature at the time of infiltration (Kloppmann *et al.*, 1998).

6.2.3.3 Strontium Isotopes: From Water–Rock Interaction to Aquifer Inter-connexions

The $^{87}Sr/^{86}Sr$ ratios are mainly used as tracers of water-rock interaction and the primary sources of Sr in groundwater are atmospheric input, dissolution of Sr-bearing minerals, and anthopogenic input (Blum, Erel & Brown, 1994; Négrel and Petelet-Giraud, 2005; Négrel, 2006).

Sr shows large content variations in the groundwaters, from $156 \mu g L^{-1}$ (ES3) to $3419 \mu g L^{-1}$ (EI2), however not proportionally to TDS ($R^2 \sim 0.32$). This reflects that part of the main source of salinity does not necessarily provide Sr to groundwaters and vice versa. Large variation are observed in the $^{87}Sr/^{86}Sr$ ratios, from 0.707597 (EI2) to 0.711847 (SIM4). The lowest $^{87}Sr/^{86}Sr$ values agree with the weathering of carbonates and evaporites as they have recorded the seawater signal at time of deposition and their weathering release the same signal in the water (Kloppmann *et al.*, 2001; Négrel and Petelet-Giraud, 2005). Contrary to that, highest $^{87}Sr/^{86}Sr$ values refer to the weathering of silicate such as granite that deliver to water $^{87}Sr/^{86}Sr$ ratios higher than 0.710 (Blum *et al.*, 1994; Négrel, 2006). The spatial variation reveals two areas of high $^{87}Sr/^{86}Sr$, one located south near the central zone of the Pyrenees, and the other one eastward near the Massif Central, certainly reflecting the weathering of granites. On the other hand, low values are observed along a transect northwest-southeast that can be related to the weathering of low $^{87}Sr/^{86}Sr$ rocks (like carbonates and/or evaporates).

The Sr isotopic ratios coupled with Sr concentrations can be used to investigate mixing of different groundwater bodies or connections between surface and groundwater. When plotted against the $^{87}Sr/^{86}Sr$ ratio, the Sr contents (or 1/Sr) enable clear identification of geochemical end-members (Figure 6.2.5). Most of the signatures of the groundwater samples in this diagram can be explained by mixtures in various proportions of at least three identified geochemical end-members (1 to 3). Theoretical mixing lines between different geochemical end-members can be calculated by considering two-component mixtures according to Equation 6.2.3. End-members are characterized as to encompass all the considered samples. The first mixing curve rely end-member 1 and 2 whose characteristics are ($^{87}Sr/^{86}Sr = 0.70725$; $Sr = 60 \mu mol L^{-1}$ and $^{87}Sr/^{86}Sr = 0.70900$; $Sr = 1.4 \mu mol L^{-1}$ respectively). The second mixing curve rely end-member 1 and 3, the characteristics for the latter are ($^{87}Sr/^{86}Sr = 0.71500$; $Sr = 6 \mu mol L^{-1}$). The third mixing curve rely end-member 2 and 3.

The end-member 1 may represent the weathering of evaporite formations. Water that experienced interaction with evaporites generally display high Sr contents (up to $350 \mu mol L^{-1}$, Négrel and Petelet-Giraud, 2005) that is reflected in the Sr content of end-member 1. As the Sr isotopc ratios in seawater are constant at any point in time and evaporite (as well as carbonate) are Rb-free (Kloppmann *et al.*, 2001; Négrel and Petelet-Giraud, 2005), the $^{87}Sr/^{86}Sr$ ratio of theses rocks reflect that of the seawater at time of deposition. The $^{87}Sr/^{86}Sr$ ratio of end-member 1, representing the evaporite, fully agrees with the value of marine evaporites from the Eocene sulphate minerals from Cenozoic evaporites located in Anatolia (Turkey) with a mean $^{87}Sr/^{86}Sr$ ratio of 0.707271 (standard deviation = 0.000604, n = 16, Palmer *et al.*, 2004). It is worth noting that this ratio also agrees with the seawater value given for the Eocene by Burke *et al.* (1982).

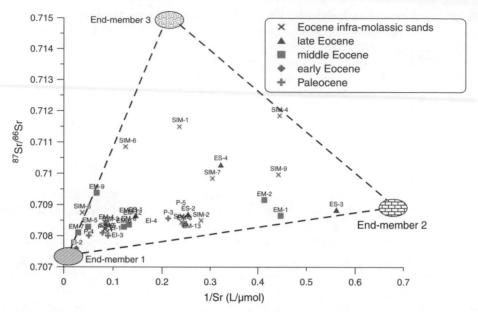

Figure 6.2.5 Plot of $^{87}Sr/^{86}Sr$ vs. $1/Sr$ ($1/\mu mol/L$) for the ground waters collected in the Eocene sand aquifers.

The end-member 2 may correspond to the weathering of carbonate as suggested by the increase in the Ca/Sr ratio from around 30–100 for end-member 1 up to 500–1300 for end-member 2. However, the Sr content for this end-member is the lowest one observed in the three end-members and its $^{87}Sr/^{86}Sr$ (0.70900) is clearly higher than that of Eocene carbonate that may display a $^{87}Sr/^{86}Sr$ ratio of around 0.7072 (Burke *et al.*, 1982). Thus this end-member may not represent the Eocene carbonate as there is no evidence of carbonate with such high $^{87}Sr/^{86}Sr$ ratio in the considered aquifers. Two samples located north of the area in the vicinity of the Gironde estuary characterize this end-member (EM1 and ES3) and an impact of seawater intrusion may be considered. However, the Ca/Sr ratio of these samples refers to carbonate weathering (700 and 1000 respectively, Négrel and Petelet-Giraud, 2005) and do not plead in favour of a relict of seawater or to direct seawater intrusion yielding to a $^{87}Sr/^{86}Sr$ ratio close to that observed for the past 8 Ma (0.7089–0.7092, Burke *et al.*, 1982). Otherwise this end-member may represent the leakage effect due to the overexploitation of the Eocene aquifer in this northern part of the basin and the origin of the dissolved Sr should be the weathering of the Miocene carbonate deposits with $^{87}Sr/^{86}Sr$ ratio around 0.7089–0.7091.

The end-member 3 with a radiogenic value clearly refers to the weathering of silicate. The highest $^{87}Sr/^{86}Sr$ ratios are observed in the southern part of the basin, near the Pyrenees and in the eastern part, on the edge of the Massif Central where the silicate rocks are present. Part of the Sr mass-balance in a crystalline environment should be controlled by Sr supplied by chemical weathering of Sr-bearing phases from the host rock. Minerals from granites that most commonly influence the Sr isotopic budget in waters are apatite, plagioclase, K-feldspar, biotite and muscovite and the $^{87}Sr/^{86}Sr$ ratio of water draining crystalline environment generally fall in the interval 0.712–0.718 (Blum *et al.*, 1994; Négrel, 2006).

The mixing proportion can be calculated for each sample on the mixing trend as well as for those lying within the triangle. Along the mixing line between the end-members 1 and 2, the proportions of end-member 2 is around 70% in sample ES3, 60% in sample EM1 and 35% in sample EM13. Along the mixing line between the end-members 2 and 3, the proportions of end-member 2 is around 80% in sample SIM4. Along the mixing line between the end-members 1 and 3, the proportions of end-member 1 is around 94% in sample EI2, 50% in sample EM7 and reach 20% in sample EM9.

6.2.3.4 Sulfur and Oxygen Isotopes of Sulfates: From Water–Rock Interaction to Biological Processes

Isotopic signature of dissolved sulfates $\delta^{34}S_{SO4}$ et $\delta^{18}O_{SO4}$ are capable of tracing sulfates sources, either from lithological origin or from anthropogenic origin. On the Adour-Garonne district, large variations are observed both for groundwater sulfate concentrations, from $0.004\,\mathrm{mmol\,L^{-1}}$ (P-3) to $10\,\mathrm{mmol\,L^{-1}}$ (SIM-8), for $\delta^{34}S_{SO4}$ signature, from $-2.7‰$ (SIM-15) to $39.8‰$ (SIM-1) and for $\delta^{18}O_{SO4}$ signature, from $2.6‰$ (SIM-4) to $16.1‰$ (SIM-8). These variations have been discussed in order to answer to the following question: 'What are the sources of dissolved sulfates in groundwater?' For dissolved sulfates concentrations higher than 1 mmol/L ($96\,\mathrm{mg\,L^{-1}}$), $\delta^{34}S_{SO4}$ vary between 12.6 and 31.3‰ (Figure 6.2.6). The mean value measured for local gypsum deposits from the upper part of the Eocene sands (mean value $= 12.72 \pm 1.2‰$, André, 2002) and isotopic signature documented for evaporite of Eocene and Paleocene ages (17 to 22‰, Claypool *et al.*, 1980), suggest that dissolution of evaporite layers is

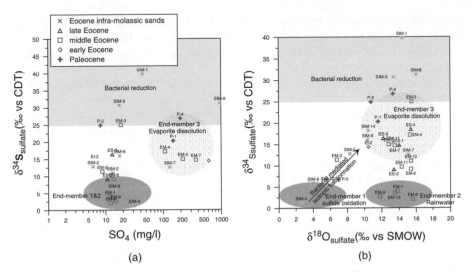

Figure 6.2.6 $\delta^{34}S$ vs. sulfate concentrations (mg/L) (a) and $\delta^{34}S$ vs. $\delta^{18}O$ of dissolved sulfates (b) in groundwaters of the Eocene aquifers.

probably one of the uppermost source of sulfates in the basin. Thus the end-member #3 on Figure 6.2.6 corresponds to dissolved sulfates coming from evaporite dissolution. Groundwater sampling points, with sulfates concentrations lower than $0.4\,\mathrm{mmol\,L^{-1}}$ ($38\,\mathrm{mg\,L^{-1}}$) and $\delta^{34}S_{SO4}$ values lower than 10‰ (Figure 6.2.6), suggest that other sources than evaporite dissolution occur in the basin. Observed concentrations and $\delta^{34}S_{SO4}$ values are both compatible with sulfates from rainwater and from oxidation of sedimentary sulfur occurring in local lithologies (André, 2002). On Figure 6.2.6, for similar $\delta^{34}S_{SO4}$ values (1.3 à 6‰), $\delta^{18}O_{SO4}$ values distinguish the sampling points: SIM-4 and SIM-9 ($\delta^{18}O_{SO4} = 2$ to 4‰) from the sampling points: EM-1, EM-8, EM-9 and EM-13 ($\delta^{18}O_{SO4} = 11$ to 16‰) and suggest that 2 end-members are prevailing (end-members #1 and #2). The expected signature of sulfates produced by sulfur oxidation is included in the range -10 to 10‰ (Clark and Fritz, 1997), and thus correspond to the end-members #1. Whereas, O-isotopic signature of end-members #2 is compatible with published $\delta^{18}O_{SO4}$ values for rainwater (10–17‰, Brenot *et al.*, 2007). As a result, the CARISMEAU project demonstrates that sulfates isotopes are relevant to identify three dominant sources of dissolved sulfates for the Eocene aquifers on the basin (evaporite dissolution, rainwater contribution and oxidation of sedimentary sulfur) and to trace their relative contribution depending on the sampling location.

Sulfate isotopic compositions $\delta^{34}S_{SO4}$ and $\delta^{18}O_{SO4}$ are also capable of tracing processes, in particular sulfate bacterial reduction which consume sulfates from the dissolved phase. This sulfates loss induces an isotopic fractionation of the remaining sulfates in the dissolved phase of groundwater. The $\delta^{34}S_{SO4}$ values higher than 25‰ (25 à 39.8‰, Figure 6.2.6) clearly testify that, dissolved sulfates in groundwater from the Eocene aquifers have undergone bacterial reduction (Nriagu *et al.*, 1991). A positive correlation between $\delta^{34}S_{SO4}$ and $\delta^{18}O_{SO4}$ values (Figure 6.2.6) for groundwater sampled in the infra-molassic sand aquifer (SIM-4, SIM-9, SIM-2, SIM-2bis and SIM-6) is observed. Because of the absence of significant increase of sulfates concentrations for these sampling points, this positive correlation can not be attributed to increasing evaporite dissolution. Bacterial reduction of dissolved sulfates is the most likely process to explain the evidenced positive correlation. Furthermore, the observed correlation factor is almost compatible with values published in the literature (Yamanaka and Kumagai, 2006).

Because of the important spatial disparities in the lithologies drained by the Eocene aquifers and the capacity of sulfate isotopes to trace water/rock interactions, this isotopic approach is able to trace the mixing of the different water reservoir for the Eocene aquifers on the Adour-Garonne district. Thus the geochemical approach, using sulfate isotopes, allows us to answer the following questions: 'Where and how is the groundwater recharge occurring? What are the connections between the different aquifers?' Nevertheless, due to bacterial reduction of dissolved sulfates, the initial isotopic signal, tracking the origin of dissolved sulfates, may be lost. It is for this reason that sulfate isotopes have been combined with Sr isotopes. The sampling points EI-2 and SIM-2, respectively in the early Eocene aquifer and in the infra-molassic sand aquifer, are located roughly at the same place. Sr and sulfate isotopic signature measured for this two sampling points display significant differences. This result clearly testifies that early Eocene aquifer and infra-molassic sand aquifer are locally disconnected in the central part of the Adour-Garonne basin.

6.2.3.5 Boron Isotopes: From Water–Rock Interaction to Heterogeneities

Boron Concentration, B Isotopes and Water–Rock Interaction

It is well known that boron is a mobile element during water/rock interaction and is essentially derived either from rock weathering or from atmospheric inputs in uncontaminated waters. Groundwater display a large range of boron concentrations between 7 and 2250 µg L^{-1} (Figure 6.2.7), in close agreement with concentrations found in pristine aquifers (Barth, 2000a,b) and some polluted aquifers (Pennisi *et al.*, 2006a) while they are lower than that observed in formation waters, brines and thermal springs and highly polluted aquifers (e.g. range between 5 and 570 mg L^{-1}, Kloppmann *et al.*, 2001; Pennisi *et al.*, 2006b; Millot *et al.*, 2007). The high boron content observed in the two surveys in groundwater like SIM8 (2550–2156 mg L^{-1}), 13 (1080–890 mg L^{-1}), P4 (820–728 mg L^{-1})..., surely reflects the predominance of evaporites in the rocks and their weathering that release dissolved boron. These high concentrations of dissolved B correlate with high concentrations of Cl, Na, agree with halite dissolution. The role of the dissolution of gypsum on the B concentration can be evidenced by the Pearson's and Spearman's R coefficients of 0.81 and 0.56 for SO$_4$ (respectively). However the

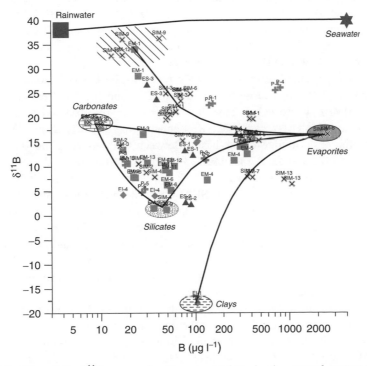

Figure 6.2.7 Plot of the δ^{11}B versus the B concentrations in the ground waters collected in the Eocene sand aquifers. End-members are illustrated (carbonates, silicates, evaporates, clays, rainwater, seawater). All mixing lines are calculated according to Equation 6.2.2.

dissolution of gypsum is not fully supported by the relationship between B, SO_4 and Ca as the Pearson's and Spearman's R coefficients are around -0.15 and -0.32 between B and Ca respectively and around 0.25 and 0.19 between SO_4 and Ca. Furthermore, evaporites dissolution as the unique process for supplying elements to the water is only supported by $2mCa + mNa$ versus $mCl + 2mSO_4$ (m refers to molarity) for some samples like SIM5, 8, P1, 4 and EM5, 7. As for the other results, a shift toward depleted $mCl + 2mSO_4$ values reflects the role of carbonate dissolution where Ca is balanced by HCO_3. Moreover there is another process that may explain the worse relationship between B, SO_4 and Ca. Indeed, many waters are oversaturated with respect to calcite, aragonite and dolomite as the simulations with PHREEQC demonstrate (Parkhurst and Appelo, 1999). As on the other hand there is no evidence of any oversaturation with respect to halite, anhydrite and gypsum, Ca can be depleted in the waters while sulfates are not controlled and thus there is no relationship between B and Ca, the role of Ca-SO_4-B bearing phases can only be seen by the relationship between B and SO_4.

Variations of the B concentrations between the two periods of collection span over a wide range. The coefficient of variation ranged between 1% (EI1) and 117% (EI4); 12 samples have a coefficient of variation lower than 10% (P1, 2, 3, 5, EI1, 3, EM1, 8, 11, SIM1, 10, 14) and 6 samples have a coefficient of variation higher than 50% (EI4, EM3, 4, 13, SIM2, 9). But however, the variation in the B concentration is generally not accompanied by a variation in the salinity. The coefficient of variation for the TDS ranged between less than 0.5% (EI1, EM12, ES1) and 45% (EM4), but only 7 samples have a coefficient of variation higher than 10%. Comparing the coefficients of variation between TDS and B concentrations highlight no relationships as the Pearson's and Spearman's R coefficients are of 0.50 and 0.28 (respectively).

Groundwater are characterized by a wide range in the boron isotopic compositions with the lowest $\delta^{11}B$ value close to -17.96% (EI1) and the most enriched one close to 36.25% (SIM9). These values cover the range generally observed in natural waters (Barth, 2000a,b; Coplen *et al.*, 2002; Kloppmann *et al.*, 2001). Low $\delta^{11}B$ values are observed on the eastern edge of the basin. Between the two sampling surveys, a $\delta^{11}B$ value of around 4.77 and 5.38% is observed in P5, located more southward in the basin. Similarly, on the most eastern edge, a $\delta^{11}B$ value ranging between 2.56 and 7.82% is observed in the samples SIM4 and SIM13. On the northern edge of the basin, the lowermost value is for the sample EI1 with a $\delta^{11}B$ of around -17.96 and -17.10% between the two sampling surveys. More southward, a set of samples shows also low $\delta^{11}B$ values (2.31 and 2.71% in ES2, 4.04 and 4.23% in EI4, 1.16 and 1.49% in EM9 . . .). The highest $\delta^{11}B$ values are observed in the north-western part of the basin (28.52 and 34.04% in EM1, 26.79 and 23.77% in ES3) and along the Pyrenean edge (36.03 and 36.25% in SIM9, 32.82 and 32.67% in SIM12).

Temporal variation in the $\delta^{11}B$ are generally low, the coefficient of variation ranges between 1 and 7% in 25 samples between the two sampling periods. Few samples exhibit a larger variation in the $\delta^{11}B$, the most important being in SIM4 with a $\delta^{11}B$ of around 7.82% in May and 2.56% in October leading to a coefficient of variation of 205%. The other large variations are for the samples EM4 (55%), SIM2 and 10 (38%). As for B concentration, there is no relationship between the variation in the $\delta^{11}B$ and the salinity of the groundwater as the Pearson's and Spearman's R coefficients are around 0.40.

No simple relationship can be observed when $\delta^{11}B$ values are plotted vs. the boron concentration in the waters (Figure 6.2.7) and poor Pearson's and Spearman's R coefficients with values lower than 0.02. This implies the existence of more than two end-members in the control of the dissolved boron in the groundwater.

Mixing Processes in the Groundwater

The weathering of rocks (carbonate, evaporites and silicates) may controls the dissolved boron in groundwater, the other end-member is related to the atmospheric inputs that can be dominant in some samples as can be seen in weathering of large portion of continents (Rose *et al.*, 2000). The B concentrations in continental carbonate rocks are typically in the order of $1 \, mg \, g^{-1}$, and never exceed $5 \, mg \, g^{-1}$ while biogenic carbonates can reach $60 \, mg \, g^{-1}$ (Hemming and Hanson, 1992). With regard to the $\delta^{11}B$, carbonates generally display values in the range $+15$ to $+25‰$ (Hemming and Hanson, 1992).

Evaporites (essentially halite and gypsum) do represent an important proportion of the rocks exposed to chemical weathering in the district, their rapid dissolution rate together with high B concentrations may potentially impact the dissolved B (Kloppmann *et al.*, 2001; Lemarchand and Gaillardet, 2006). Halite minerals generally display variable B concentrations, from very low values ($0.5 \, \mu g \, g^{-1}$, Liu *et al.*, 2000; Kloppmann *et al.*, 2001) up to few tenth of $\mu g \, g^{-1}$ (Liu *et al.*, 2000). Mg-sulfates, gypsum and borates have higher B concentrations, in the range 2 – few thousands of $mg \, g^{-1}$ (Kloppmann *et al.*, 2001). The $\delta^{11}B$ in evaporites range from $+15$ in halite up to $+35‰$ in Mg-sulfate minerals and increase in salts minerals as a function of the degree of evaporation of brines (Vengosh *et al.*, 1992; Liu *et al.*, 2000).

The role of silicate in controlling the $\delta^{11}B$ was investigated more through the values of the water rather than that of rocks as few studies gave $\delta^{11}B$ of whole rock. Some data refer granite and give a $\delta^{11}B$ from -3.2 to $+6.8‰$ while basalts display low $\delta^{11}B$ from around 0 up to $-10‰$. Water interacting with silicate rocks generally have low $\delta^{11}B$ in the range -10–$+10‰$ as numerous studies have shown (Barth, 2000b; Pennisi *et al.*, 2000; Casanova *et al.*, 2002; Négrel *et al.*, 2002; Lemarchand and Gaillardet, 2006; Min *et al.*, 2007; Millot *et al.*, 2007).

As stated above, plotting the data on a $\delta^{11}B$ and [B] mixing diagram reveals a large scatter in the groundwaters either for the isotope composition or for the B concentration. This scatter reflects well the different lithologies whose weathering control the dissolved boron. At least 3 mixing lines, basically reflecting water–rock interaction, between the carbonate, silicate and evaporite end-members can be calculated using Equation 6.2.2. The end-members representing the lithologies are constrained by the extreme points SIM8 for evaporite, SIM15 and EM2 for carbonate and EM9-SIM4 for silicate. The choice of these points as end members is corroborated by the nature of the sedimentary rock that constitutes the aquifer system. SIM8 is a 1410 m depth borehole in which carbonate and gypsum have been identified, EM2 is only made of calcareous deposits while SIM15 is made of carbonate and sands, SIM4 is a 130 m depth borehole in which the last screened 50 m are made of clays, gravels and sands and the screened part of the EM9 borehole (345 m depth) is made of Eocene sands. The mixing field defined by these three end-members encompasses numerous points with some of them plotting along the mixing lines reflecting binary mixing. This can be illustrated by EM3 (252 m depth borehole), that plots along the mixing line between the carbonate and evaporite end-members, which shows 99% of boron originating from the carbonate weathering as the screened part (125–252 m) is made of carbonate and sandstone. Sample P3, that plots near the mixing line between the carbonate and silicate end-members, agrees with this mixing line as it is a 483 m depth borehole made of carbonate, marls, sands and clayey marls.

Another mixing line can be calculated between the evaporite end-member and the most depleted point EI1 with a $\delta^{11}B$ of around $-17‰$. The most depleted point can be related to the interaction with clay minerals. A series of mechanisms like ion exchange and adsorption-desorption may influence the concentration and isotopic composition of boron in groundwater, particularly in part of sedimentary aquifer with an important clay matrix. The dissolved B as trigonal $B(OH)_3$ and tetrahedral borate anions can be fixed on clay minerals through adsorption reactions (Pennisi *et al.*, 2006a,b) leading to a ^{11}B depletion in adsorbed B and a corresponding enrichment in the adsorbed boron. This induces an increase of the $\delta^{11}B$ in waters when adsorption processes prevail which is not observed in the point EI1. The unique way to explain the high depleted value in EI1 is to consider desorption processes as this process led to preferential released of ^{11}B that induces a depleted $\delta^{11}B$ in waters. In the desorption processes, the $^{11}B/^{10}B$ ratio in the dissolved and adsorbed B is supposed to decrease continuously if a continuous isotopic re-equilibration takes place between both. The $\delta^{11}B$ will decrease until a steady state according to the concentration and isotopic composition of B dissolved in recharging water. The $^{11}B/^{10}B$ ratio will continue to decrease as in Rayleigh-type processes if no B is added to the system (i.e. no recharge). The possible desorption processes in EI1 is corroborated by the nature of the sedimentary rock that constitutes the aquifer system. Over the 140 m of the borehole in which the last 30 m are screened, the sedimentary deposits are made of sandy clays. Along the mixing line between SIM8 (evaporite) and EI1 (clays), the sample SIM 7 plot with a proportion of 90% of the boron derived from the clays end-member. This is corroborated by the nature of the sedimentary rock that constitutes the aquifer system (anhydrite-gypsum and clayey marls). Considering that SIM8 (evaporite) and EI1 (clays) may not be the pure end-members, the sample SIM13 can be the results of such mixing between evaporite and clays.

A last mixing line can be calculated between the evaporite end-member and the most enriched points SIM9, 12 and EM1. This most enriched group of points reflects the rainwater input with low B concentration and low to intermediate Cl concentration (0.35–1.8 mmol L^{-1}). The $\delta^{11}B$ of around $+32.67–+36.25‰$ correlates well with the characterization of rainwater near the coast (Millot *et al.*, in preparation; Chetelat *et al.*, 2005) which is illustrated by the mixing line between the seawater and the rainwater. The location of the rainwater along this mixing line depends on the proportion of sea salts. Considering SIM9, 12 and EM1 as reflecting the recharge for the B systematic, the mixing line with the evaporite end-member can be calculated as example but all other mixing with carbonate and silicate should also be considered. As a result, the mixing line with the evaporite do not encompasses most of the samples, the dispersion to the left of the points can be due to the influence of the carbonate and/or silicate end-members. The samples P1 and SIM1, plotting near this curve, are mainly made of carbonates and dolomite with few sandy deposits. This is not in agreement with the location of these samples along the mixing line with evaporite but when combined with the high Cl concentration, this suggests that the water has interacted with evaporite deposits. However, this interaction is not predominant in the mixing as the proportion of recharge is around 88–90% for SIM1 and P1 respectively.

6.2.4 HOW TO CHOOSE THE ISOTOPE TOOL TO ANSWER A QUESTION

One of the CARISMEAU aims is to apply numerous isotope systematics on one water body (in the sense of the WFD) in order to reinforce the knowledge, particularly with regards to heterogeneities and interconnections between the aquifers. For that, we have selected common isotopic methods with $\delta^{18}O$ and $\delta^{2}H$ of the water molecule and $\delta^{34}S_{SO4}$ and $\delta^{18}O_{SO4}$; innovative isotopic method with strontium isotopes ($^{87}Sr/^{86}Sr$) and potential isotopic methods with boron ($\delta^{11}B$) and lithium ($\delta^{7}Li$) isotopes as well as U series (Li and U are not illustrated here). A result of this project is, owing the increase in the knowledge, to affine the criterion that allows one (or more than one) isotope systematic to be chosen, according to questioning that arises.

The *stable isotopes of the water molecule* ($\delta^{18}O$ and $\delta^{2}H$) allowed to trace the origin of the aquifer recharge, either in space (defining the recharge zone) and time as well as to highlight spatial heterogeneities between aquifers and inside one aquifer. Even though stable isotopes are not a tracer of the water ages, they can be used to constraint the recharge time, particularly in the context of deep sedimentary confined aquifers. They thus showed a recharge in some part of the aquifers under a colder climate (e.g. $>10\,000\,y$).

The *strontium ($^{87}Sr/^{86}Sr$) and sulfur isotopes* ($\delta^{34}S_{SO4}$ and $\delta^{18}O_{SO4}$) are able to trace the origin of dissolved Sr and SO_4, either from natural origin (rainwater, water rock interaction) or from anthropogenic ones (mainly fertilizers). Thus they help in answering the questions: From where do dissolved Sr and SO_4 originate in groundwater? Is there spatial and/or temporal variation in the source(s) of the dissolved elements? We observed contrasted signature in the groundwaters according to the weathered lithologies in the aquifer systems and/or the anthropogenic inputs. Mixing between different water (e.g. the end-members) with $^{87}Sr/^{86}Sr$ and $\delta^{34}S_{SO4}$ and $\delta^{18}O_{SO4}$ isotopes signatures that differ led to a final signature that depends on the respective contribution of each end-member and thus $^{87}Sr/^{86}Sr$ and $\delta^{34}S_{SO4}$ and $\delta^{18}O_{SO4}$ isotopes allows the proportions of the mixing to be calculated. Therefore, they help in answering to the questions: Where does the groundwater come from? How is the groundwater being recharged? What are the relationship between the different aquifer levels? A better estimate of the groundwater circulation results from their investigation.

The *sulfur and oxygen isotopes* ($\delta^{34}S_{SO4}$ and $\delta^{18}O_{SO4}$) also allow us to trace the bacterial reduction processes of dissolved sulfates. Thus they allow us to answer the questions: Is there any sulphate loss through bacterial reduction processes? Is the media anaerobic? For a similar initial isotopic signature, reflecting the sources of dissolved sulfates, the bacterial reduction of sulfates may yield to contrasted isotope signature in the groundwater. On one hand, this process may allow us to identify mixing between aquifers with different signature but, on the other hand, this processes hides the initial signature of the dissolved sulphate and thus may also make it difficult to identify mixing between different aquifers.

The *boron isotopes* allowed us to identify mixing between groundwater that have interacted with various lithologies and to confirm this mixing by coupling with the observed lithologies in the boreholes. Contrary to the strontium isotopes, for which the seawater signal has been variable over geologic time, that of boron remained identical. Thus the isotope signature of the evaporite and carbonate lithologies are well constrained and numerous groundwater in the district are explained, with regard to B isotopes, by a mixing of boron originating from the weathering of these two rock types. Similarly, the boron isotope signature of water interacting with a silicate is less variable than that of strontium isotopes. Here again, boron isotopes highlight mixing between the carbonate and silicate end-members. Lastly, boron isotopes illustrated the impact of the recharge zone, in complement to stable isotopes of the water molecule and the role of some specific part of the district (clayey rich deposits) on the water signature, allowing a precise characterisation of these very specific zones.

6.2.5 VARIABILITY DURING A HYDROLOGIC CYCLE: A WAY TO MANAGE THE GROUNDWATER RESOURCES

The chemical variability during the hydrologic cycle was tested in the Eocene sands aquifer water body. Two sampling surveys were carried in two distinct hydrological conditions (high flow in May 2006 and low flow in October 2006). As a general rule, at the water body scale, there is no chemical (and isotopic) variation between the two periods considered, reflecting homogeneous water masses without significant water contribution with distinct chemical characteristics. This argument is true for each point sampled in the various aquifers through the existing bore-holes (42 accesses at the water body scale). Among all the sampling points, only one presents distinct chemical and isotopic signatures in contrasted hydrological conditions. This point (SIM-4) is located at the eastern border of the water body (Figure 6.2.2) potentially in the recharge area. It presents clearly distinct $\delta^{18}O$ and δ^2H signature probably reflecting the water contribution of distinct hydrological compartments of the aquifer rather than a direct input of 'new' water as the stable isotopes clearly suggest 'old' water recharged under a colder climate than that prevails nowadays. Strontium isotopes also present a distinct signature during the low flow period with a higher $^{87}Sr/^{86}Sr$ value that can be related to a contribution of water that have interacted with silicates, in agreement with the hypothesis of a water originating from another hydrologic compartment. This is also attested by the boron isotopes.

This kind of result is fundamental to design and/or revise groundwater monitoring networks as required by the Water Framework directive. Indeed, the groundwater monitoring plays an essential role in the implementation of the WFD and the Groundwater Daughter Directive. Member States are required to establish programs that 'provide a coherent and comprehensive overview of groundwater chemical status', can 'detect presence of long-term ... upward trends in pollutants' and 'provide a reliable assessment of quantitative status'. The groundwater monitoring networks of course differ according to the variability of hydrogeological settings notably. Under the Common Implementation Strategy (CIS), the understanding of the groundwater system is called a 'conceptual model' (see Chapter 2.2 of this book for further details). This refers to both natural characteristics (structure, background levels, flow path,....) and pressures on the groundwater body.

According to the CIS guidance document no. 15 on 'Groundwater monitoring', the design of groundwater monitoring networks shall be supported by the use of a conceptual model. Figure 6.2.8 summarizes the links between the content of the conceptual model and the monitoring (European Commission, 2003, 2006).

As illustrated by the CARISMEAU research project, isotopic tools can help improving and feeding conceptual models. When designing chemical and quantitative monitoring networks and answering the two following questions, isotopic tools bring answers:

- **Where?** Groundwater bodies are large-scale and heterogeneous systems. In the specific case studied by the CARISMEAU project, isotopes have contributed to a better understanding of leakages between the different aquifers that make the whole groundwater body. Furthermore, the project has allowed highlighting different compartments characterised by different background levels. And finally, it has contributed to the identification of young and old waters. In other words, it has been possible to precise the recharge area.
  →This will help defining how much monitoring sites are necessary to build a representative monitoring network of the groundwater body and where exactly they should be implemented.

- **When?** For chemical aspects as well as for quantitative aspects, understanding the seasonal and inter-annual variations of the system is a key issue to define the frequency of sampling or of the water-table measure. In the case of the Eocene/Paleocene

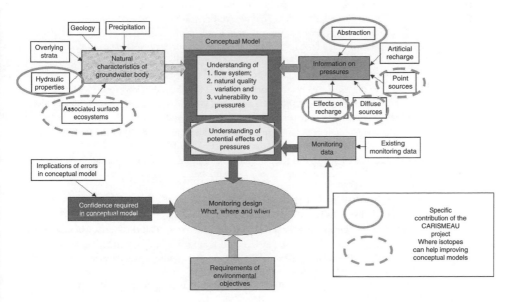

Figure 6.2.8 Link between the conceptual model/understanding and monitoring (from European Commission, 2006)

groundwater bodies of South-West France, isotopes have shown that there were no significant variations of the groundwater bodies' characteristics between the high and low-water levels.

→ As a conclusion, yearly samples and measures of the water table are sufficient in the case of the CARISMEAU groundwater bodies.

The WFD is a cyclic process and in the case of CARISMEAU groundwater bodies, information brought by isotopes will help improving and optimizing the monitoring network in 2012. Very briefly, isotopes have shown that it was not necessary to collect more than one sample a year to have a representative monitoring of the Eocene/Palaeocene groundwater bodies. Due to the complexity of the system, efforts should be put on increasing the number of monitoring stations, both spatially and at different depths.

REFERENCES

Aggarwal P.K., Gat J.R. and Froehlich K.F.O. 2005. *Isotopes in the Water Cycle: Past, Present and Future of a Developing Science*. Springer Ed., Dordrecht.

Allègre C.J., 1987. Isotope geodynamics. *Earth and Planetary Science Letters*, 86, 175–203.

Allègre C.J., 2005. *Géologie Isotopique*. Belin Ed., Paris.

André L., 2002. Contribution de la géochimie à la connaissance des écoulements souterrains profonds. Application à l'aquifère des Sables Infra-Molassiques du Bassin Aquitain, Thèse, Université de Bordeaux 3.

André L., Franceschi M., Pouchan P. and Atteia, O., 2005. Using geochemical data and modelling to enhance the understanding of groundwater flow in a regional deep aquifer, Aquitaine Basin, south-west of France. *Journal of Hydrology* 305, 40–62.

Barth S.R. 1993. Boron isotope variations in nature: a synthesis. *Geol. Rundsch*, 82, 640–1.

Barth S.R. 2000a. Stable isotope geochemistry of sediments-hosted groundwater from a Late-Proterozoic–Early Mesozoic section in Central Europe, *Journal of Hydrology* 235, 72–87.

Barth S.R. 2000b. Geochemical and boron, oxygen and hydrogen isotopic constraints on the origin of salinity in groundwaters from the crystalline basement of the Alpine Foreland. *Applied Geochemistry* 15, 937–52.

Blum J.D., Erel, Y. and Brown, K., 1994. $^{87}Sr/^{86}Sr$ ratios of Sierra Nevada stream waters: Implications for relative mineral weathering rates. *Geochimica et Cosmochimica Acta*, 58 5019–25.

Bowen R., 1988. *Isotopes in the Earth Sciences*. Elsevier Applied Science Ed., New York.

Brenot A., Carignan J. and France-Lanord C., 2007. Geological and land use controls on $\delta^{34}S$ and $\delta^{18}O$ of river dissolved sulfates: the Moselle river basin, France. *Chemical Geology* 244, 25–41.

Burke W.H., Denison R.E., Hetherington E.A., Koepnick R.B., Nelson H.F. and Otto J.B., 1982. Variation of seawater $^{87}Sr/^{86}Sr$ throughout Phanerozoic time. *Geology*, 10 516–19.

Casanova J., Négrel Ph., Petelet-Giraud E. and Kloppmann W. 2002. The evolution of boron isotopic signature of groundwaters through silicate weathering. In *6th International Symposium on the Geochemistry of the Earth's Surface, Hawaii*, vol. 6, 7–12.

Chetelat B. Gaillardet J., Freydier R. and Négrel Ph., 2005. Boron isotopes in precipitation: Experimental constraints and field evidence from French Guiana. *Earth and Planetary Science Letters* 235, 16–30.

Clark I.D. and Fritz P., 1997. *Environmental Isotopes in Hydrology*. CRC Press/Lewis Publishers, Boca Raton, FL.

Claypool G.E., Holser W.T. and Kaplan I.R., 1980. The age curve of sulfur and oxygen isotopes in marine sulfates and their mutual interpretation. *Chemical Geology* 28, 199–260.

Coplen T.B., Hopple J.A., Böhlke J.K., *et al*. 2002. Compilation of minimum and maximum isotope ratios of selected elements in naturally occurring terrestrial materials and reagents. U.S. Geological Survey, Water-Resources Investigations, Report 01-4222.

Craig 1961. H. Isotopic variations in meteoric waters. *Science* 133, 1702–3.

European Commission, 2000. Directive 2000/60/EC of the European Parliament and of the Council of 23 October 2000 establishing a framework for Community action in the field of water policy, Official Journal of the European Communities L 327, 22.12.2000, p. 1.

European Commission, 2003. Monitoring under the Water Framework Directive, CIS Guidance Document N°7, European Commission, Brussels.

European Commission, 2006. Groundwater Monitoring, CIS Guidance Document N°15, European Commission, Brussels.

Faure G. 1986. *Principles of Isotope Geology*. John Wiley & Sons Ltd, Chichester.

Fritz P. and Fontes J.C. 1980. *Handbook of Environmental Isotope Geochemistry*, Vol 1 to 5. Elsevier Ed., Amsterdam.

Hemming N.G. and Hanson G.N., 1992. Boron isotopic composition and concentration in modern marine carbonate. *Geochim. Cosmochim. Acta* 56, pp. 537–43.

Kloppmann W. 2003. Les nouveaux isotopes dans les sciences de l'eau. *L'actualité chimique*, **267**, 44–7.

Kloppmann W., Dever L. and Edmunds W.M., 1998. Residence time of Chalk groundwaters in the Paris Basin and the North German Basin: a geochemical approach. *Applied Geochemistry*, **13**(5): 593–606.

Kloppmann W., Négrel Ph., Casanova J., Klinge H., Schelkes K. and Guerrot C., 2001. Halite dissolution derived brines in the vicinity of a Permian salt dome (N German Basin). Evidence from boron, strontium, oxygen and hydrogen isotopes. *Geochim. Cosmochim. Acta*. **65**, 4087–4101.

Lemarchand D. and Gaillardet J. 2006. Transient features of the erosion of shales in the Mackenzie basin (Canada), evidences from boron isotopes. *Earth and Planetary Science Letters* **245**, 174–89.

Liu W.G., Xiao Y.K., Peng Z.C., An Z.S. and He X.X. 2000. Boron concentration and isotopic composition of halite from experiments and salt lakes in the Qaidam Basin. *Geochimica et Cosmochimica Acta* **64**, 2177–83.

Millot R., Négrel Ph. and Petelet-Giraud E. 2007. Multi-isotopic (Li, B, Sr, Nd) approach for geothermal reservoir characterization in the Limagne Basin (Massif Central, France). *Applied Geochemistry* **22**, 2307–25.

Millot R., Guerrot C., Négrel Ph. and Petelet-Giraud E. Li, B and stable isotopes characterisation in rainwater in France, in preparation.

Min, M., Peng, X., Zhou, X., Qiao, H., Wang, J., and Zhang, L., 2007. Hydrochemistry and isotope compositions of groundwater from the Shihongtan sandstone-hosted uranium deposit, Xinjiang, NW China. *Journal of Geochemical Exploration* 93((2), 91–108.

Mook W.G. 2001. Environmental Isotopes in the hydrological cycle. Principles and applications IHP-V Technical Documents in Hydrology, N° 39. UNESCO – IAEA, available at www.naweb.iaea.org.

Négrel Ph. 2006. Water-granite interaction: clues from strontium, neodymium and rare earth elements in saprolite, sediments, soils, surface and mineralized waters. *Applied Geochemistry*, **21**, 1432–54.

Négrel Ph. and Petelet-Giraud, E. 2005. Strontium isotope as tracers of groundwater-induced floods: the Somme case study (France). *Journal of Hydrology*, **305**, 99–119.

Négrel, Ph., Petelet-Giraud, E., Kloppmann, W. and Casanova, J., 2002. Boron isotope signatures in the coastal groundwaters of French Guiana. *Water Resources Research* 38(11), 1262, DOI: 10.1029/2002WR001299.

Nriagu J.O, Rees C.E. and Mekhtiyeva V.L., 1991. Hydrosphère – Ground water. In: Krouse, H.R, Grinenko V.A. (eds), *Stables Isotopes: Natural and Anthropogenic Sulphur in the environment*. SCOPE 43. John Wiley & Sons Ltd, Chichester: 229–42.

Palmer M.R., Helvaci C. and Fallick A.E., 2004. Sulphur, sulphate oxygen and strontium isotope composition of Cenozoic Turkish evaporites. *Chemical Geology*, **209** (3-4): 341–56.

Parkhurst D.L. and Appelo C.A.J., 1999. User's guide to PHREEQC (version 2) – a computer program for speciation, batchreaction, one-dimensional transport, and inverse geochemical calculations: U.S. Geological Survey Water-Resources Investigations. Report 99-4259, 312 p.

Pennisi M., Leeman W.P., Tonarini S., Pennisi A. and Nabelek P. 2000. Boron, Sr, O, and H isotope geochemistry of groundwaters from Mt. Etna (Sicily) - hydrologic implications. *Geochimica et Cosmochimica Acta* **64**, 961–74.

Pennisi M., Bianchini G., Muti A., Kloppmann W. and Gonfiantini, R. 2006a. Behaviour of boron and strontium isotopes in groundwater–aquifer interactions in the Cornia Plain (Tuscany, Italy). *Applied Geochemistry* **21**, 1169–83.

Pennisi M., Gonfiantini R. Grassi S. and Squarci P. 2006b. The utilization of boron and strontium isotopes for the assessment of boron contamination of the Cecina River alluvial aquifer (central-western Tuscany, Italy). *Applied Geochemistry* **21**, 643–55.

Rose E.F., Chaussidon M. and France-Lanord C., 2000. Fractionation of boron isotopes during erosion processes: the example of Himalayan rivers. *Geochimica et Cosmochimica Acta* **64**, 397–408.

Vengosh A., Starisnky A., Kolodny Y., Chivas A.R. and Raab M. 1992 Boron isotope variations during fractional evaporation of sea water: New constraints on the marine vs. non marine debate. *Geology* **20**, 799–802.

Vengosh A., Heumann K.G., Juraske S. and Kasher R. 1994. Boron isotope application for tracing sources of contamination in ground water, *Environ. Sci. Technol.* **28**, 1968–74.

Vengosh A., Chivas A.R., Starinsky A., Kolodny Y., Baozhen Z. and Pengxi Z., 1995. Chemical and boron isotope compositions of non-marine brines from the Qaidam Basin, Qinghai, China. *Chem. Geol.* **120**, 135–54.

Yamanaka M. and Kumagai Y., 2006. Sulfur isotope constraint on the provenance of salinity in a confined aquifer system of the southwestern Nobi Plain, central Japan. Journal *of Hydrology* **325**, 35–55.

Part 7
Associating External Stakeholders

7.1

Groundwater Teaching at University Level in Spain

Emilio Custodio[1,3], Antoni Gurguí[2,3] and Eduard Batista[3]

[1] *Technical University of Catalonia (UPC), Department of Geotechnics, Barcelona, Spain*
[2] *Nuclear Security Council (CSN)*
[3] *Fundación Centro Internacional de Hidrología Subterránea (FCIHS), Barcelona, Spain*

7.1.1   INTRODUCTION

In relation to the importance of the resource, groundwater training received little attention until recent times. First specific lectures on groundwater appeared probably in the second half of the nineteenth century in France, England and Germany. Some of them were oriented to explain observations and to obtain new resources of freshwater, especially after the success of the artesian wells in the Paris basin. Other lectures were directed to groundwater hydraulics to solve drainage problems and to obtain water for public supply, and relied on the 1856 Darcy's Law and the early works of Thiem and Forchheimer. In the United States an important boost in teaching groundwater appeared after the progress of the Geological Survey, leaded by Meinzer (Biswas, 1970).

In Spain, in spite of the potential importance of groundwater to solve the problems of a country with extensive semi-arid areas (Custodio, 2008), groundwater teaching was

Groundwater Monitoring Edited by Philippe Quevauviller, Anne-Marie Fouillac, Johannes Grath and Rob Ward
© 2009 John Wiley & Sons, Ltd

formally included into university teaching quite late, in the decade of 1960. Previously, it had been only considered as part of other subjects in the degrees of geology, natural sciences, and civil and mining engineering, often reduced to a few lectures by a non-specialist.

Worldwide deficiencies in hydrological training to solve development goals have been recognised as critical for the sustainable use of groundwater. They prompted UNESCO to include hydrological (including hydrogeological) training as a relevant issue of the International Hydrological Decade (IHD) 1965–1975. UNESCO recommended all countries vehemently to start training and education programmes as soon as possible, filling the wide gap in knowledge existing in all but a few places. Since then, efforts in this direction have continued. Regular courses have been created in Delft Technical University (The Netherlands) and organised in South America (Porto Alegre, Brazil), later in various places in Argentina, and now in Uruguay, etc. They will not be considered further, as this chapter will focus on the Spanish experience.

UNESCO produced a series of documents on the progress of this initiative (UNESCO, 1974) and diffused papers on needs and course syllabi (Mostertman, 1970).

7.1.2   THE SPANISH EXPERIENCE

Following the original recommendations of UNESCO's IHD, the then created Spanish Committee for the IHD recommended the creation of courses with an international orientation, mostly to try to fill national needs and those of Ibero-American countries. Three courses were started:

- The International Course on General and Applied Hydrology in the Hydrographic Studies Centre of the Experimentation Centre for Public Works (CEDEX), an autonomous body then ascribed to the Public Works Ministry. It started in 1966 and still exists. It is not linked to any university, although for some time was integrated into a Section of the National Council for Scientific Research. The Course deals almost exclusively with surface water hydrology, with a preference for river water use and regulation. Later on, a course on irrigation was created mostly directed to irrigation technology and the use of surface water in agriculture.

- The Course on Hydrogeology Noel Llopis (CHNL) in the Faculty of Geology of the Madrid's Complutense University. It started in 1967 and was directed mainly – but not exclusively – to geologists. It closed in 1999. Average engagement was about 25 students.

- The International Course on Groundwater Hydrology (CIHS), initially in the Industrial Engineering School and afterwards in the Civil Engineering School, Technical University of Catalonia, Barcelona. It has a managing board, formed by different public and private organisations. It started in 1967 and is working with different extensions, as it will be described later. Average engagement is around 25 graduate students, with strict admission rules. Most participants are geologists and engineers, but not exclusively.

The CHNL and especially the CIHS have had a clear impact on Spain and Ibero-America by providing not only professionals but also hydrogeology professors in a series of universities (Llamas, 1970). Although the number of teaching hours in some of the courses offered in them is often small, global knowledge on groundwater has greatly increased. In fact, Spain has one of the three most numerous active groups of the International Association of Hydrogeologists (see Chapter 7.2 of this book), and has been the seed of national groups in many other countries.

In Spain, the growth of specialised training has been fast (Fornés and Senderos, 2002; MIMAN, 2000). Currently, there are about 20 universities offering some course on groundwater. The most active ones are those of Barcelona (Technical University of Catalonia), Valencia (Technical University of Valencia), Cartagena (Technical University of Cartagena), Almería, Granada, Madrid (Complutense University) and A Coruña.

In Ibero-America, active professors on hydrogeology graduated from the Spanish courses can be found in Mar del Plata, Antofagasta, Montevideo, Porto Alegre, Río de Janeiro, Belo Horizonte, Piura, and many other places. There are also professors in Lisbon and Aveiro (Portugal), and Udine and Catania (Italy). This is a satisfactory result to the call from UNESCO, where the groundwater specialist at the International Hydrological Programme is also an old student from the CIHS.

In the early editions the participation of foreign students was fostered and supported by the Spanish International Cooperation Programme, but it ceased due to policy changes. This has been a serious limitation for possible participants, but this has not halted their participation through other international sources or through research projects. Erratic policies and short-time vision at government level have been a permanent handicap.

7.1.3   THE INTERNATIONAL COURSE ON GROUNDWATER HYDROLOGY

The International Course on Groundwater Hydrology was created in 1966 and started in 1967 following the call of UNESCO. It was conceived as a six month, half-day, intensive course addressed to graduate students and professionals with a high level university degree. It is taught in Castilian (Spanish). The managing board was formed by public and private entities which compromise to subsidise about half of the running costs, the remaining coming from registration fees. This board has suffered some changes along time, but at least four different entities from the public and private sector have always been in it. What is now the Technical University of Catalonia (UPC) provided facilities for teaching and adhered soon to the Course Board. In order to solve managing rigidities of administrative and financial issues, in 1987 the CIHS was incorporated into an Association, which evolved in 1991 to the Foundation 'International Centre for Groundwater Hydrology' (FCIHS), whose main activities were to manage the CIHS and carry out activities and studies related to hydrogeology not provided by others. The foundation has a Board of Sponsors controlled by the Government of Catalonia under non-profit (charity) organisation rules, with yearly audits. This gives to the FCIHS full legal entity, and capacity to carry out its goals: groundwater training, R&D, publications, meetings, etc.

The CIHS is the oldest running post-graduate course of any public university in Spain (Batista *et al*., 1994, 2002), which is quite a surprise considering that groundwater is far from a fashionable topic in the public eyes. The students who complete the course satisfactorily receive a Certificate of Studies from the FCIHS and an official postgraduate degree of the UPC. This last diploma does not add legal competencies to those corresponding to each participant's graduate degree, but is actually highly prized by the civil society. Their existence was recognised early in the hydrogeological work (Davis, 1969), and through UNESCO's reports.

The CIHS has maintained its format along these 43 years, although adapting its contents to the new developments arising in this field (Custodio, 1994). Current contents are explained in Box 7.1.1. The CIHS also offered a two-month intensive course in English for developing countries in 1982, sponsored by the Ministry of Foreign Affairs of Spain), and an intensive course for water managing staff persons of the public administration, in 1977. Other activities along this line were considered in different moments, as well as a version of the CIHS in English, but the necessary resources to carry them out were not found.

BOX 7.1.1 Contents of the International Course on Groundwater

	Effective hours						
Topic	Lectures	Workshop and tests	Field work	Visits	Total	Personal work	Total general
Water cycle basics	10	5	–	10	25	30	55
Groundwater hydraulics	45	25	24	10	104	140	244
Hydrogeochemistry	25	20	8	5	58	90	148
Hydrogeological surveys and study	35	25	16	–	76	120	196
Applied aspects	15	5	–	5	25	40	65
Isotopes in hydrogeology	10	5	–	5	20	30	50
Well and borehole construction	20	15	8	–	43	70	113
Groundwater quality	25	15	–	5	45	80	125
Groundwater models	15	10	–	–	25	50	75
Social aspects of groundwater	40	15	–	–	55	110	165
Personal report	–	10	40	–	50	150	200
	240	150	96	40	526	910	1436

The course is currently a mix of topics on groundwater evaluation, development and management.

The Teaching Commission and the professor staff are specialists working in water aspects dealing with groundwater study, management, administration and development, both from public and private entities. The holistic vision helps in the effective training.

The main teaching tool is the book by Custodio and Llamas (1976/1983) and reprinted several times, which is now also available in Italian. This is complemented by papers to expand some areas, introduce new ones, or update parts. This book has been a key element to CIHS success, and is also a reference text in other courses taught in Castilian (Spanish), Portuguese or Italian.

The course is currently a mix of topics on groundwater evaluation, development and management.

The duration is of six months, or 180 days, five afternoons per week with about 15 non-working days besides Sundays and most Saturdays, and a series of mornings for visits and field work. Considering that a normal academic year for an average student contains 60 ETCS credits, or about 400 h of classroom, the course is equivalent to almost a full academic course that can be evaluated as 55 to 60 ETCS credits. The European Transfer Credit System (ETCS) has a ratio of total work to classroom work of 3.7.

Apart from the board responsible to oversee the activities of the Foundation, the CIHS has its own Teaching Committee, which elaborates each year's contents, choose the professors and evaluate the progress of individual students. Typically, each edition has 10 permanent professors and about 60 invited professors, coming not only from university departments but also from private companies and government bodies and agencies. Although such diversity has to be managed very carefully to assure the adequate coordination of the contents being taught, it also assures a wide overview of the different perspectives and interests involved in the whole water cycle: economy, environment, public interest, etc. This three-tier concept: university–industry–administration, which permeates the whole structure of the Foundation up its Board, is one of the keys of the continued success of the CIHS, proven every year by the strong demand of students aiming to enrol in the CIHS.

The course completion requires a field assignment resulting in a hydrogeological report on a specific problem of a concrete location. Even though this assignment has a training objective and is directed by one of the permanent professors, it is evaluated as if it were a professional study. This has resulted in more than 500 reports – mostly in Catalonia but also in the whole Spain and elsewhere – which today constitute a unique body of knowledge and data, spanning more than 40 years.

After six months, most students reported to be simply exhausted, as shown in the poll which is their very last task to complete the course just before the closing ceremony. However, after some time and professional experience, the general view is very positive: they feel that they have learned a lot, and that what they have learned has been useful in their work, no matter whether they have gone to the private or public sectors, or continued in the academia, or else. This explains that the CIHS can be proud to have produced more than 1000 new professionals highly skilled in groundwater presently working worldwide (29% of the students come from Ibero-America, 7% from the rest of Europe and 2% from the rest of the world).

In the present text, however, it may be more interesting to analyse how teaching has evolved and what are the main problems resulting from the CIHS experience. The next section will refer to the necessary adaptations of the contents to the development of water sciences and technologies and the evolution of problems associated with groundwater use and management worldwide. But from the teaching experience, is there any conclusion regarding how people learn?

The CIHS is almost as old as the transistor, meaning that, in the first editions, computers were only some exotic concept. Groundwater models were exclusively analog ones (viscous fluids, electrical, etc.). Nowadays, a click with the mouse lets anybody download in a few seconds a sophisticated multiphase transport code running on his personal laptop and resulting in spectacular 3-D views based on standard GIS software and data. In parallel with those powerful tools, teaching must have become easier and more efficient, or is that not the case?

Experience shows that this is not the case. First of all, the knowledge base that a hydrogeologist must master is bigger every day. Initially, he was expected to be able to find the resource and to exploit it. Today, he has to produce sound conceptual models, has to deal with a broad variety of pollution problems, takes care of environmental impacts, and will soon be asked to address the impact of greenhouse warming at local scale and what to do about it. Nevertheless, none of it can be tackled, no matter how powerful the tools, unless there is a previous understanding of the concepts and the basic principles involved in groundwater flow and transport. This has nothing to do with an accumulation of information, but with a clear understanding of how things work in nature. This, in turn, requires a solid knowledge and comprehension of the basic concepts of mathematics, physics, chemistry, geology, isotope science, and environmental science.

Then things like the internet are paradoxically a mixed blessing. They are revolutionary if they do not distract from the need to be able to visualise and understand what is happening in nature, and even to be able to advance an approximate estimation – orders of magnitude – of the parameters involved with a simple pencil and paper envelope. This is not an exaggeration, we have seen people using powerful mathematical models who do not know what Darcy's law means. Recently, in Barcelona, a few disasters related to the construction of new metro lines and a high speed train occurred, resulting in damage and tremendous overcosts and delays, mostly as a consequence of an inadequate understanding of groundwater principles in one of the best studied aquifers in the world (the Llobregat Delta).

Therefore, now and in the future, the biggest challenge in groundwater teaching will continue to be able to provide the adequate conceptual background for hydrogeologists to enable them to use the power of new technologies. Otherwise, there is the danger of producing highly skilled pilots of ultra sophisticated vehicles who do not know what they carry, where they are, where they come from, or even where they have to go.

The CIHS is not the only contribution of the Foundation to the development of groundwater sciences. In order to catch with development of science and technology, jointly with the Department of Geo–Engineering of the UPC (initially with the Nuclear Engineering Laboratory) different doctoral thesis have been written and doctoral courses were included in the graduate offer. This has been also a source of groundwater studies, benefiting mostly Catalonia, but also other Regions of Spain and abroad.

In 1989 the Master on Groundwater Hydrology started, designed as an extension of the CIHS with complementary studies and an additional thesis. This is one of the Masters offered by the UPC that often leads to complete a doctoral degree. This is now under revision to adapt it to the so called 'Bolonia' directives for the European Superior Teaching Space.

The shortage of grants to students, especially for those from other countries, mostly Ibero-America, fostered the creation of an 'on line' version of the CIHS. This new experience started in 2002. It has been quite successful, with a growing number of engagements. The 'on line' course is based on the platform developed by the CIMNE (International Centre for Numerical Methods in Engineering) of the UPC, although the course contents have been heavily reworked to adapt them to the interactive character required by the on-line work of the student. This is a tremendous effort, but it soon became obvious that a simple translation of the texts and presentations used in the CIHS would be inadequate for people learning alone – often in after-work hours – in front of a computer screen. Such effort has required its own organisation in the FCIHS. Students in the on-line version of the CIHS have to report regularly to a centre near the residence of the student. There are about 12 associated centres in Spain, Portugal, Argentina, Chile, Perú, Uruguay, Brazil, Colombia and Mexico, and the number of centres (that may be several per country) is growing, always depending on a pre-existing university structure offering groundwater teaching. Not surprisingly, many of these Centres include advantaged CIHS alumni in their staff.

7.1.4   EVOLUTION OF KNOWLEDGE NEEDS

The CIHS started in parallel with the intensive groundwater development in many countries. Intensive groundwater use is at most 60–70 years old in the western United States and near New York, and also near Barcelona. It was not until the 1960s in the arid and semi-arid zones of more or less developed countries, including Spain and Mexico, when groundwater began to be a very significant freshwater source. In other countries, such as India and China, it became a fast growing activity only 20–30 years ago. All these countries had to face, and are facing, the shortage of groundwater specialists and the lack of understanding of politicians, government officials and the mass media.

In relatively humid countries the main use of groundwater has been supplying towns and rural settlements, where in most cases water quality and its protection are the main concern. But in the countries where agricultural irrigation is important, not only for survival but as an important economic source of income, water quantity became the main issue, with situations in which groundwater development was so intense that groundwater reserves were being depleted (mined).

These were the circumstances under which the CIHS started, in a country with large semi-arid, intensively irrigated areas, mostly concerned by water quantity, and with medium-level technology. These conditions coincide with those of many areas of Ibero-America, which together with the common language (Portuguese is also a close language to Castilian) greatly favoured the interest for professionals and graduates from these countries. Should the course be taught in French and/or in English, it would attract participants from the southern area of the Mediterranean, Sub-Saharan Africa, Middle East and South Asia.

From the development stages of groundwater resources, at the beginning, other topics gradually appeared. Quality issues became more and more important, especially related to intensive use of fertilisers and pesticides in agriculture, areas with intensive stockbreeding, urban areas and industrial settlements, and, in some locations, even mining activities. More recently, quality protection and aquifer restoration are issues with a strong demand for specialists.

Large civil works, including dense urban housing, high buildings, etc. appeared also as a new issue demanding water specialists to address the problems related to groundwater existence in the subsoil, beyond the normal field of geotechnical experts. This opened a new field of interest.

The intensive use of groundwater (Llamas and Custodio, 2003) is deeply related with the role of groundwater in nature, such as river base flow, lake and wetland water supply and phreatophite surface area. The modification of the water cycle through groundwater development may produce significant ecological changes to be considered, studied and corrected. This is a bunch of new issues for specialists. There is also the social aspect of sustainable development of natural resources compatible with the environment. But this development is carried out silently (Fornés *et al.*, 2005) by very numerous initiatives, especially when irrigation is significant, and this is decided individually, without regulation – or poorly effective regulation – in front of governmental institutions not used to this issue, understaffed, with untrained personnel, and/or without the resources to cope with the new problems which they do not master, and which develop silently but strongly, bottom-up. Under this situation, groundwater may become a new form of the 'tragedy of the commons'. How to control these situations to preserve the social benefits from groundwater development and also the environment is a complex combination of governmental and civil institutions with the co-responsibility of stakeholders. This is a set of completely new issues beyond what has been considered classical groundwater hydrology or hydrogeology. It demands new capabilities not in the normal geological, environmental and engineering teaching. Here, other areas of knowledge must be incorporated, such as economy, sociology, law, public administration and policy management.

The current concern of possible future global (including climate) change is a very recent issue regarding groundwater, still in its early stages and very incipiently addressed in teaching. A main aspect is that of uncertainty, which seems that cannot be reduced beyond a certain limit (Roe and Buker, 2007).

All this opens new ways and needs for hydrogeological training, which cannot be addressed in a single course, although the main principles have to be included in some way. The solution requires a general purpose training in basics, with later focus on the different issues in branches using specialised subjects. This needs resources – or high fees – and professors who need quite a long time to be trained.

7.1.5 TAKING INTO ACCOUNT THE WATER FRAMEWORK DIRECTIVE NEEDS

The European Water Framework Directive (WFD, Directive 2000/60/CE) introduces new thinking about water resources protection in a multinational conglomerate, as is Europe. Applying the subsidiarity principle (Custodio, 2003) most efforts of control, regulation and administration are the task of member states – or the regions where they are defined. The Directive aims at preserving and restoring the common heritage and the fair-play between the parts to avoid one taking advantage over the other by unfair concurrence and transferring costs to the community or to future generations. Thus, the WFD refers mainly to environmental quality, other issues being the consequence or conditioning this. This legal framework provides a fully new point of view that needs to understand the orientation from the beginning, and to establish teaching and training along these lines. Concepts on ecology, systemics, commonwealth and sociology are to be explained, jointly with hydrogeological principles and basics. Intensifications in topics on the aspects of interest for each student should come later. All this needs redrafting teaching contents and goals, both individual matters and speciality courses, perhaps by adding the necessary lectures.

The spectrum of graduates enlarges the classical geological and engineering backgrounds, to include biologists, environmentalists and social sciences graduates. This implies that basics have to adapted, perhaps in two stages, one common to all of them, and later an extension to those following pathways closer to hydrogeology.

Using the European Credit Transfer System (ETCS) and referring to a person who has achieved a graduate level of 240 ETCS credits, the adaptation means about 30 ETCS credits for completing previous knowledge, 30 ETCS credits of oriented basic knowledge, and 60 ETCS credits of specialized teaching, plus a thesis, field work and in-project training.

7.1.6   TOOLS TO DIFFUSE HYDROGEOLOGICAL KNOWLEDGE

Tools to diffuse hydrogeological knowledge are those directed to schools and social centres. There have been different efforts since at least twenty years ago in various institutions, which are not to be reviewed here. Spain has tried to contribute to these tools by producing posters and other means. Most of the effort in groundwater has been carried out by the Geological Survey of Spain (IGME), but with limited diffusion of the produced material. This is a complex issue that needs the full cooperation of other institutions in charge of school teaching and social information.

A recent new intent started in 1998 with the production of a book directed to secondary teaching professors (López Geta *et al*., 2001), a series of posters, and an interactive video film starred by a water drop (Ramos, 2007). The results are still uncertain but encouraging.

The International Association of Hydrogeologists has a Working Group on Training, which has produced one primer, whose diffusion is poor, and needs new boost under some dynamic fostering group. Other primers on groundwater exist such as that of Price (2004), Younger (2007), Franke *et al*. (2000; 2002) and USGS (1999). This seems a correct way but it takes time to see their efficacy. Most efforts should be directed to primary and secondary schools, expecting that the new generation will be more conscious on the important role of groundwater, and the need of its sustainable use under an ecological point of view. The impact will not come, probably, until the second generation.

As a final word, it should be realised that the real success of the WFD will not only be through its enforcement (effective implementation), fines and investments, but also by its dedication to the diffusion of knowledge to the new generations.

REFERENCES

Batista E., Custodio E. and Valverde M., 1994. The Barcelona's International Course on Groundwater Hydrology: experience of 28 years and fitting process to a changing background of students. UNESCO's Intern. Workshop on Postgraduate Hydrology Education. Czeck Nat. Comm. for IHD. Praha.

Batista E., Valverde M. and Burgos R., 2002. La formación de especialistas por el Curso Internacional de Hidrología Subterránea: 36 años de experiencia (1967–2002). Aguas Subterráneas y Desarrollo Humano. XXXII Congress Intern. Assoc. Hydrogeologists–ALHSUD. Mar del Plata, Argentina.

Biswas A. K., 1970. *History of Hydrogeology*. North–Holland Publ., Amsterdam: 1–336.

Custodio E., 1994. Tendencias en la docencia e investigación hidrogeológica en España. La Contribución Académica al Conocimiento y Aprovechamiento de las Aguas Subterráneas en España. *Royal Acad. Sci. Mat., Fis., Nat*. (Spain). Madrid, 1: 127–48.

Custodio E., 2003. Tasks of EuroGeoSurveys under the subsidiarity principle. *Geologisches Jahrbuch, Bundesanstalt für Geowissenchaften und Rohrstoffe*, Hannover, G10: 216–44.

Custodio E., 2008. History of hydrogeology in Spain. History of Hydrogeology (ed. J. Matter). *Intern. Assoc. Hydrogeologists, Contributions to Hydrogeology Series*. (submitted).

Custodio E. and Llamas M. R., 1976/1983. *Hidrología Subterránea*. Ed. Omega, Barcelona, 2 vols.: 1–2350.

Davis S. N., 1969. Ground–water training in Barcelona. Spain. *Ground Water*, 7(6): 21–23.

Fornés J. M. and Senderos A., 2002. Las aguas subterráneas en la enseñanza española. III Congreso Ibérico. Sevilla.

Fornés J. M., de la Hera A. and Llamas M. R., 2005. The silent revolution in groundwater intensive use and its influence in Spain. *Water Policy*, 7(3): 253–68.

Franke O. L., Reilly T. E., Haefner R. J. and Simmons D. L., 1990, 1992, 2000, 2002. Study guide for a beginning course in groundwater hydrology. Part 1 and Part 2. U.S. Geological Survey Open File Report 90–183, Reston VA: 1–180; 1–128.

Llamas M. R., 1970. La formación de expertos en aguas subterráneas. *Agua. Barcelona*. Jul–Agost: 4–22.

Llamas M. R. and Custodio E. (eds)., 2003. *Intensive Use of Groundwater, Challenges and Opportunities*. Balkema Publ., Lisse, NL.: 387–414.

López Geta J. A., Fornés J. M., Ramos G. and Villarroya F., 2001. *Las aguas subterráneas, un recurso natural del subsuelo*. Instituto Geológico y Minero de España/Fundación Marcelino Botín, Madrid: 1–94 (an English version was produced later on).

MIMAM, 2000. Libro blanco del agua en España. Secretaría de Aguas y Costas, Ministerio de Medio Ambiente. Madrid.

Mostertman, L. J., 1970. Manpower requirements, training research, technical assistance policy. UN Panel of Experts of Water. Buenos Aires. ESA/RT/AC.1/29.

Price M., 2004. *Introducing Groundwater*. Routledge, UK: 1–304.

Ramos G., 2007. Educational material on hydrogeology and groundwater produced by the Geological Survey of Spain. In: *The Global Importance of Groundwater in the 25th Century*. The National Ground Water Association Press. Westerville, Ohio: 297–302.

Roe G. H. and Baker M. B., 2007. Why is climate sensitivity so unpredictable? *Science*, 318: 629–32.

UNESCO, 1974. The progress of hydrological education since the inception of the International Hydrological Decade. SC/WS/581. UNESCO, París: 1–39.

USGS, 1999. *Ground Water*. U.S. Geological Survey General Interest Publ., Reston VA.

Younger P. L., 2007. *Groundwater in the Environment, An Introduction*. Blackwell Publ., Oxford: 1–317.

7.2

Factoring in Expertise: International Scientific Networks – Roles and Benefits[*]

Philippe Quevauviller[1], Alice Aureli[2], Stephen Foster[3], Patrice Christmann[4] and Neno Kukuric[5]

[1] *Vrije Universiteit Brussel (VUB), Department of Water Engineering, Brussels, Belgium*
[2] *UNESCO – IHP, Paris, France*
[3] *c/o IAH, Kenilworth (Warks), United Kingdom*
[4] *EuroGeoSurveys, Brussels, Belgium*
[5] *IGRAC, TNO Princetonlaan, The Netherlands*

[*] The views expressed in this chapter are purely those of the author and may not in any circumstances be regarded as stating an official position of the European Commission.

Groundwater Monitoring Edited by Philippe Quevauviller, Anne-Marie Fouillac, Johannes Grath and Rob Ward
© 2009 John Wiley & Sons, Ltd

7.2.1   NEED FOR STAKEHOLDER'S INVOLVEMENT IN THE ENVIRONMENTAL POLICY PROCESS

The development of environmental policies is a complex process, which mixes legal requirements with issues of technical feasibility, scientific knowledge and socio-economic aspects and which requires intensive multi-stakeholder consultations. In this context, the consideration of scientific progress and the access to technical information represent key aspects for the design of new policies and the review of existing ones (Quevauviller *et al.*, 2005). On-going discussions highlight the increasing needs for an operational 'science-policy interfacing' as exemplified in the case of EU water policy (Quevauviller, 2007).

Within the European Union, this consideration is fully embedded into the Sixth Environmental Action programme which stipulates, namely, that 'sound scientific knowledge and economic assessments, reliable and up-to-date environmental data and information, and the use of indicators will underpin the drawing-up, implementation and evaluation of environmental policy' (European Commission, 2001). This requires, therefore, that scientific inputs should constantly feed the environmental policy process. This integration also involves various players, namely the scientific and policy-making communities but also representatives from industry, agriculture, NGOs etc. (Quevauviller *et al.*, 2007).

An example of successful multi-stakeholder consultation and participation concerns the implementation of the Water Framework Directive (2000/60/EC). In this context, a Common Implementation Strategy (CIS) has been agreed with the Member States, Candidate and Associate countries and stakeholder organisations and is operational since 2001 (European Commission, 2003). In this framework, various topics are under discussion by experts from EU Member States, industry, agriculture, scientists etc. with the aim to gather and share knowledge and concern on WFD relevant issues, as examined from different perspectives. This approach, albeit time-consuming, has considerably enhanced the knowledge and common interpretation of the key provisions of the WFD, and it has been considered as a very powerful tool for sharing good practices and an example of good gouvernance (Figure 7.2.1).

Consultation loop

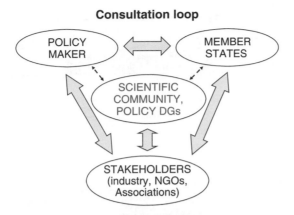

Figure 7.2.1   Consultation process involving stakeholders (from Quevauviller, 2006).

7.2.2   EU GROUNDWATER POLICY AND RELATED IMPLEMENTATION EXPERT GROUPS

7.2.2.1   The Common Implementation Strategy of the WFD

As already mentioned in the introductory part of this chapter, it has become clear, soon after the WFD adoption, that the successful implementation of the Directive will be, at the least, equally as challenging and ambitious for all countries, institutions and stakeholders involved. Therefore, a strategic document establishing a Common Implementation Strategy (CIS) for the Water Framework Directive (WFD) has been developed and finally agreed by the European Union's Water Directors under the Swedish Presidency in 2001 (European Commission, 2003). Despite the fact that the full responsibility of the individual Member States for implementing the WFD was recognised, a broad consensus existed among the Water Directors of the Member States, Norway and the Commission that the European joint partnership was necessary in order to:

- develop a common understanding and approaches;
- elaborate informal technical guidance including best practice examples;
- share experiences and resources;
- avoid duplication of efforts;
- limit the risk of bad application.

Furthermore, the Water Directors stressed the necessity to involve stakeholders, NGOs and the research community in this joint process as well as to enable the participation of Candidate Countries in order to facilitate the cohesion process. Following the decision of the Water Directors, a comprehensive and ambitious work

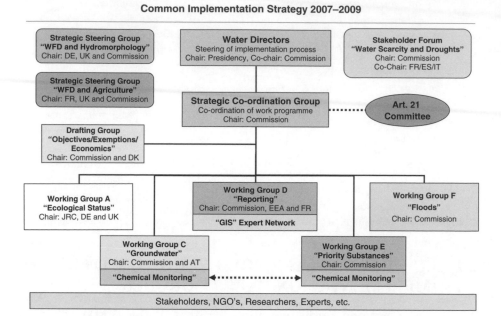

Figure 7.2.2   CIS operational diagram for the period 2007–2009.

programme was started of which the first phase, including ten Working Groups and three Expert Advisory Forum groups (EAF), was completed at the end of 2003 and led to the availability of fourteen Guidance Documents which are publicly available (in the form of CD-ROM and on Internet on the WFD *europa* website, see http://ec.europa.eu/environment/water/water-framework/index_en.html). The second phase of the CIS (2003–2004) involved four working groups, namely on ecological status (WG A), Economics and Pilot River Basins (WG B), Groundwater Body Characterisation and Monitoring (WG C) and Reporting (WG D), as well as two EAF groups, of which the discussions focused on the developing policies linked to the WFD (i.e. Priority Substances Directive, and revision of the Reporting Directive). These groups were re-conducted in the third phase (2005–2006), and this process is now continued under new mandates for the period 2007–2009 (see Figure 7.2.2), which is detailed with regard to groundwater in the section below.

7.2.2.2   The CIS Groundwater Working Group (WG C)

The CIS Groundwater Working Group (C) aims both to clarify groundwater issues that are covered by the WFD and prepare the development of technical guidance documents and exchange best practices on several issues in the light of the orientations of the newly adopted Groundwater Directive (European Commission, 2006). The Commission/DG ENV chairs the WG C which is co-chaired by Austria. The Working Group is composed of representatives of EU Member States, Associated and Candidate countries, industrial and scientific stakeholders, and NGO representatives (around 80 members in total). Plenary meetings are opened to all participants, while ad-hoc activities are operated by groups of around 15–20 participants which develop documents that are scrutinised by the plenary group.

The focus in the period 2003–2006 has been on the development of technical reports and guidance documents primarily focusing on the issues covered by the WFD, namely monitoring, prevent/limit measures and groundwater protected areas. In addition, a specific activity will concern exchange of views on groundwater management in the Mediterranean area (linked to the EU Water Initiative). Activities of the WG were conceived with the view of collecting targeted data and information, avoiding duplication with existing guidance documents and ensuring an efficient use of available data and information. Series of workshops have been held in 2003–2004, which led to three technical reports gathering Member State's practices in the field of groundwater risk assessment, monitoring and programmes of measures (Quevauviller *et al.*, 2007).

The orientations in 2005–2006 concerned the drafting of guidance documents on groundwater monitoring, protected areas and measures to prevent/limit pollutant introduction into groundwater, which have been endorsed by the EU Water Directors in 2007 (CIS Guidance documents 15, 16 and 17, see http://ec.europa.eu/environment/water/water-framework/groundwater/scienc_tec/cis/index_en.htm).

The main orientations of the 2007–2009 mandate of the WG C have been discussed at the occasion of the Groundwater Conference in Vienna on 22–23 June 2006 and through an email consultation of all WG C members. The main aims and objectives are to pursue exchanges in support of the implementation of the new Groundwater Directive along the CIS principles, focusing in particular on:

- best practices related to groundwater programmes of measures, including measures related to diffuse sources of pollution and megasites;

- common methodology for the establishment of groundwater threshold values;

- compliance, status and trend assessment;

- recommendations for integrated risk assessment, including conceptual modelling.

The WG C work programme 2007–2009 consists in three core activities led or co-led by Member States or Stakeholder Organisations, which develop their work programme as described in activity sheets. The activities (drafting or exchanges of good practices) are undertaken with selected WG participants (groups of ideally 20–25 participants) willing to actively contribute to the drafting of documents and to participate in ad-hoc meetings (possibly organised by the activity leaders). The progress of the activities is reported and discussed at plenary meetings of the WG C held twice a year and organised under the EU Presidency umbrella.

7.2.3   EUROPEAN NETWORK OF GEOLOGICAL SURVEYS – EUROGEOSURVEYS

7.2.3.1   Introduction

EuroGeoSurveys, the Association of the Geological Surveys of Europe (www. eurogeosurveys.org) is an European NGO legally registered in France, with its office in Brussels, Belgium. Founded in 1995, it had 33 members on 1 January 2008, it is the successor of two pre-existing organisations. The Western European Geological Surveys forum (WEGS), a yearly informal gathering of Geological Surveys Directors started its activities in 1971. In 1992 it was succeeded by the the Forum of European Geological Surveys (FOREGS), an equally informal set-up, opening up participation to the Geological Surveys of the Central and Eastern European Countries further to the political transition in that region of Europe.

EuroGeoSurveys members are the national Geological Surveys of European countries, including the Geological Survey Organisations from 26 EU Member States (only Malta is not a member). Autonomous regional Geological Surveys, existing in various European countries (Belgium, Germany, Italy, Spain) can participate in its technical activities. Its membership comprises over 20,000 persons working in, or in support of, almost any conceivable thematic domain of modern public geology: climate change impact assessment and modelling; CO_2 storage in geological formations; energy and mineral resources; environmental management; geohazards; geographic information processing and dissemination; geology and public health; geophysics; soil science; subsurface space use; water resources management.

The statutory objectives of EuroGeoSurveys are:

- to jointly address European issues of common interest;

- to promote the contribution of geosciences to European Union affairs and action programs;

- to publish, or see its Members and Partners publishing technical advice for the European Union institutions;

- to provide a permanent network between the Geological Surveys of Europe and a common, but not unique, gateway to each of the member Surveys and their national networks.

EuroGeoSurveys is entirely financed by its members on the basis of multi-tiered membership fee system, ensuring the independence and balance of its positions.

While individual Geological Surveys operate according to differing national remits and economic models data acquisition, processing and modelling on water resources is a very important part of their activity, with a strong emphasis on groundwater. Only about on-third of all EuroGeoSurveys members have activities related to surface water (hydrology).

The development of the European geological information and knowledge infrastructure is a core objective of EuroGeoSurveys. Through its experts Working Groups EuroGeoSurveys contributes to the EU policy making process for instance by means of participation of stakeholder working groups set up by the European Commission, such as the above Water Framework Directive Common Implementation Strategy Groundwater Working Group (WG C), the formulation of opinions on EU policies and legislation, the submission to Members of the European Parliament of suggestions for amendments to European Commission's legislative proposals and the participation to public consultations.

EuroGeoSurveys maintains a large website[2] providing news and information on events of relevance to Geological Surveys and persons interested by the application of geosciences to society. From its 'about' menu it is possible via a map service to navigate to the website of its individual members, several of them offering extensive web-based information services on water resources. Its download section offers the possibility to download the recently published International Hydrogeological Map of Europe at 1:5,000,000 scale compiled by W. Struckmeier, from the German Federal Institute for Geociences and Natural Resources (BGR). EuroGeoSurveys sponsored its publication.

It has good relations with other sectoral organisations depicted in this chapter, such as the International Association of Hydrogeologists (IAH) and the International Groundwater Resources Assessment Centre – (IGRAC). Several leading groundwater experts from EU Geological Surveys also play key roles in these organisations.

7.2.3.2   European Geological Survey Organisations and Groundwater Management

In almost every European country, geological Surveys provide the public data, information and expertise needed for sustainable groundwater use. Data and information is provided as hydrogeological maps and/or digital information layers in raster or vector formats, including information such as the nature and the spatial distribution of the hydraulic properties of geological formations (aquifers, aquitards, aquicludes, main hydrogeological parameters), of recharge areas, of climatological data (atmospheric recharge), of hydrodynamic parameters (run-off, transmissivity, groundwater flow directions and rates), of hydrogeochemical parameters (major and trace elements), of production wells and well fields or of piezometers. The lack of a common pan-European or, better, global hydrogeology relevant extension of the Geographic Mark-up Language (GML) remains a considerable hindrance for the development of at least schematic interoperability between the various national and regional digital hydrogeological information systems that so far are developed on the basis of differing national/regional data models. This is an issue that needs to be addressed by the European Commission, especially in view of the management of the numerous high-value cross border aquifers.

A wide range of information can be attached to each well, such as the borehole specifications, a hydrogeological profile of the well, pumping tests results, groundwater heads, hydrochemical and hydroclimatological data. Some geological surveys produce monthly reports on the groundwater table variations of their country's main groundwater bodies.

Several European Geological Surveys also have a long history of cooperation with developing countries on various aspects of groundwater use and management, for instance

---

[2] www.eurogeosurveys.org.

on the development of national groundwater databases and GIS, modelling, prospecting for groundwater bodies and selection of production borehole sites.

Finally, European Geological Survey Organisations could play a greater science-policy interface role in support of the Water Framework and of the Groundwater Directives. In this role they could provide Member States water resources using/management bodies with advice on the technical aspects of groundwater resources management and assist the European Commission in its role of providing guidance and support to the Member States in the process of implementing the Directives. However this would require a clear mandate and funding. None of them was available at the time of writing (July 2008).

7.2.3.3   European Geological Surveys and Groundwater Research

The activity of Geological Surveys frequently involves an applied research component, on both national/regional and EU levels. The research projects database available from the EuroGeoSurveys' website[3] provides information on over 200 EU funded applied research projects to which at least one Geological Survey participated, including 58 related to water. Most of these projects were implemented under the 5th (1998–2002) and 6th (2002–2006) EU Research and Technological Development Framework Programmes. The water related projects involved 13 EuroGeoSurveys member surveys, along a wide range of partners from academia, research institutes and industry. For 15 out of these 58 projects an EuroGeoSurveys member was the project coordinator. A database providing a short summary of each project and a link to its website is available for download from the EuroGeoSurveys website: http://www.eurogeosurveys.org/assets/files/research/EGS_Research_database_20.06.07.xls

Research topics cover issues such as groundwater management through a better understanding of the river-sediment-soil-groundwater system as a whole; the study of natural attenuation and of the phenomena at the soils-groundwater interface; quantitative and qualitative monitoring technologies; groundwater geochemistry; the determination of natural background values and of legal threshold values; pesticide risk prediction and the assessment of their fate in soils and groundwater; biotechnologies for the decontamination of polluted groundwater; efficient resource use and water conservation; integrated water resources management at the river-basin scale and of cross-border groundwater bodies; spatial digital hydrogeological data management and the development of Internet based services; groundwater and geochemical modelling; the assessment of groundwater contribution to flooding mechanisms.

EuroGeoSurveys geochemistry working group is also engaged in the development of a first-ever pan-European harmonised geochemical assessment of groundwater based on a sampling of commercial bottled water. This survey, based on about 1500–2000 sampling points will become available in 2009. Although this is a very low sampling density, it will most likely show some significative regional differences in the groundwater chemical composition, and hopefully trigger support for a later higher resolution coverage.

---

[3] http://www.eurogeosurveys.org/assets/files/research/EGS_Research_database_20.06.07.xls.

7.2.3.4 EuroGeoSurveys Position on EU Groundwater Management Policy

EuroGeoSurveys praises the progress jointly made by the EU Member States and the European Commission in developing and implementing a common European groundwater policy. The Water Framework Directive (2000/60/EC) that came into force on 22/12/2000 required (Article 5 and Annex 2) the Member States to characterise their surface and groundwater bodies and deliver the related information to the European Commission not later than 22/12/2004. As the groundwater bodies characterisation was essentially devolved to Geological Surveys a pan-European groundwater workshop of the European Geological Surveys leading groundwater experts, coorganised by BGR and EuroGeoSurveys was organised in Berlin on October 25 and 26, 2005 to compare the information resulting from the characterisation work. The conclusions of the workshop, given in Box 7.2.1, remain valid.

BOX 7.2.1 Conclusions of the BGR- EuroGeoSurveys Groundwater workshop, October 2005

- Groundwater bodies (GWBs) should be addressed as consistent groundwater entities within subsurface systems. They must be:

  o As uniform and homogenous as possible with regard to hydrogeology (structure), flow conditions (hydraulics), properties (quality);

  o Consistent with connected surface water bodies concerning hydraulic dependencies, flow systems (recharge-discharge, age).

- The delineation of GWBs needs a conceptual model and a clear understanding of hydraulic boundary conditions in order to support the description and the delineation of agreed management units, to allow monitoring requirements and to allow the management of the quantitative status; to foster planning and operation of remediation and/or mitigation measures.

- GWBs should be addressed everywhere in Europe based on the same features, i.e. hydrogeological (not necessarily aquifer) boundaries, surface water-groundwater interaction; groundwater highs (divides); groundwater lows (rivers, depressions).

- Harmonised information systems about GWBs are required to document GWBs consistently and up-to-date, and to allow coherent views on transboundary GWBs.

- The hydrogeological community contributes relevant knowledge, expertise, information and data appropriate for political and management decisions.

- The Berlin workshop:

  o Recognised the need for harmonising data in order to foster the interoperability of data and information;

  o Suggested a common groundwater typology for Europe;

- o Required to include more complementary data (e.g. on land use, discharges, regional information on water use);

- o Stressed that monitoring and assessment must be addressed together;

- o Recommended that the goal of monitoring should be included more tightly into monitoring programmes, to address the dilemma between compliance versus uses and ecological status (calling for different methods and data sets and requiring the development of appropriate assessment methods);

- o Underlined that monitoring of groundwater should be deeper intertwined with other domains (surface water, human activities) because drivers and pressures are not sufficiently taken into account (changes in land-cover, agricultural activities) and the linkage with surface water is insufficient

The Berlin workshop suggested that European Geological surveys be given a mandate to:

- Develop a better understanding the flow and residence time in groundwater flow systems (particularly in large and deep flow systems);

- Consider the damage of aquifers from over-pumping and subsequent compaction;

- Take a preventive, long-term, view to avoid pollution of important aquifer systems;

- Transboundary aquifers need to be identified in their natural setting without regarding their current status of use;

- Anomalous Groundwater systems (hydrothermal; in areas of volcanic activity or of ore deposits, in closed basins) need to be investigated;

- Research is needed for the assessment of hitherto unmonitored aquiifers (comparable to the Prediction in Ungauged Basins programme) in view of defining the (natural) background composition of groundwater and assessing the natural potential of groundwater.

7.2.4 THE INTERNATIONAL HYDROLOGICAL PROGRAMME OF UNESCO

7.2.4.1 Introduction

The International Hydrological Programme (IHP) was established by the UNESCO General Conference during its 18th Session in 1974 and is the only intergovernmental scientific programme of the UN system devoted to hydrology, water resources management, education and capacity building. The members of the UNESCO IHP Intergovernmental Council are nominated by their governments and elected by the General Conference of UNESCO on a rotational basis. The Bureau attached to the IHP Intergovernmental Council coordinates the work of the Council between sessions. The programme is thus a political process driven by sovereign states, implemented in six-year phases. The IHP Council meets every two years to provide directives to IHP enabling it to adapt to a rapidly changing world. The IHP Secretariat is located at UNESCO Headquarters in Paris, and it works closely with over 165 National Committees assigned by their governments and focal points to contribute to the implementation of the programme.

Regional cooperation is an important aspect of this programme and the UNESCO regional cluster offices and national offices coordinate regional and local activities. The programme receives certain financial resources from UNESCO and operates in close cooperation with UN WATER and other UN System agencies and programmes. Member States may apply for additional support for specific regional and local activities through the UNESCO participation programme. They may also work with the Secretariat to develop project proposals for funding from donor agencies, such as the Global Environmental Facility (GEF), the United Nations Development Programme (UNDP), the World Bank, the European Community, and private foundations. IHP also benefits from generous financial and technical contributions from Member States. IHP National Committees play a critical role in the programme's implementation by organizing national and regional activities and producing national reports. UNESCO is host to the Secretariat of the UN World Water Assessment Programme (UNWWAP) to which IHP provides a valuable technical contribution. IHP takes a proactive role in UN system-wide initiatives such as the United Nations Development Assistance Framework (UNDAF), and the UN Country Programme Action Plan.

IHP operates as a global series of networks that includes connections with the UNESCO water-related Institutes and Centres, the UNESCO-IHE Institute for Water Education and UNESCO water-related Chairs. The UNESCO water portal serves as a viable tool in enhancing dissemination of information and knowledge (http://www.unesco.org/water). Most of the IHP publications can be downloaded for free from the IHP web site. UNESCO IHP also gives a prize for studies on water resources in arid and semiarid zones, awarded biannually to scientists and institutions working to provide effective tools to combat water scarcity.

7.2.4.2 IHPVII – Water Dependencies: Systems under Stress and Societal Responses

The seventh phase of IHP (2008–2013) is entitled: 'Water Dependencies: Systems under Stress and Societal Responses'. Since its inception, the different phases of IHP have operated under the three pillars of 'hydrological science, surface and groundwater resources assessment and management, education and capacity building'.

In the seventh phase of IHP, headline tasks are as follows:

- integrating interdependencies of water sciences and policy making through research and education, underpinned by culture and communication;

- understanding water interdependencies in physical, biological and social environments;

- promoting participatory decision making in interdependent water-related health, food and energy systems and security in a changing world.

Building upon the previous three pillars and focusing on the demands arising from global changes, creating partnerships and initiatives for greater synergies, the IHP-Phase VII maintains its comparative advantage in promoting and leading international hydrological research for enhanced water management towards meeting the UN Millennium Development Goals on Environmental Sustainability, Water Supply, Sanitation, Food Security and Poverty Alleviation. It adds value to localised research and experience by providing a policy-relevant context and harvesting the knowledge of researchers, educators, practitioners, and policy-makers so as to maximise the value of scientific outcomes and engender confidence in innovation and reform. It provides a solid scientific underpinning for the UN Decade of Water for Life and the UNESCO led UN Decade on Education for Sustainable Development.

IHP VII is subdivided into five themes, including 22 focal areas:

- Theme 1 – Adapting to the impacts of global changes on river basins and aquifer systems

- Theme 2 – Strengthening water governance for sustainability

- Theme 3 – Ecohydrology for sustainability

- Theme 4 – Water and life support systems

- Theme 5 – Water education for sustainable development

However Member States agree that amongst the working priorities of IHP-VII, particular focus should be given to the sustainable management of groundwater resources.

7.2.4.3   IHPVII – Managing Groundwater Systems' Response to Global Changes

While 'integrated watershed and aquifer dynamics' was one of IHP's themes during Phase VI (2002–2007), UNESCO has recognised that there is still insufficient attention being given to the long term management of groundwater resources and its scientific underpinnings. With this in mind, the accelerated use of groundwater resources in the absence of effective long-range management and governance policies has already caused serious local problems in many regions of the world. To a large extent, this reflects the existing lack of scientific knowledge regarding aquifers and lack of investment in developing appropriate groundwater resource management strategies.

In finding solutions to water scarcity, IHP pays due attention to the role that aquifers can play. The large storage capacity of groundwater resources, many of which are transboundary, if properly managed, can play a crucial role in supporting adaptation measures for coping with impacts from climate variability, global changes, hydrological extremes and natural disasters.

As such, IHP-VII will continue to consolidate knowledge on groundwater resources and their sound management on all levels; continental, regional and local, through to small island developing states. This includes formulating science-based policies and principles, preparing appropriate regulations to curb over-exploitation, developing technologies and policy instruments. Closer attention is being paid to transboundary aquifers, non-renewable groundwater resources, enhancement of aquifer recharge, and adaptation measures to climate variability, groundwater quality, groundwater protection, groundwater-dependent ecosystems and urban groundwater management.

On a regional level, the activities of IHP-VII take into consideration the integrated management of groundwater aquifers and surface water basins, combining natural and man-induced processes at various scales and time. Establishment of common principles for the development, protection and management of transboundary aquifers and improvement of the scientific, socio-economic, legal, institutional and environmental issues surrounding the management of shared aquifers is a high priority for IHP-VII, implemented within the framework of the ISARM project.

Until the year 2000, no regional or global estimation for the number of transboundary aquifers (TBAs) actually existed. The IHP Intergovernmental Council responded to this knowledge gap at its 14th Session held in June of that year by adopting a resolution to launch a worldwide inventory and assessment of transboundary aquifers; the UNESCO International Shared Aquifer Resources Management (ISARM) project. The aim of the project is to work together with Member States to identify the transboundary aquifers in each continent, support countries in the assessment of these aquifers and formulate recommendations on their management. The ISARM project therefore considers the scientific-hydrogeological, socio-economic, environmental, legal and institutional aspects related to the management of transboundary aquifers.

ISARM operates in coordination with Member States and various intergovernmental and governmental organisations as well as international partners. Studies conducted across the different continents show that the number of transboundary aquifers recorded since the launch of the ISARM project is comparable to that of international river basins. This number is set to increase in future years due to more detailed investigations that are planned to be carried out by IHP in Asia and Africa. At present, 68 transboundary aquifers have been identified in the Americas, 29 in South America (18 in Central America, 17 in North America and four in the Caribbean), 48 in South Eastern Europe, 10 in Western Asia and 39 in Africa. The ISARM Asia inventory is currently in its initial phase but only 12 transboundary aquifers have been identified to date. In Europe, UNESCO IHP is working closely with the United Nations Economic Commission for Europe (UNECE) in assessing groundwater resources and transboundary aquifers located in Eastern European countries and SEE.

It is expected that under the supervision of IHP, the Global Groundwater Information System (GGIS; at www.igrac.nl), established at the UNESCO and WMO International Groundwater Resources Assessment Center (IGRAC), will collect and compile a global set of data. In addition, the World Hydrological Map (WHYMAP) project led by UNESCO and BGR provides an essential contribution with its mapping programme presenting an overview of the global groundwater resources and cartography of most of the transboundary aquifer systems inventoried to date.

Inventories, maps and data compiled under the guidance of the IHP ISARM project demonstrate that transboundary aquifers are a worldwide phenomenon with some extending over distances of several hundred kilometers. These systems play a crucial role in providing drinking water supplies as well as water for irrigation, terrestrial and coastal ecosystems, and the socio-economic development of many regions of the world. TBAs occupy a significant percentage (>15%) of the earth's total land surface. Projects are currently ongoing for several individual transboundary aquifers at different locations across the world.

However, although the scientific principles involved in the sound management of transboundary aquifers are well known and understood by groundwater specialists, there is need for an international legal instrument to deal comprehensively on a global level with TBAs by addressing their specific characteristics. In 2002, in order to redress this issue, the UN International Law Commission (UNILC) in charge of the codification of international laws included the topic of Shared Natural Resources in its programme. Since 2003, UNESCO IHP and the UNILC have been working in close cooperation in the preparation of an international legal instrument for the management and use of transboundary aquifers. This collaboration resulted in a set of articles on The Law of Transboundary Aquifers (UNILC Report, July 2008) that was presented at the 63rd Session of the UN General Assembly in October 2008 and annexed to the resolution adopted by the General Assembly on 11 December 2008. The UN General Assembly resolution represents a milestone in the international recognition of the crucial function of transboundary aquifers for man and the environment.

Amongst the different IHP-VII projects on groundwater resources it is worthwhile mentioning:

- GRAPHIC project, aimed at investigating interactions between climate change and groundwater resources;

- GWES project, aimed at establishing guidelines and an international network to address issues related to groundwater resources management in emergency situations.

7.2.5   IAH – THE WORLDWIDE PROFESSIONAL COMMUNITY FOR GROUNDWATER

7.2.5.1   Introduction

The IAH (International Association of Hydrogeologists) is the worldwide professional NGO for groundwater specialists (www.iah.org) – which reached its 50th Anniversary in 2006. Its membership come from a variety of disciplines but share the common goal of working, directly or indirectly, to develop groundwater resources sustainably and to utilise the subsurface environment safely for engineering construction and waste disposal – thus contributing to the IAH mission of 'furthering the understanding, management and protection of groundwater in the interest of humankind and the environment'. Since 1972 the IAH has been recognised by ICSU as the 'international scientific association competent in groundwater matters'.

Following prior discussions on the newly-emerging field of groundwater science, the IAH was founded at the International Geological Congress of 1956 in Mexico City, with the election of a first small Council largely of European origin, and has subsequently grown to be:

- a worldwide association with approaching 4000 members in more than 140 countries;

- proud of, but not constrained by, its geoscience roots and capable of addressing all facets of groundwater science, engineering, resource management and protection;

- a professional network embracing academics, researchers, regulators and consultants;

- a forum for professional contact at international and national level, both for its members and the wider environmental and engineering community interested in groundwater.

IAH is financed mainly by its members and supporters, without government, trade or industrial allegiances, and since 2000 has had the legal status of an incorporated company and registered charity. It is managed by a Council of 4 Executive Members and 8 Regional Vice-Presidents, who are elected to serve a 4-year term. In more than 40 countries members have also organised themselves into National Chapters to act as a focal point for debate of groundwater issues at national level. Perhaps the most direct way of capturing the essence of the role that IAH plays is to allow some of the members 'to speak for themselves' (Box 7.2.2).

BOX 7.2.2 What IAH members say

In the IAH 50th Anniversary Year (2006) a comprehensive survey of member opinion on its role, performance and priorities was carried out. One question asked was: 'What would you say is the single most important reason why you belong to IAH?' and a selection of responses is given below:

- *IAH has opened a window for me to other groundwater specialists around the world and given me personal contact and research discussion with them.*

- *I see IAH as the global leader in dissemination of groundwater understanding and technology for water management and involvement as an avenue for remaining in touch with new developments.*

- *I like a global perspective in my work and my life, and IAH is science-based and, unlike some other international organisations, appears not to be overly political or business-centred.*

- *I share with IAH the belief that we must move toward sustainable groundwater use in our own country and also help the developing world to progress in a responsible and sustainable manner.*

IAH was one of the first international professional NGOs to promote membership in low-income countries by a tiered fee structure and a sponsored membership scheme (cross-subsidised by individual members and national groups in the high-income countries) – at present it has over 500 members in low-income countries of which about 35% are sponsored. Bearing in mind that groundwater plays such a critical role in the provision of basic water supplies, and thus has great importance in socio-economic development in the developing world, in 1985 IAH founded the 'Burdon Commission' to ensure the needs of these countries were taken fully into account in its programmes and publications. During 2005–15 the priority for its volunteer work and charitable expenditure is Sub-Saharan Africa with the objective of disseminating groundwater know-how, together with closer cooperation with UNICEF and international NGOs working in basic water-supply and sanitation provision – thus providing practical support to the achievement of the UN Millenium Development Goals.

7.2.5.2   IAH as an Advocate for Groundwater Policy Formulation

The IAH harnesses the research, knowledge, expertise and enthusiasm of its members to promote full consideration of groundwater in integrated water resources management – and puts special emphasis on the complex, but intimate, relationships with both agricultural and urban land-use. It plays an important role in speaking-up for groundwater as a human and environmental resource in world and regional water policy dialogue, championing its proper use and protection – this especially because groundwater issues are not always well understood by the general public, policy makers and even some professionals in related 'water disciplines'.

IAH considers that historically the policy base for groundwater has been generally weak and poorly understood, especially as regards governance provisions and institutional arrangements for resource management and protection. It has thus taken on significant advocacy and advisory role in this regard at global level through being an active member of *UN-Water, the WWC and the GWP* – and has promoted major sessions at the 3rd World Water Forum (Japan–March 2003), jointly with UNESCO-IHP, World Bank GW-MATE and other interested organisations, and at the 4th World Water Forum (Mexico–March 2006), with World Bank-GEF and WWC).

The EC Water Framework Directive (2000) and Groundwater Protection Directive (2006) put groundwater conservation for drinking water-supply and ecosystem health in the European political spotlight. IAH, which has a total of over 1500 members and 17 National Chapters in the EU countries, warmly welcomed this initiative and is making a special effort to support and facilitate its development – through both the active membership of the *EC-DGE Groundwater Working Group* (which meets on a six-monthly basis in the country holding the EU Presidency) and the organisation of parallel symposia or workshops on a topic of sub-regional interest complimentary to the meeting and work of the Group.

7.2.5.3   IAH as a Scientific Forum and Publisher

The scientific, and science-policy interface, activities of IAH are mobilised by three separate, but interrelated, types of activity:

- international congresses and regional or national conferences and workshops;
- the work of specialist commissions and working groups;
- publication of a regular journal and two periodic book series.

Major international meetings are held regularly – with the IAH 34th, 35th and 36th Congresses having been held respectively in Beijing-China (October 2006), Lisbon-Portugal (September 2007) and Toyama-Japan (October 2009) (Box 7.2.3 gives a broader idea of the geographical trajectory). As appropriate IAH cooperates with other organisations (such as IAHS, IAEG, IWA, etc) in the organisation of events, and commonly receives co-sponsorship from various UN agencies. It also participates in the IGC's and (through its National Chapters and Commissions) organises or sponsors at least 20 other conferences and workshops every year.

BOX 7.2.3 Geographical distribution of IAH Congresses

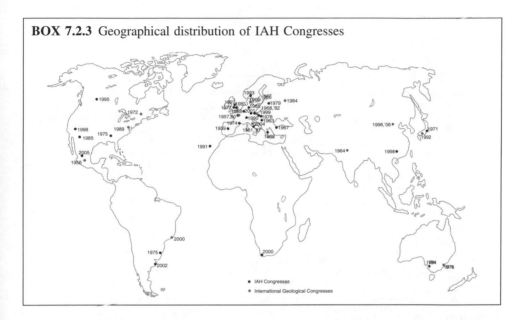

The development and application of hydrogeoscience, through the interaction between experts, is one of the longest-standing roles of IAH – this has been carried out by standing commissions and more specific working groups, many of which have enjoyed strong support from national geoscience and environmental agencies around the world. Some IAH Commissions (Box 7.2.4) have a long history of activity (notably Hydrogeological Maps founded in 1959, Mineral & Thermal Waters and Karst Hydrogeology both founded in 1968).

During the 1980s IAH developed a role as the major *UNESCO-IHP Partner* for groundwater – a partnership which has enabled its members to pool their expertise and provide state-of-the-art syntheses for widespread dissemination and application. There has subsequently been a degree of convergence of successive UNESCO-IHP themes and IAH Commissions (eg. on Groundwater Protection, Internationally-Shared Aquifer Management, Managed Aquifer Recharge and Groundwater & Climate Change) and broadening of IAH cooperation with other UN agencies (IAEA, UNEP, UN-FAO, WHO, etc.).

IAH is also playing a key role in trying to ensure the information base for groundwater and hydrogeological terrain characteristics is secured and amplified through its strong long-term inputs to UNESCO-WHYMAP and support for the establishment of IGRAC.

BOX 7.2.4 IAH commissions and working groups

TOPIC AREA Conveners	SCOPE & STATUS
Hydrogeological Maps Willi Struckmeier	established in 1959 and has coordinated the publication of the 30-sheet map of Europe, produced the world standard legend and is now developing (with UNESCOsupport) WHYMAP, the hydrogeological map of the world.
Hydrogeology of Karst (vacant)	long-standing group that has published extensively and meets regularly at IAH sponsored karst conferences.
Mineral & Thermal Waters Werner Balderer	founded in 1968 and meeting annually in classic areas of mineralised and thermal groundwaters
Groundwater Protection Jaroslav Vrba	has addressed both point source and diffuse pollution issues and their management through publications and conferences and made a major contribution to UNESCO-IHP
Urban Hydrogeology Ken Howard	has several regional and thematic working groups and has run numerous workshops.
Hydrogeology of Hard Rocks Jiri Krasny	established in 1993 and has a number of active regional working groups, which organise workshops
Transboundary Aquifer Management Shaminder Puri	collaborates with UNESCO and many other international agencies dealing with shared water resources to tackle the technical, legal and institutional aspects of managing these resources.
Managing Aquifer Recharge Peter Dillon	established to encourage the development and adoption of improved practices for the management of aquifer recharge
Groundwater & Climate Change Nico Goldscheider & Ian Holman	to address the understanding of implications of climate variability for groundwater
Groundwater Dependent Ecosystems Christine Colvin	established to provide a forum for multi-disciplinary studies on the environmental impacts of groundwater development.
Aquifer Dynamics & Coastal Zone Management Giovanni Barrocu	seeks to co-ordinate various groups working on the management of coastal zone aquifers and control of saline intrusion
Groundwater Management Partnerships (convenors to be nominated)	to distil experience on the institutional aspects of groundwater resource management and groundwater source protection

☐ very long standing   ☐ well established   ☐ recently formed and forming

The premier IAH publications are:

- Hydrogeology Journal (published by Springer) – which reaches its 16th volume in 2008 and has recently doubled its publication frequency (to 8 times per year) and trebled its number of pages (to 1400 per year)

- International Contributions to Hydrogeology and Selected Papers on Hydrogeology (published by Taylor & Francis) – periodically producing both text-book reference works of long shelf-life and volumes on more topical themes.

And, of course, IAH maintains an extensive website with information on recent developments, up-coming activities and information sources of interest to all those working with or concerned about groundwater resources – in addition to the procedure for becoming a Member and receiving fuller benefits (www.iah.org). It also publishes and distributes a hard-copy newsletter 'IAH News & Information' (3 times per year) summarising the latest information on the website.

7.2.6 INTERNATIONAL GROUNDWATER RESOURCES ASSESSMENT CENTRE

7.2.6.1 IGRAC in Brief

International Groundwater Resources Assessment Centre (IGRAC) is founded to facilitate and promote global sharing of information and knowledge required for sustainable groundwater resources development and management. Accordingly, IGRAC main activities are meant to contribute to:

- groundwater (meta)data, information and knowledge management;

- understanding of groundwater processes and patterns on regional and global scales;

- management of transboundary aquifers;

- global monitoring of groundwater resources;

- promotion of groundwater use in sustainable water resources management.

IGRAC is a non-profit centre that operates under auspices of UNESCO and the World Meteorological Organisation (WMO). The centre is hosted by Netherlands Organisation for Applied Sciences TNO (in Utrecht, The Netherlands) and receives a financial support from the Government of the Netherlands.

The following sections provide a short overview of IGRAC activities, presented according to their contributions to the issues listed above. More attention is paid to activities that are particularly interesting for the subject of this book, namely assessment and monitoring of groundwater quality. Additionally, Section 7.2.6.6 (IGRAC and Promotion of Groundwater Use) also addresses a role of IGRAC as an international networking organisation.

7.2.6.2 Groundwater Information Management

As a global groundwater information centre, IGRAC has a task to increase international availability of groundwater-related information and knowledge. This is a complex undertaking because the information made available by IGRAC ought to be relevant, consistent, documented, structured and easy to search through. Nevertheless, the first of all information needs to be collected and most often processed (or assessed) as well (Figure 7.2.3).

Figure 7.2.3 IGRAC logo.

IGRAC is not collecting data and it is not developing a global mega-database. The centre collects rather metadata on various sources of information, linking those in a powerful global information network. Next to metadata, only the information and knowledge generated by IGRAC assessments are stored in IGRAC databases.

IGRAC databases are connected in the Global Groundwater Information System (GGIS). The system is meant for various categories of stakeholders, including both professionals and the general public. It is simple to use and completely publicly accessible. It leads the user from aggregated, global information (Global Overview) via related information sources (Meta Information Module) towards a direct information exchange (Collaborative Environment). Global Groundwater Monitoring Network (GGMN) is the youngest and the most demanding GGIS module. Its purpose is to monitor a global change of groundwater resources (see Section 7.2.6.4).

The GGIS Global Overview contains a world map of countries and a set of aggregated groundwater-related attributes for each of the countries. The attribute set contain more than seventy attributes, divided in several categories, such as aquifer characteristics, groundwater quantity, groundwater quality, groundwater development and groundwater problems. For example, Figure 7.2.4 shows reported cases of pollution from domestic sewage worldwide. In some cases, the sources of information used for the overview (i.e. documents, websites, other databases) are available and directly accessible. If not, MIM contains contact data of information owners and/or providers. Eventually, the Collaborative Environment can be used to facilitate collaboration established with related persons and organisations.

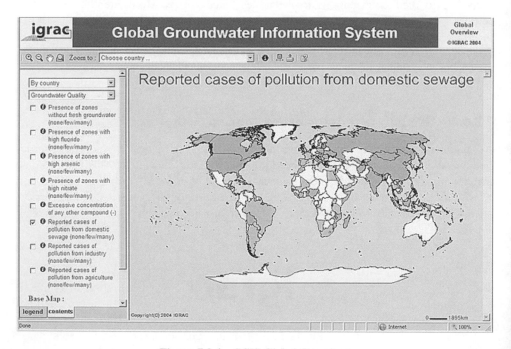

Figure 7.2.4   GGIS Global Overview.

IGRAC is making information available primarily through its own on-line services, including GGIS, but also ISARM portal (see Section 7.2.6.4), GGMN, Guidelines and protocols database (see Section 7.2.6.4) and other web-based systems, databases and services. Additionally, IGRAC provides assistance to countries/organisations to set up their own information services, such as NAMIS (National Meta Information System on Groundwater in Yemen, see www.igrac.net) and AEGOS (African-European Georesource Observation System, together with EuroGeoSurvey).

7.2.6.3 Groundwater Assessment

Global groundwater assessment is a core activity of IGRAC. This assessment is necessary for better understanding of groundwater processes and patterns on regional and global scales.

Quite often, IGRAC assessment activities are a contribution to an international programme such as WWAP (World Water Assessment Programme), GWES (Groundwater in Emergency Situations) and GRAPHIC (Groundwater Resources under the Pressure of Humanity and Climate change). In other cases, initiative to address a certain topic is taken by IGRAC, based on analysis of global knowledge gaps and demands. For instance, artificial recharge methodologies have been thoroughly elaborated, yielding clear overviews, classifications, illustrative examples and a database with numerous case studies (see www.igrac.net).

All the assessment activities mentioned above either contain a groundwater quality component or are related to groundwater quality. Additionally, IGRAC has set up a programme dedicated to mapping of hazardous substances in groundwater on a global scale. The first mapping step has resulted in the global groundwater quality maps as made available in GGIS (see Section 7.2.6.2). These maps provide a general indication and allow for intercomparison, but are not well suited to small-scale water resources management. In order to move from the global indication to refined spatial distribution, IGRAC conducted the assessment of (probability of excessive occurrence of) arsenic and fluoride worldwide (Vasak *et al.*, 2006). The assessment (see Figure 7.2.5) consisted of the following major steps:

1. determination of locations for reported country cases;

2. gaining knowledge on sources and geochemical behaviour;

3. selection of proxy information as an indicator of probability of occurrence.

Once the occurrence of contaminated water is assessed, suitable removal techniques need to be determined and applied as soon as possible. Therefore, IGRAC has made an analysis and comparison of removal technologies in terms of efficiency, costs and required technology (the report is available on www.igrac.net).

Also related to groundwater quality is a mapping and assessment of brackish and saline groundwater worldwide that is on-going activity at IGRAC. The first outcomes of the mapping are expected some time before the end of 2008.

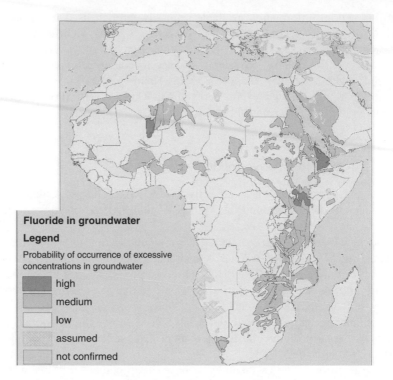

Figure 7.2.5 Arsenic in Africa.

7.2.6.4 Transboundary Aquifers

The management of transboundary aquifers is one of the major contemporary international challenges. It asks for a comprehensive approach, combining the hydrogeological, but also institutional, socio-economical and legal aspect of the issue.

In principle, a hydrogeological assessment of transboundary aquifer does not differ from the assessment of a random aquifer within the national borders. However, involved countries ('aquifer states') need to agree about the assessment methodology, starting from a basic definition of the aquifer. The main relevant hydrogeological definitions are now included in the draft articles of the first low on transboundary aquifers that is prepared by UN ILC (International Law Commission), with a modest IGRAC contribution.

The success of a characterisation of any aquifer relies heavily on availability and quality of related data. For internationally shared aquifers, however, the harmonisation of data across the border plays an equally important role; if two data sets cannot be mutually compared (and further processed), they are not much of use. Besides, these data need to be made accessible internationally, which brings up the issue of information management. Essentially, data harmonisation and information management are technical activities related to harmonisation of formats, classifications, terminologies, reference systems and reference levels, software and hardware specifics, etc. Yet, they are very much determined by political, organisational, legal, cultural and economical situation and agenda (Figure 7.2.6).

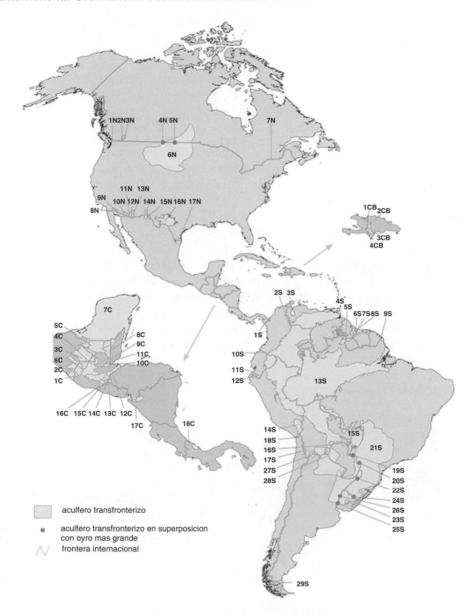

Figure 7.2.6 Transboundary aquifers of Americas.

The need for an independent groundwater centre in the assessment of transboundary aquifers is very obvious. Most of IGRAC transboundary activities are conducted under the framework of ISARM (Internationally Shared Aquifer Resources Management) programme. As a part of the ISARM core group, IGRAC is involved in the development of the programme, also the course material, the web-portal (www.isarm.net), etc. The major IGRAC transboundary activity is, however, assessment of transboundary aquifers in various parts of the world (including Americas, Southern Africa, South-East Europe and other regions). IGRAC is also involved in a number of GEF (Global Environment Facility) projects or project proposals that deal with transboundary waters (IW-LEARN, DIKTAS, IW-Science, TWAP).

7.2.6.5   Global Monitoring

A few years ago, IGRAC has conducted a comprehensive inventory of guideline and protocols for groundwater data acquisition and monitoring. An on-line database contains information on about 480 relevant guidelines and protocols. Many of them are related to groundwater quality, including both field and lab test manuals.

An international working group (set up by IGRAC) has produced a Guideline on groundwater monitoring for general reference purposes, which is also available through the IGRAC portal.

Advised and supported by Global Groundwater Monitoring Group, IGRAC is in a process of establishing a Global Groundwater Monitoring Network (GGMN). The aim is to use monitored data for a periodic assessment of the global groundwater resources (Kukuric & Vermooten, 2007). IGRAC does not intend to create a new, separate global network of monitoring wells. Likewise, no redesign of existing groundwater monitoring networks should be expected. The GGMN is a network of networks that uses information from existing networks in order to represent a change of groundwater resources at the scale relevant for the regional and global assessment. Thanks to its unique position as a UNESCO/WMO centre, IGRAC is contacting various organisations, associations and

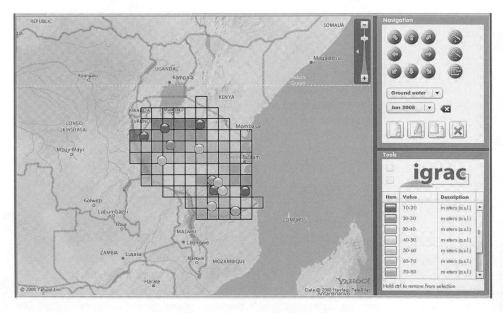

Figure 7.2.7   Groundwater Monitoring Network.

programmes that are either involved in monitoring activities or in position to support those. Again, the involvement of an non-profit, independent centre becomes very important for the success of this kind of initiatives.

The selection of the groundwater quality variables (to be monitor as a part of GGMN) has still to be made (including the protocols, methods, thresholds, interpretation, etc). The experience gained in implementation of European Water Framework Directive can be very helpful in this process.

7.2.6.6   IGRAC and Promotion of Groundwater Use

IGRAC is founded in order to promote groundwater use in sustainable groundwater management. The centre is accomplishing its mission through the activities presented in the previous four sections (7.2.6.2–7.2.6.5). Furthermore, IGRAC provides an 'ABC of groundwater management' in a form of information, advice and guidelines, this mainly for the groundwater specialists and less for the policymakers and general public. General information on groundwater can be found at various places on internet (e.g. www.groundwater.org) and there are also groundwater publications meant for the broad public. However, more should be done in promoting groundwater among policymakers.

According to IGRAC experience, groundwater-related information needs to be translated to policymakers in terms of problems and solutions, and in context of groundwater interaction with its environment. In practice, that means (among others) strengthening the role of invisible groundwater in Integrated Water Resources Management (IWRM). EU Water Framework Directive (WFD) and some other programmes contribute to integration of groundwater in water resources management, with much more to be done. IGRAC is trying to establish cooperation with large catchment or river–based organisations and programmes, convinced in the strength of integration.

Translation of groundwater impact should go beyond the surface water and IWRM. The links between groundwater and ecosystems, water supply, food production, energy production, industry and tourism need to be presented more clearly and disseminated widely.

IGRAC strongly believes in the participative approach, where the owners and users of information form an interactive network. Examples of successful participative initiatives that include general public are a precipitation monitoring network www.rainlog.org and a worldwide monitoring initiative www.worldwatermonitoringday.org. Response of the groundwater specialists to the global initiatives is more enthusiasts when that additional effort is planned and budgeted (such as contribution to WFD or WWAP in some governmental organisations). Individual contribution to GGIS Global Overview has been quite limited. On contrary, response to IGRAC activities dedicated to a certain theme or a problem (such as the arsenic and fluoride analysis and the monitoring guidelines development) has been more than satisfactory.

A participative, networking role of IGRAC asks for the appropriate internet-based tools and services. The user interfaces should be simple and the tools fully interactive. Besides, data and information remain most often at the source, being shared, combined and processed in common portals. In this way owners and provides retain the full control over data they make accessible to others. Finally, IGRAC is developing tools (such as the GGMN application) that can be used as both: a web-based application and a desktop tool (i.e. without internet connection).

A digital infrastructure has been improving rapidly all over the world, extending the possibilities for international participative activities. Nevertheless, face-to-face contacts remain essential for the intensive brainstorming and the confidence building. No wander, because whatever IGRAC does is eventually about the people.

7.2.7   ROLES AND BENEFITS OF THE VARIOUS NETWORKS – PERSPECTIVES FOR FUTURE COLLABORATIONS

The efficiency of the international or EU regulatory framework for groundwater protection closely depends upon a successful technical implementation of various milestones (e.g. risk analysis, monitoring, programmes of measures etc.) which are part of larger river basin management principles. This in turn requires an efficient participatory approach and operational multisectoral and multidisciplinary collaborations. The EU Groundwater Working Group works along these lines with a focus on implementing the Groundwater Directive in the best way. Other international associations or expert groups have different scope and aims, as illustrated in this chapter, but most of them are tailored on expertise and represent an invaluable source of knowledge and advice for decision-makers. This chapter gives different examples about such expert groups (the list is not exhaustive). It highlights the high potential for international cooperation and synergies, which should led to a drastic improvement of groundwater protection from the regional to the worldwide scale.

REFERENCES

European Commission (2001) 6th Environment Action Plan 2001–2010.

European Commission (2003) Common Implementation Strategy for the Water Framework Directive, European Communities.

European Commission (2006) Directive 2006/118/EC of the European Parliament and of the Council of 12 December 2006 on the protection of groundwater against pollution and deterioration, *Official Journal of the European Union*, L372, p. 19.

European Commission (2007a) *Groundwater monitoring*, CIS Guidance No. 15, Common Implementation Strategy of the WFD.

European Commission (2007b) *Groundwater in Drinking Water Protected Areas*, CIS Guidance No. 16, Common Implementation Strategy of the WFD.

European Commission (2007c) *Prevent & Limit*, CIS Guidance No. 17, Common Implementation Strategy of the WFD.

European Commission (2007d) Mandate of the Working Group C 'Groundwater', Common Implementation Strategy of the WFD.

European Groundwater Conference 2006, Proceedings, Vienna, 22–23 June 2006.

GGIS, Global Groundwater Information System; Global Overview, IGRAC 2003-2008, www.igrac.nl.

Kukuric, N. and Vermooten (2007) 'Global Monitoring of Groundwater Resources', International Association of Hydrogeologists, IAH Congress, Lisbon, Portugal, 2007.

Quevauviller Ph., ed. (2007) *Groundwater Science and Policy – An International Overview*, RSC Publishing, Cambridge.

Quevauviller Ph., Balabanis P., Fragakis C., *et al*. (2005) *Environ. Sci. Pol.*, **8**, 203.

Quevauviller Ph., Grath J. and Scheidleder A. (2007) In: *Groundwater Science and Policy – An International Overview*, Quevauviller Ed., RSC Publishing, Cambridge.

Vasak S., Brunt, R. and J. Griffioen (2006), Mapping of hazardous substances in groundwater on a global scale. In: *Mapping of Hazardous Substances in Groundwater, IAH-Symposium Arsenic*, NNC-IAH Publication No. 5.

7.3

Communication of Groundwater Realities Based on Assessment and Monitoring Data

Juan Grima[1], Enrique Chacón[2], Bruno Ballesteros[3], Ramiro Rodríguez[4] and Juan Ángel Mejía[5]

[1,3] *Instituto Geológico y Minero de España (IGME), Valencia, Spain*
[2] *Universidad Politécnica de Madrid, Madrid, Spain*
[4] *Instituto de Geofísica UNAM, Circuito Institutos Delegación Coyoacán, DF Mexico*
[5] *Instituto de Ecología de Guanajuato, Guanajuato, México*

7.3.1   INTRODUCTION

Reality is a quite complex issue to be communicated in a simple way. To better understand physical systems, an abstraction of reality must be done. That is the reason why models

---

Groundwater Monitoring Edited by Philippe Quevauviller, Anne-Marie Fouillac, Johannes Grath and Rob Ward
© 2009 John Wiley & Sons, Ltd

are needed. They are in fact a simplified representation of some real world phenomena (Fetter, 1994). Monitoring networks are then designed and constructed with the aim of feeding data to models that allow simulation of groundwater flow and contaminant transport, and assess about the efficacy and design of remediation systems.

From this perspective, if there is a lack of understanding of natural processes, the monitoring networks will provide erroneous conclusions about what happens in the groundwater body. In addition, monitoring objectives must be defined for an efficient data collection at the early stages of investigation.

To extract information from data some work must be done. The raw data must be elaborated to draw conclusions about ongoing processes in the subsurface. Detection of trends, for example, calls for the application of standardized statistical techniques in order to provide sound scientific conclusions about the existence of such trends. Data must be in some way certified or checked the used methodology for collected or created the data. Data from different origin must be homogenised to avoid erroneous conclusions

7.3.2   KEY ELEMENTS FOR GETTING THE WHOLE PICTURE

7.3.2.1   Conceptual Model

It is very important, prior to communicating the groundwater realities to the stakeholders that a good conceptual model of the groundwater body at risk has been made, as the selection of an appropriate monitoring network (Sandoval and Almeida, 2006) and an adequate monitoring program (Figure 7.3.1). Findings based on a wrong conceptual model or on no model at all can produce results in an opposite direction than expected.

Groundwater bodies are lodged in permeable rocks of different geological composition. That set of geological units conform the aquifers or aquifer systems. Aquifer conceptualization must be done looking for a better understanding of water bodies. The misconception of underground rivers or water veins is well accepted for rural communities even for some stakeholders. The transcription of aquifer to manageable, understandable models implies the integration of data collected or created in diverse institutions and with multiple purposes. A reliable model must be based on geological information extracted from wells, exploratory drillings, geophysical data and geological mapping.

To translate the physical model to a functional model, the local and regional hydrodynamic must be incorporated. Groundwater flows obey dynamic laws. Preferential directions are controlled by the piezometric gradient. The functionality of a groundwater body is defined by local and regional hydrologic balances.

The contribution of regional flows to local groundwater bodies occurs is some areas. To demonstrate and prove its existence and the estimation of the compromise volume represent a challenge for hydrogeologists.

The role of big rivers or lakes and reservoirs in recharge mechanisms is other phenomenon to consider in the establishment of functional models. The use of stable isotopes as deuterium and oxygen 18 can help to define if those surface water bodies are hydraulically connected with groundwater bodies (see Chapter 6.2 of this book).

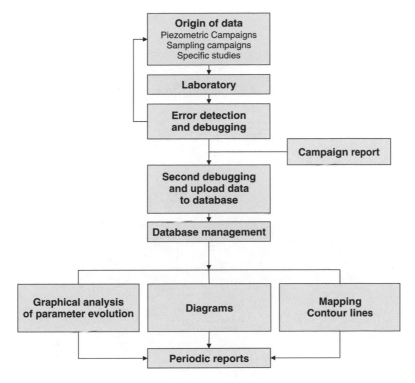

Figure 7.3.1   Flow of information from monitoring networks (from Bouwer *et al*. 1988).

This interaction is very important in water supply policies based on Integrated Water Resources Management.

Groundwater quality is the more relevant aspect for users and stakeholders related to water supply. Series of chemical data must be analyzed to calculate background values of potential contaminants. A complex functional model included not only water composition also potential surface and underground sources. Representative data are collected in monitoring networks.

A conceptual model results from the integration of the physical and functional models. A conceptual model strong defined can be handled in a mathematical computational model. This modelling process offers realistic scenarios for management and preservation of volume and quality. All parameters and boundary conditions can be easily got from a solid conceptual model.

7.3.2.2   Heterogeneity of Groundwater Bodies

Groundwater bodies are complex systems presenting a high variability not only in geometry and hydrodynamic characteristics of its geological media, but in physical-chemical properties (Ballesteros *et al*., 2001), what involves multiple scenarios and behaviour in relation to its flow pattern and quality.

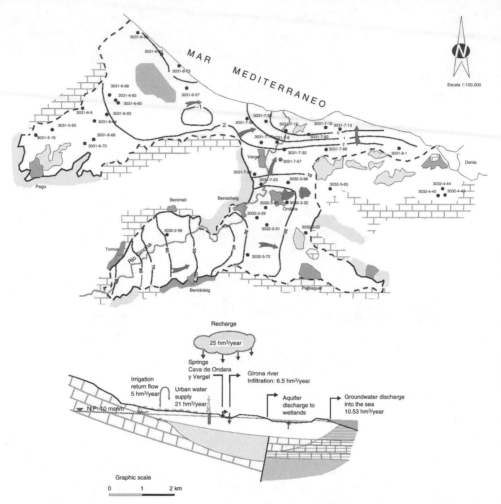

Figure 7.3.2 Conceptual model and heterogeneities of an aquifer at the Mediterranean coast (Ballesteros *et al*., 2001).

To define particularities and heterogeneities of groundwater systems, mainly of rocks confirming aquifer systems, monitoring networks are a key factor as well as information coming from well logs. Nevertheless, a proper design, according to the typology and characteristics of investigated systems (Wendland *et al*., 2006) is call for. Again, the importance of building a conceptual model coming from a detailed investigation is highlighted (Figure 7.3.2).

It must be clear that homogeneity and heterogeneity of a system are concepts closely related to the scale of the investigation. When talking about groundwater, the same groundwater body can be considered as homogeneous at large scale and heterogeneous at the small one. Among other considerations, the subject will depend on the dimensions and extension of the groundwater body. In short, the scale of investigation must be considered at the very beginning of work.

From that consideration four basic aspects of aquifers must be dealt with: groundwater body geometry, aquifer parameters, pressure head regime and hydro geochemical features of groundwater. Particularities of each system will be the driving force for the design and dimension of monitoring networks.

Geometry is the first factor to be considered due to its relevance. There is a wide range of geologic formations, from uniform morphology with no variations at spatial scale (like tabular formations) to complex ones (where processes of deformation and fracturing of previous structures are dominant). The study of the latter can require the division in several sectors and subsequently more dense monitoring networks could be needed.

Changes in hydrodynamic parameters, namely hydraulic conductivity and storage will affect the flow pattern and the hydraulic gradient and, consequently, the potentiometric surface and its evolution. Hydrodynamic regime of groundwater bodies depend upon endogenous and exogenous characteristics of geological formations. In detritic aquifers the permeability is due to intergranular porosity, while in fractured rock aquifers is due to water-bearing fractures (where dissolution processes occur) and their interconnection. Within the former, the flow regime is laminar and homogeneous having a slower and more predictable movement. Within the latter the flow is faster, trough preferential channels and turbulent flow can happen when channels get a given proportion. There are, then, obvious implications depending on the type of permeability of the groundwater body being investigated.

When studying heterogeneities of groundwater bodies, one key element is the hydrostatic pressure of groundwater within an aquifer. In unconfined aquifers, the hydrostatic pressure is defined by the water table, whereas in confined aquifers, the water level will depend upon the pressure head (Villanueva and Iglesias, 1984). In nature, groundwater bodies can range from simple aquifer formations to very complex multilayer aquifer-aquitard systems. As it can be easily understood, in the latter ones more precise measurements are needed and the correct assignation of measures taken to identified aquifer layers is of crucial importance.

Hydrogeochemistry is another source of variation within groundwater bodies, as interaction processes between water bearing rocks and the water itself generate different water quality sectors, and this fact will determine the natural evolution of every particular system. The effect of interactions with hydraulically connected aquifers as well as surface water is to be considered when looking for a detailed description of heterogeneities. This factor is particularly important when dealing with coastal aquifers, where a freshwater and saltwater interface exists. Interactions between both types of waters and geological formations must be accounted for, as seawater intrusion is a dynamic and three-dimensional process that originates water quality variations at horizontal and vertical scale. That implies more dense monitoring networks than required for continental aquifers. Obviously, pollution processes are contributor factors to spatial and temporal heterogeneities within groundwater bodies.

7.3.2.3 Receivers of the Information

Not only policy makers are the target of the communication process, but water user associations, Academia, water research and water service institutions, representatives of waste water treatment plants, watershed and planning authorities, Non Governmental organisations (NGOs), Public health departments, Ministries of Environment, Water an Agriculture and civil society. Specific mechanisms must be implemented in order to enhance the communication procedure at different levels: national, regional and local.

7.3.3   ANALYSIS AND ASSESSMENT OF DATA

7.3.3.1   Objectives

When running monitoring networks, analytical measurements are made, and each observation provides a piece of additional information. The primary goal for groundwater sampling is to obtain representative samples and field parameters. The analysis of these data is a process in which the information collected is looked at and summarized.

As mentioned before, nature is very complex, so a model has to be developed (that rests on a number of assumptions) to reproduce processes occurring in groundwater. The underlying structure is then examined and hypothesis tests to check assumptions are delineated, prior to getting conclusions about ongoing processes.

As an example, determination of trends is required to assess if there is an increasing concentration of pollutants within a given groundwater body. As stated by the Groundwater Directive, criteria should be established for the identification of any significant and sustained upward trends in pollutant concentrations and for the definition of the starting point for trend reversal. Member States should, where possible, use statistical procedures, provided they comply with international standards and contribute to the comparability of results of monitoring between Member States over long periods.

7.3.3.2   Methods of Analysis

Heterogeneity of hydrogeological environment is due to a great number of factors, and it is present to a bigger or smaller extent in every groundwater body. Consequently, for groundwater body monitoring, it is required a sufficient level of information to be available, provided that previous studies at the appropriate scale has been done. Designed monitoring networks are then representative of ongoing hydrodynamic and hydro chemical processes. The degree of knowledge of the investigated systems must be as much as close to reality as possible within a context of a reasonable cost benefit framework. As it is easily understandable, the requirements related to spatial distribution and frequency of measurements will depend upon the complexity of the system under study.

First thing to keep in mind when analyzing monitoring data, i.e. contaminant concentration, is that measurement are not exact values, but the obtained value is composed of the measurement itself, the margin error and the confidence level. That is, we could assert that the concentration of a specific substance in a given groundwater body is 27 µg plus or minus 0.1 µg with 95% level of confidence (Helsel and Hirsch, 2002). It clearly indicates the necessity of incorporating uncertainty in data analysis, and calls for the application of tiered approaches. At first stages, when data sets are small, the associated uncertainty will be high, being reduced the margin error as new data become available. As stated by the BRIDGE Project (Müller *et al*., 2006), 'A tiered approach allows the effort to be proportional to the risk involved (greater risks greater effort)'. Thus a tiered approach supports a practical way forward and cost effectiveness.

After collecting experimental data mathematical procedures can be used to make statistical decisions. In the case that the distribution of the parameter to be analyzed is assumed to follow a known probability distribution, the inferential statistical method is called a parametric model. Non-parametric methods, on the other hand, are referred to as distribution free methods, it is to say that no assumption is made about a given probability distribution from which it is supposed the data are drawn.

In order to make a statistical decision, a hypothesis test must be designed. It will answer, for example, the question if we are able to reject the existence of an increasing concentration of contaminant in a groundwater body with the experimental data alone. The significance level (in this case probability of rejecting there is no trend when in fact there is not) must be established by the decision maker, and should be based on an integrated evaluation of both the objectives of the analysis and the current phase of the stepwise approach.

Linear regression is one of the most popular parametric methods to analyze hydro-geological information. One of the most important applications of ordinary least squares is the hypothesis test for the slope of the regression line (check if it is significantly different from zero). The Pearson correlation coefficient is a parametric statistic, and it means that if the assumptions underlying the model are met, it is the most powerful statistical approach for linear correlation. Nevertheless, when the assumptions are found to be false, the use of non parametric statistics is advised.

When linear regression is applied to a set of environmental data, one of the most frequent questions we are interested in is if the slope of the regression line is significantly different from zero. It means that we need a hypothesis test with null hypothesis of slope equal to zero. A number of assumptions are then required (Figure 7.3.3), like homocedasticity and normality and independence of residuals.

To what extent a parametric model can be used when normality assumptions are not met (like normality of residuals) is the key point of the statistical analysis (Helsel and Hirsch, 2002). If departure from normality is not very high, parametric tests are robust, what means that they will provide right conclusions even if the underlying distribution is not normal.

On the other hand, there is the general opinion that non parametric methods are less efficient, due to the fact that they discard data values or information. This is only right when the hypotheses of parametric tests are known to be true, which is very unusual in most environmental data sets, especially in those of small size.

Software environment for statistical computing and graphics is freely available on the Internet under the GNU General Public Licence (Hornik, 2008) and pre-compiled binary versions are provided for Microsoft Windows, Mac OS X, and several Linux and other Unix-like operating systems.

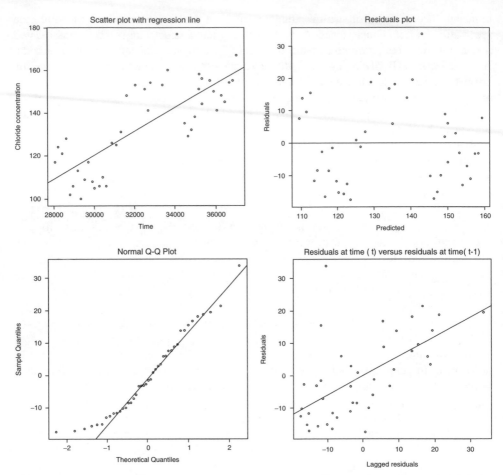

Figure 7.3.3 Plots needed when constructing a regression model. Graphics generated with R for chloride concentration in a coastal aquifer.

7.3.3.3 Limitations and Misperceptions

As mentioned before, reality is very complex, and models developed to conceptualize aquifers or groundwater bodies are just a simplification of the physical reality and they are based on a number of assumptions. Therefore, it is important to test whether the models chosen are consistent with available data and recognize the inherent limitations of any model.

When dealing with groundwater, an important question to be accounted for is the response time-lag of groundwater bodies (Figure 7.3.4). Human activities may affect groundwater and associated ecosystems in a number of ways. Sometimes, there is a direct relationship between an action and its consequences on the environment. It is the case of groundwater exploitation and the change in natural groundwater regime.

Nevertheless, it can take several decades (even millennia) for a pollutant before reaching water table (López-Geta *et al*., 2006), due to the response time-lags to anthropogenic

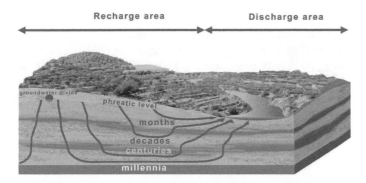

Figure 7.3.4 Response time-lag of groundwater bodies (from López *et al.*, 2006).

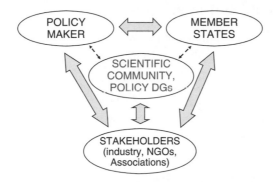

Figure 7.3.5 The consultation loop (from Quevauviller, 2006).

input. This inertia of aquifers in its response to external influences can delay the detection of problems or the effects of management for a long time.

Recharge is a key factor in aquifer characterization. Its relevance is given by volume and time. Users and stakeholders consider rainfall and infiltration from surface water bodies as rivers, lagoons or reservoirs the main source of available water for recharge. A general misperception is recharge temporality. Only in cases where water table is very shallow (some dozens of meters) and/or surface rocks are very permeable transit times are very short (some days). In general surface water arrives to aquifers after long periods of time, years to decades or even more. Stakeholders must catch the importance of residence times and the fact that the defacement between recharge and abstraction can cause aquifer storage decreasing (Figure 7.3.5).

Another factor to be considered is the natural variation of the chemical state of groundwater (Custodio and Llamas, 1976). As the water table rises or fall and temperatures and recharge rate vary, the composition of groundwater changes. This effect may play an important role in shallow groundwater.

When analyzing environmental data some misapplications of statistical methods are made by the water resources community, including some hydrogeologists. One of the most common errors is to introduce the raw data into commercial software and obtain a value for the square of the Pearson's r correlation coefficient (r reflects the degree of linear relationship between two variables and it ranges from $+1$ to -1).

The coefficient of determination, R^2, is the proportion of variability that is accounted for by a statistical model. It is incorrectly assumed that values of R^2 close to 1 are indicative of a good regression model, when it can be far from reality. Another limitation is due to the fact most of the commercial software assumes that the data set is large enough to run tests based on large sample approximations.

7.3.4   COMMUNICATION PROCESS AND INTERACTION WITH STAKEHOLDERS

7.3.4.1   Policy and Legal Framework

As stated by the European Commission (Communication from the Commission to the European Parliament and the Council – addressing the challenge of water scarcity and droughts in the European Union), in order to be fully effective, policy action on water needs to be based on high-quality knowledge and information on the extent of the challenge. In addition, in order to achieve the Lisbon aim of a knowledge-based society, closed links are foreseen to be established between the European area of lifelong learning and the European research area.

According to the Water Framework Directive,

> to ensure the participation of the general public including users of water in the establishment and updating of river basin management plans, it is necessary to provide proper information of planned measures and to report on progress with their implementation with a view to the involvement of the general public before final decisions on the necessary measures are adopted.

To better address the diverse and complex challenges that science and policy making communities are facing related to the communication process, an efficient interface to handle information exchange and promote coordination mechanisms must be developed (Quevauviller, 2007). In fact, to tackle this issue, the European Commission is developing a knowledge-based approach to an operational science-policy interface linked to WISE (Water Information System for Europe), (http://www.wise-rtd.info/).

7.3.4.2   Effective Implementation of Water Policies through Participative Approaches

Participation is a process through which stakeholders influence and share control over development initiatives and the decisions and resources which affect them (World Bank). It is widely accepted that holistic and participatory processes are needed for an effective implementation of water policies (Garduño *et al*., 1995). The position of an actor depends upon its perception about challenges and opportunities of the groundwater environment.

Consultation and establishment of partnerships among stakeholders (see Figure 7.2.1 in Chapter 7.2), on the other hand, appear like the cornerstone of any participatory approach. In such a scenario it is of crucial importance to extract as much information as possible from monitoring networks. Stakeholder's conclusions can then be based on realities more than in opinions by means of the information process.

The role of stakeholders is of growing importance due to the challenges water resources are facing and they are going to in the near future and due to some other factors, like policy changes (decentralization) or even climate change. Presentation of technical results is then foreseen to the group of stakeholders, taking into account that such a group is highly heterogeneous and non technical in orientation.

According to the Common Implementation Strategy for the Water Framework Directive, presentation and communication of the rather technical analysis to stakeholders has led in many cases to misperceptions. In particular, a highly aggregated and non-differentiated presentation of the results did not lead to an understanding of the key issues that might need to be address in the water management.

It is necessary to remember one more time that no model can provide absolute evidence of a process, due to its inherent limitations (Fetter, 1994). In fact, what can be provided to stakeholders in many cases is the probability that something is happening, with a confidence interval for the null hypothesis. In this context, it is important to highlight the important of flexible and adaptive management of natural resources due to uncertainties related to ecosystem management and the changing conditions of natural systems.

Although it has been a recurring issue in water resources protection strategies, it is of crucial importance to raise awareness of stakeholders and civil society of the importance of the risk for health due to groundwater contamination. Pollution prevention is one of the pillars of environment protection, and can be considered as important or even more than remediation.

7.3.4.3 Turning Data into Useful Information

When data are organized and presented properly, they become information. This information is used to make decisions that will impact on the management of the groundwater body. Some steps are then required to turn the raw data into guiding principles for stakeholders and decision makers. Communication is a two-way exchange of information, and it must include groundwater threats, like pollution or deplenishment. A distinction must be made between scientific data and scientific information. The former is the quantitative information obtained from our observations, while the latter includes not only the data themselves, but a description of what was done, how it was done and why it has been done.

While monitoring data is a collection of quantitative information, it is far more effective to make a graphical presentation of data, because charts convey ideas about the data that would not be perceptible if they were displayed on a table. Some times conclusions can not be drawn by just visualizing the data, and then numerical summary methods are needed.

7.3.5 DISCUSSION AND CONCLUSIONS

Due to the complexity of natural systems a simple one-size-fit-alls policy is not enough, but varying management approaches to suit specific situations are needed.

According to the Groundwater Directive, the assessment of data will be based on statistical methods. It is important to have in mind that the objective of statistics is not to make assumptions about the value of a specific parameter, but to build confidence intervals about how uncertainty is behind our findings.

Before data are analyzed, a good conceptual understanding of what really happens down surface is needed, so a model must be built. Design of monitoring networks based on a wrong conceptual model can be the source of inconsistent monitoring networks investment strategies that will no provide additional reliable scientific knowledge of the groundwater body.

In addition, possible sources of uncertainty (like data below the limit of quantification) must be analyzed and considered within our model. In order to reduce the associated uncertainty it is needed not only to build a conceptual model and check the assumptions, but to implement quality assurance and quality control plans. Data quality objectives describe the overall level of uncertainty due to monitoring. When communicating groundwater realities inherent limitations of the model must be clearly explained.

Monitoring is a procedure to obtain understanding of groundwater characteristics through statistical sampling. Monitoring objectives must be defined at the very beginning of the process for an efficient data collection, and harmonization is needed in order to get comparable results at the monitoring stations. The driving force for monitoring networks design is the particularity of the groundwater body.

The size of many environmental data sets is small, what implies a high degree of uncertainty, and calls for the application of tiered approaches, allowing the effort to be proportional to the risk involved. The concept of risk implies that there is an element of likelihood and uncertainty that must be communicated to the stakeholders.

Taking the previous considerations into account, the use of standardized methods like linear regression must be checked to avoid providing conclusions when the assumed hypothesis are not true, what could be very misleading. The use of non parametric approaches should not be discarded in that case.

A number of misconceptions are still present in the water community, including some hydrogeologists. For example, no conclusions should be provided based on the application of commercial software alone, but expert judgement and local knowledge is needed. In this respect it is worth to mention that free software for statistical computing and graphics is available on the Internet, which provides a wide variety of statistical and graphical techniques, like the R statistical package.

Finally, the process of communication is a two-way exchange, so conclusions must be clearly explained and the technical information must be summarized. One key objective in groundwater communication strategies must be focussed on providing environmental information, including uncertainty and risk, and how these issues are related to the precautionary principle. When communicating groundwater realities to stakeholders, emphasis should be put on the uncertainty related to the management of ecosystems and the dynamism of watershed systems, and the advantages of adaptive approaches for both management and policy making.

REFERENCES

Ballesteros B. J., Rodríguez L., López J., *et al*., 2001. Análisis y ordenación de recursos hídricos de la Marina Alta (Alicante). Alternativas y Directrices. Vol II. Evaluación de recursos subterráneos. Instituto Geológico y Minero de España-Diputación Provincial de Alicante. Fondo documental IGME. 179 pp. Alicante

Custodio E. and Llamas M. R., 1976. *Hidrología Subterránea*. Ed. Omega, 2 vol. Barcelona.

European Commission, 2001. Making a European area of lifelong learning a reality.

Fetter C. W., 1994. *Applied Hydrogeology*. Charles E. Merrill. Pub. Co. 3rd ed., Prentice-Hall.

Garduño H. *et al*., 1995. Stakeholder Participation in Groundwater Management: mobilizing and sustaining aquifer management organizations, GW Mate, Briefing Note Series, Note 6, 2006. N. Wilson, *Soil Water and Ground Water Sampling*, CRC Press, Boca Raton, Florida.

Helsel D. R. and Hirsch R. M., 2002. Techniques of Water-Resources Investigations of the United States Geological Survey. Book 4, Hydrologic Analysis and Interpretation. Statistical Methods in Water Resources. *Environmental Science and Pollution Research International*, **14**(5): 297–307.

Hornik K., 2008. Frequently Asked Questions on R. {ISBN} 3-900051-08-9. http://CRAN.R-project.org/doc/FAQ/R-FAQ.html

López-Geta J. A., Fornés J. M., Ramos G. and Villaroya F., 2006. *Groundwater and Natural Underground Resource*. IGME and Fundación Marcelino Botín.

Müller D., Fouillac A. M., Hart A. and Quevauviller Ph., 2006. BRIDGE – Background Criteria for the Identification of Groundwater Threshold Values. European Groundwater Conference. Ed. Umweltbundesamt GMBH. Vienna

Quevauviller Ph., 2007. Water protection against pollution. Conceptual framework for a science-policy interface. *Environ Sci Pollut Res Int*. **14** (5): 297–307.

Sandoval R. and Almeida R., 2006. Public Policies for Urban Wastewater Treatment in Guanajuato Mexico. In *Water Quality Management in Las Americas*, eds. Biswas A., Tortajada C., Braga B. and Rodriguez D. Springer Edit. 147–65.

Villanueva M. and Iglesias A., 1984. Pozos y acuíferos. Técnicas de evaluación mediante ensayos de bombeo. Publicaciones del Instituto Geológico y Minero de España. pp 426. Madrid.

Wendland F., Blum A., Coetsiers M., *et al*., 2006. Aquifer typologies: A practical framework for an overview about major groundwater composition on a European scale. European Groundwater Conference. Ed. Umweltbundesamt GMBH. Vienna.

Index

Note: Page numbers in *italic* refer to figures; those in **bold** to tables.

Groundwater Monitoring Edited by Philippe Quevauviller, Anne-Marie Fouillac, Johannes Grath and Rob Ward
© 2009 John Wiley & Sons, Ltd